Cross-Cultural and Cross-Disciplinary Perspectives in Social Gerontology

Tannistha Samanta
Editor

Cross-Cultural and Cross-Disciplinary Perspectives in Social Gerontology

Springer

Editor
Tannistha Samanta
Department of Humanities and
Social Sciences
Indian Institute of Technology
Gandhinagar, Gujarat
India

ISBN 978-981-10-1653-0 ISBN 978-981-10-1654-7 (eBook)
DOI 10.1007/978-981-10-1654-7

Library of Congress Control Number: 2016943812

Printed on acid-free paper

This Springer imprint is published by Springer Nature
The registered company is Springer Science+Business Media Singapore Pte Ltd.

Foreword

In the last decade or so, the development in collecting large-scale, longitudinal aging datasets around the globe has been nothing short of spectacular. For example, we have witnessed the ongoing growth of the sibling studies of Health and Retirement Study (HRS, US) in multiple continents, now including the Survey of Health, Ageing, and Retirement in Europe (SHARE), the Chinese Health and Retirement Longitudinal Survey (CHARLS), and the Mexican Health and Aging Survey (MHAS). At the same time, the World Health Organization (WHO) study on global AGEing and adult health (SAGE) is expanding to African countries such as Ghana and South Africa—relative newcomers to the global aging process. Such an impressive growth in data availability certainly provides unprecedented opportunities to gerontological researchers. However, it is important to keep in mind that the expansion of such data does not automatically lead to a growth in knowledge. Instead, the way in which data are utilized and interpreted are crucial to scholarly advances in the field of aging research.

This book edited by Dr. Tannistha Samanta is both timely and significant. It reminds us that it is no time for complacency even when a large body of empirical observations have been accumulated. While the book evidently acknowledges and values empirical studies utilizing large survey data, the book goes beyond the positivistic paradigm by also including extensive anthropological, ethnographic and phenomenological body of work in different chapters. The integration of work across different disciplines is a clear strength of the book, giving the readers not only a clear picture of how social contract between generations operates at the societal level but also a sense of everyday experience of aging in diverse settings.

The second notable feature of the book is that it reminds us theory building has become more important and should be actively incorporated into empirical research. Contrasting the abundant attention to data collection effort, theory development has remained a weak spot for social gerontology. As Dr. Samanta has pointedly noted, "theorizing without empirical grounding is unproductive and mute as sophisticated empirical analysis without theoretical propositions." The various chapters in the book address the urgent need in gerontological scholarship for integrating dominant theoretical frameworks together. From stress coping models to feminist theories of care, from a social psychological approach in understanding ageism to the life course principles, these different research lenses lucidly illustrate the usefulness of interdisciplinary research and how they can collectively inform policy interventions and initiatives.

Finally, what I appreciate tremendously about the book is how it puts *culture* front and central rather than "backgrounding" it in the scientific inquiry. It is well known that some of the most applied theories in gerontological research, such as the convoy model or cumulative advantage/disadvantage theory, have originated in Western countries. Thus, the test of these theoretical frameworks in the global south in and itself is much needed. Moreover, the chapters in the book extend beyond mere replication and are explicit in addressing the cultural limits of some of the dominant theoretical frameworks in various settings. As a result, readers get a much nuanced and multifaceted understanding of how a diverse set of cultural concepts, such as norms, attitudes, and belief, operate at different levels.

Feinian Chen
Professor, Department of Sociology
University of Maryland, College Park
College Park, MD

Acknowledgements

The idea of this volume was born at the workshop in the summer of 2013, supported by the Indian Institute of Technology, Gandhinagar. I am thankful to the Director Prof. Sudhir Jain for his openness and encouragement to support the "Research Initiative on Social Gerontology" project that made this workshop possible. I would like to thank the participants of the workshop for their lively engagement and astute observations which allowed the book to take a rather unconventional goal of generating a unique interdisciplinary body of scholarship that is empirical, theoretically rooted and culturally diverse. This journey would not have been accomplished without the enormous support of the contributors who have been both friends and colleagues and have worked with me patiently to bring this project to fruition.

I am grateful to Prof. Leela Visaria and Dr. G. Giridhar for their insightful feedback on the earlier draft of the manuscript. I would like to express gratitude to my colleagues at the discipline of humanities and social sciences, who have graciously stepped in and have allowed me time to complete this book project. Very special thanks to my colleague, Rita Kothari, for being a generous friend and an exacting critic. My doctoral students, Jagriti Gangopadhyay and Anusmita Devi, offered cheerful assistance in organizing the summer workshop and later, some editing help in preparing the manuscript.

Finally, this volume has greatly benefitted through many conversations (and disagreements) over coffee and dinners with my husband, Kaushik. Long-distance phone calls with my parents and sister have contributed to certain thoughts about the social process of aging that the papers in this volume collectively address. Preparing this volume has been a journey of new learnings, new collaborations and exploring newer ways of understanding aging. I hope this volume will be a useful resource for future efforts in social gerontology.

Contents

Part I Frameworks and Integrative Approaches

1 **Bridging the Gap: Theory and Research in Social Gerontology** 3
Tannistha Samanta

2 **Who Will Care for the Elder Caregiver? Outlining Theoretical Approaches and Future Research Questions** 23
Subharati Ghosh, Benjamin Capistrant and Greta Friedemann-Sánchez

3 **Social Psychology and Gerontology: Integrating Theory to Explain and Intervene in Age Discrimination Towards Older People in Europe** 45
Sibila Marques, Christin-Melanie Vauclair, Hannah J. Swift, Christopher Bratt, Maria Luísa Lima and Dominic Abrams

Part II Culture, Contexts and Aging

4 **Aging and Subjectivity: Ethnography, Experience and Cultural Context** 69
Jason Danely

5 **Neoliberalism and Resilience Among Older Yoruba People in a Semiurban Community, South West Nigeria** 85
Ojo Melvin Agunbiade and Akanni Ibukun Akinyemi

6 **Social Capital, Interrupted: Sociological Reflections from Old Age Homes in Ahmedabad, India** 109
Tannistha Samanta and Jagriti Gangopadhyay

7 **Engaging the Future as Ethnographic Object: Japan's Aging Society Crisis, Ontogenesis and Cybernetics** 125
Ender Ricart

Part III Health and Well Being

8 **Antecedents of Subjective Wellbeing Among Older Adults in Kerala** 143
S. Irudaya Rajan, Anusmita Devi, Tannistha Samanta and S. Sunitha

9 **Early Life Critical Transitions in the Relationship Between Current Life Stressors and Depressive Symptoms Among Community-Residing Older Lebanese Adults** 159
May H. Aydin, Abla Mehio Sibai and Anthony Rizk

10 **Analysis of Perceived Health Status Among Elderly in India: Gender and Positional Objectivity** 177
Zakir Husain and Dona Ghosh

11 **The Relationship Between Spatial Activity and Wellbeing-Related Data Among Healthy Older Adults: An Exploratory Geographic and Psychological Analysis** 203
Michal Isaacson, Hans-Werner Wahl, Noam Shoval, Frank Oswald and Gail Auslander

12 **The Changes of Disability-Free Life Expectancy and Inter-Generation Support for the Elderly in China: 2005–2010** 221
Wei Guo

13 **Elderly Inpatient Care Utilization and Financing in India: Is There a Gender Difference?** 245
William Joe, Abhishek Kumar and Udaya Shankar Mishra

Index 271

Editor and Contributors

About the Editor

Tannistha Samanta is Assistant Professor of Humanities and Social Sciences, Indian Institute of Technology, Gandhinagar. Samanta received her Ph.D. from the Department of Sociology at the University of Maryland, College Park, MD. Samanta specializes in blending both quantitative and interpretive approaches to study intergenerational relationships, health, social capital and aging focusing primarily on India, Email: tannistha@iitgn.ac.in.

Contributors

Dominic Abrams is Professor of Social Psychology and Director of the Centre for the Study of Group Processes at the University of Kent, UK. He is an expert on intergroup relations. He is codirector and founder of EURAGE and is coeditor of the journal *Group Processes and Intergroup Relations*, Email: D.Abrams@kent.ac.uk.

Ojo Melvin Agunbiade is Lecturer in the Department of Sociology and Anthropology, Obafemi Awolowo University, Ile-Ife, Nigeria. Melvin teaches courses on sociological theory, social research methods and sociology of health. His research interests include sexual health, aging, youth, gender and society, traditional medicine and development. He has published in reputable journals, Email: oagunbiade@oauife.edu.ng, ojomelvin@yahoo.com.

Akanni Ibukun Akinyemi is Associate Professor of Demography and Social Statistics, Obafemi Awolowo University, Ile-Ife, Nigeria. His research interests include demography of aging, migration, family demography and reproductive health. He teaches research methods at undergraduate and postgraduate levels. He is versatile in data analysis, Email: maydin@nsf.gov.

Gail Auslander is Zena Harman Professor, Paul Baerwald School of Social Work and Social Welfare, the Hebrew University of Jerusalem. Her research focuses on psychosocial aspects of health and illness, including the role of family caregiving to elderly people in the community and in-patient settings.

May H. Aydin is Director of the Information and Technology Services Program at the National Center for Science and Engineering Statistics, National Science Foundation, USA. Her interest in aging research includes life course trauma, stress, war and conflict, late-life psychological wellbeing, and resiliency among aging populations, Email: maydin@nsf.gov.

Christopher Bratt was awarded a Ph.D. in sociology in 1992 from the University of Oslo, Norway, and works in the fields of sociology and psychology, with intergroup relations being his main interest. Bratt also holds a strong interest in quantitative methods. He is a founder member of the EURAGE group (www.eurage.com), Email: Christopher.Bratt@gmail.com.

Benjamin Capistrant is Assistant Professor at the University of Minnesota School of Public Health, USA, where he previously held the Fesler-Lampert Chair in Aging Studies. He was a postdoctoral scholar at the University of North Carolina—Chapel Hill, and completed his graduate training in social epidemiology at Harvard University, Email: bcapistr@umn.edu.

Jason Danely is Senior Lecturer of Anthropology at Oxford Brookes University, and author of *Aging and Loss: Mourning and Maturity in Contemporary Japan* (2014, Rutgers University Press). Jason is currently conducting cross-cultural research on the lived experiences of family caregivers of older adults in Japan and the UK, Email: jason.danely@gmail.com.

Anusmita Devi is a doctoral candidate in the Department of Humanities and Social Sciences, Indian Institute of Technology, Gandhinagar, India and holds a Masters' degree in Sociology from Jawaharlal Nehru University (JNU), New Delhi. Currently, as part of her dissertation project, she is examining the cultural representation of aging bodies and embodiment, social constructionism and sexism in popular Indian media, Email: anusmita.devi@iitgn.ac.in.

Greta Friedemann-Sánchez a medical and economic anthropologist, is Associate Professor at the Hubert H. Humphrey School of Public Affairs, University of Minnesota, USA. She aims to understand how and why gender equity or inequity is generated at homes through the interlacing of family bargaining dynamics, policy and cultural norms. Her primary research areas include intimate partner violence and unpaid caregiving labour, Email: gfs@umn.edu.

Jagriti Gangopadhyay is a doctoral candidate in the Humanities and Social Sciences Department at the Indian Institute of Technology Gandhinagar (IITGN). Her research interests include family sociology, aging, intergenerational relationships and grand-parenting. Jagriti holds an MA in Sociology (Jadavpur University, Calcutta) and an M.Phil in Planning & Development from the Indian Institute of Technology Bombay, Mumbai, Email: jagriti.gangopadhyay@iitgn.ac.in.

Dona Ghosh is an institute-sponsored research fellow undertaking her doctoral research on gender disparities in health of the aged at the Department of Humanities and Social Sciences, Indian Institute of Technology Kharagpur, Email: ghoshdona.1707@gmail.com.

Subharati Ghosh is Assistant Professor of Social Work at the Tata Institute of Social Sciences, Mumbai, India. She specializes in the study of disability and caregiving across the life course. She has conducted several studies on caregiving and disability in the USA and India. Her work aims to understand the impact of caregiving on caregivers' physical, social, psychological and economic wellbeing, with the aim to develop services and inform policy to help caregivers, Email: subharati.ghosh@tiss.edu.

Wei Guo, Ph.D. is Associate Professor of Demography and Social Policy at Nanjing University, China. A demographer and a sociologist by training, Dr. Guo's research interests are in the areas of social gerontology, adolescent health and development, and social demography. He has published over 20 articles in mainstream English and Chinese refereed journals from 2008 to the present. He is also a regular reviewer for several Chinese and English academic journals, Email: weiguo@nju.edu.cn.

Zakir Husain is Associate Professor in Department of Humanities and Social Sciences, Indian Institute of Technology Kharagpur. His research interests are aging, gender and socio-religious discrimination, education and health. His work has been published extensively in journals like *Social Indicators Research*, *Journal of International Development* and *Journal of Biosocial Science*, Email: dzhusain@gmail.com.

Michal Isaacson is Assistant Professor at the Gerontology Department, University of Haifa, Israel. Trained as an Urban Geographer, Michal is interested in how people both consume and create urban space. Her research focuses on spatial aspects of later life with a special interest in the mobility of older adults.

William Joe is Assistant Professor at the Population Research Centre, Institute of Economic Growth, Delhi, and Bernard Lown Visiting Scholar, Harvard School of Public Health, Boston. He has a Ph.D. in Economics from Jawaharlal Nehru University with research interests in health economics, demography and development economics, Email: william@iegindia.org.

Abhishek Kumar is a research analyst at the Institute of Economic Growth, Delhi. He is currently working on various projects related to population health. He has a Masters' degree in economics from the Gokhale Institute of Politics and Economics, Pune. His research interest lies in the field of human development and health economics, Email: abhishekkmr48@gmail.com.

Maria Luísa Lima is Professor in Psychology at ISCTE-IUL. She is an expert in the social psychology of health with a focus on environment, health risks and aging, and the author of several publications. She is codirector and founder of EURAGE (www.eurage.com), Email: Luisa.Lima@iscte.pt.

Sibila Marques is a researcher at the Centre for Social Psychological Research and Social Intervention at the university ISCTE-IUL (University Institute of Lisbon), Portugal. She holds a Ph.D. in Social Psychology and has several publications in the gerontological field. She is a founder member of the European Research Group on Attitudes to Age (www.eurage.com), Email: sibila.marques@iscte.pt.

Udaya Shankar Mishra is Professor and a statistician/demographer at the Centre for Development Studies, Thiruvananthapuram, India. His research contributions are broadly in the areas of aging, health, nutrition as well as population policy and programme evaluation. His current research interests include measurement issues in health and equity focus in evaluation of outcomes, Email: mishra@cds.ac.in.

Frank Oswald is Professor for Interdisciplinary Ageing Research at the Goethe University Frankfurt, Germany. He is a psychologist by training and co-author of several articles in the area of Environmental Gerontology, addressing issues of p-e transaction, contexts of adult development, housing and relocation.

S. Irudaya Rajan is Professor at the Centre for Development Studies, Thiruvananthapuram, India. He has published extensively in national and international journals on social, economic, health and demographic facets of aging over the past three decades. He initiated the first longitudinal Aging Survey in Kerala since 2004 and entered its fifth wave in 2016, Email: rajancds@gmail.com.

Ender Ricart, Ph.D. is at the Battelle Memorial Institute, USA. Ender received a Ph.D. in Anthropology from the University of Chicago. She has held fellowship appointments with Fulbright and the Japan Society for the Promotion of Science (JSPS), and visiting research positions at Tokyo Metropolitan Institute of Gerontology in Japan and the East-West Center in Washington, D.C, Email: ricart@battelle.org.

Anthony Rizk is a research assistant in social epidemiology and medical anthropology at the American University of Beirut, Lebanon. He has been involved in a number of research projects on aging, gerontology, knowledge translation, war and conflict, sexuality and migration, Email: anthony.rizk@outlook.com.

Noam Shoval is Professor at the Department of Geography, the Hebrew University of Jerusalem, Israel. His research focuses on the implementation of tracking technologies in various areas of research such as tourism, urban studies and medicine.

Abla Mehio Sibai is Professor of Epidemiology at the Faculty of Health Sciences, American University of Beirut and co-founder of the Centre of Studies on Aging in Lebanon. Her interest in aging research includes several interrelated clusters: social and psychological dimensions of aging, disease and disability, and knowledge translation.

S. Sunitha is Research Associate at the Centre for Development Studies, Thiruvananthapuram, India. She has a Ph.D. in Demography from the University of Kerala. She has been working closely with S. Irudaya Rajan on several projects on aging and has published articles in edited volumes as well as journals regarding different demographic aspects of aging, Email: sunithasyam@gmail.com.

Hannah J. Swift is a research fellow at the Centre for the Study of Group Processes, University of Kent, UK. Her research focuses mainly on ageism and healthy aging, and has been published in academic journals and reports for the UK government. She is a founder member of EURAGE (www.eurage.com), Email: H.J.Swift@kent.ac.uk.

Christin-Melanie Vauclair is a researcher at ISCTE-IUL, Centre for Psychological Research and Social Intervention. She holds a Ph.D. in Cross-Cultural Psychology and her research focuses on cultural comparative research in regard to ageism. She is a founder member of the European Research Group on Attitudes to Age (www.eurage.com; www.melanie-vauclair.com), Email: Melanie.Vauclair@iscte.pt.

Hans-Werner Wahl is Professor of Psychological Aging Research at the Institute of Psychology, Heidelberg University, Germany. His research activities include the understanding of the role of physical-technological environments for aging well, adaptational processes in the context of age-related chronic functional loss, processes of awareness of aging from a lifespan perspective, and intervention research.

List of Figures

Figure 2.1 Dependency ratio. *Source* United Nations (2014). Probabilistic population projects based on the world population prospects: the 2012 revision. Population Division, DESA/ST?/ESA/SER.A/353. http://esa.unc.edu/unpd/ppp 24

Figure 6.1 Frequency of visits: actual and expected. *Source* Based on authors' interviews with the study participants, summer 2014 118

Figure 8.1 Subjective wellbeing index for 2004 and 2013. *Source* Kerala Aging Surveys (2004, 2013). 150

Figure 8.2 Subjective wellbeing Index by gender (2004 and 2013 compared). *Source* Kerala Aging Surveys (2004, 2013) 152

Figure 8.3 **a**, **b** Subjective wellbeing and perception of physical health (2004 and 2013 compared). *Source* Kerala Aging Surveys (2004, 2013) 153

Figure 10.1 Explaining gender differences in actual health status. 182

Figure 10.2 Age specific sex ratio-census 2011. *Source* Office of the Registrar General and Census Commissioner of India (2011). Five year age group data (Table C14), accessed at http://www.censusindia.gov.in/2011census/C-series/C-14.html on 17 March 2016 193

Figure 12.1 SOCSIM simulation results and the results by the census: population age structure. *Data source:* calculations by aggregated data of previous censuses and SOCSIM micro-simulation data 226

Figure 12.2 Proportion of Chinese elderly living free of disability in 2005 and 2010 (male versus female). *Data source:* 1 % national population sample survey in 2005 and the sixth national population census in 2010 (similarly hereinafter) 227

Figure 12.3 Proportion of Chinese elderly living free of disability in 2005 and 2010 (urban versus rural) 228
Figure 12.4 Changes of life expectancy of Chinese elderly between 2005 and 2010 (male versus female) 229
Figure 12.5 Changes of disability-free life expectancy of Chinese elderly between 2005 and 2010 (male versus female) 230
Figure 12.6 Changes of life expectancy of Chinese elderly between 2005 and 2010 (urban versus rural) 231
Figure 12.7 Changes of disability-free life expectancy of Chinese elderly between 2005 and 2010 (urban versus rural). 232
Figure 12.8 Changes of the proportion of DFLE in LE of Chinese elderly between 2005 and 2010 (male versus female) 233
Figure 12.9 Changes of the proportion of DFLE in LE of Chinese elderly between 2005 and 2010 (urban versus rural). 233
Figure 12.10 The increasing speed of LE and DFLE of Chinese elderly between 2005 and 2010 234
Figure 12.11 The increasing speed of LE and DFLE of Chinese elderly between 2005 and 2010 235
Figure 12.12 The average number of surviving children and sons of Chinese elderly at age X. *Data source:* calculations by SOCSIM micro-simulation data 236
Figure 13.1 A stylized model of elderly health care and household utility 252
Figure 13.2 Concentration curve for elderly inpatient care, India 2004 257
Figure 13.3 Marginal effects for using borrowings by intersectional groups (*Note* the marginal effects are based on econometric results reported in Table 12.5. The groups are formed by intersecting three prominent determinants viz. household income (poorer and richer households), gender (male and female) and relation with household head (self head or others). The intersection of these determinants yields eight mutually exclusive subgroups, namely: (1) poorer-female-head (PFH); (2) poorer-female-other (PFO); (3) richer-female-head (RFH); (4) richer-female-others (RFO); (5) poorer-male-head (PMH); (6) poorer-male-others (PMO); (7) richer-male-head (RMH); (8) richer-male-others (RMO) 267

List of Tables

Table 2.1 Old age dependency ratio (# aged 65+ per 100 aged 15–64): 2000–2050 36
Table 5.1 Socio-demographic profile of the interviewed elderly men and women 93
Table 8.1 Selected descriptive statistics of older adults by levels of subjective wellbeing index (% reported). . . . 151
Table 8.2 SUBI by household size of the older adults 154
Table 8.3 SUBI by older adults living with grandchildren 154
Table 9.1 Age distribution of the study sample across time (1975–2003). . . . 165
Table 9.2 Descriptive characteristics of the study sample (n = 490) 166
Table 9.3 Ordinary least square regression main effects for occurrence of critical life course transitions on depressive symptomatology 167
Table 9.4 Ordinary least square regression main effects for timing and developmental life stage of critical life course transitions on depressive symptomatology 168
Table 9.5 Interaction effects for off-time critical life transitions and total number of life event stressors (TLE) on depressive symptomatology 169
Table 9.6 Interaction effects of critical life transition and type of recent life event stressor on depressive symptomatology 170
Table 9.7 Interaction effects of critical life transition and type of recent life event stressor on depressive symptomatology 171
Table 10.1 Percentage of people in good health across socio-economic correlates—by gender, place of residence and time 188
Table 10.2 Logistic regression results for reported health status—by place of residence 190
Table 10.3 Logit regression results for females of good health in different zones in India 192
Table 10.4 Inconsistency between actual and perceived health status 194

Table 10.5 Gender differences in consistency of responses by place of residence . . . 195
Table 10.6 Results of logit models of response bias on female dummy . . . 196
Table 10.7 Sample profile of elderly population in rural areas . . . 198
Table 10.8 Demographic profile of elderly population in urban areas. . . . 199
Table 10.9 Results of tests for gender differences in sample proportions for inconsistent responses . . . 200
Table 11.1 Sample description including differentiation according to gender . . . 208
Table 11.2 Distribution of spatial variables including differentiation according to gender. . . . 210
Table 11.3 Wellbeing indicators including differentiation according to gender . . . 210
Table 11.4 Spearman correlations between reported life satisfaction and spatial activity . . . 211
Table 11.5 Correlations between perceived health and spatial activity . . . 213
Table 11.6 Correlations between perceived physical functioning and spatial activity . . . 213
Table 11.7 Correlations between positive affect and spatial activity . . . 214
Table 11.8 Correlations between negative affect and spatial activity . . . 214
Table 12.1 Disability-free life expectancy of the Chinese elderly in 2005 (male) . . . 239
Table 12.2 Disability-free life expectancy of the Chinese elderly in 2005 (female). . . . 239
Table 12.3 Disability-free life expectancy of the Chinese elderly in 2010 (male) . . . 240
Table 12.4 Disability-free life expectancy of the Chinese elderly in 2010 (female). . . . 240
Table 12.5 Disability-free life expectancy of the Chinese elderly in 2005 (urban) . . . 240
Table 12.6 Disability-free life expectancy of the Chinese elderly in 2005 (rural) . . . 241
Table 12.7 Disability-free life expectancy of the Chinese elderly in 2010 (urban) . . . 241
Table 12.8 Disability-free life expectancy of the Chinese elderly in 2010 (rural) . . . 241
Table 13.1 Percentage elderly reporting inpatient care utilization, India 2004 . . . 256
Table 13.2 Probit selection model results for elderly hospitalization: selection equation . . . 257

Table 13.3 Probit selection model results for private sector hospitalization and hospitalization expenditure for elderly: marginal effects from outcome equation 260
Table 13.4 Sources of financing used by households for elderly inpatient care, India 2004. 261
Table 13.5 Probit selection model results for sources of financing for elderly hospitalization: marginal effects from outcome equations 263

Part I
Frameworks and Integrative Approaches

Chapter 1
Bridging the Gap: Theory and Research in Social Gerontology

Tannistha Samanta

Abstract Social gerontology's growth as a distinctive discipline has remained contested with continuing debates on gerontology's limited research focus, disciplinary boundaries and inadequate theoretical development subsequently constraining the cumulative knowledge building of the discipline. Given this intellectual background, in this introduction, I assert the goal of this volume: This volume, based on original research drawing from different disciplines and theoretical orientations, acknowledges the pursuit of a common gerontological imagination by foregrounding social gerontology as an integrative discipline. A related effort has been to emphasize the importance of cultural diversity and plasticity in understanding the social process of age and aging. This introduction maps how scholars in the volume have examined the sociological question of inequality and its intersection with age, gender, health, family and social relations. In the process, the studies in this volume have highlighted the unique historical, institutional and social systems that govern the subjective experience of aging in diverse contexts globally. Specifically, societies in transition including India, Lebanon, Nigeria, Japan, China, Israel and the Europe are studied while connecting the micro social experience of aging (loneliness, wellbeing, discrimination, relationships and resilience) with the larger temporal and political contexts. In this introductory chapter, I show ways how this exercise has generated an intellectual capital that reformulates linkages between aging research and policy in innovative ways.

Keywords Social gerontology · Theoretical development · Culture · Contexts · Data · Global south

T. Samanta (✉)
Humanities and Social Sciences, Indian Institute of Technology, Gandhinagar, Palaj, Gujarat, India
e-mail: tannistha@iitgn.ac.in

T. Samanta (ed.), *Cross-Cultural and Cross-Disciplinary Perspectives in Social Gerontology*, DOI 10.1007/978-981-10-1654-7_1

1.1 Theory and the Route to Emancipatory Knowledge

> He who loves practice without theory is like the sailor who boards ship without a rudder and compass and never knows where he may cast.
>
> The supreme misfortune is when theory is unconnected to performance
>
> —Leonard da Vinci

The above lines capture the rationale and intent of this volume. It has been widely acknowledged that published works on social gerontology have often neglected the intellectually fertile link between theory and empirical research (Bengtson et al. 1997; Hendricks et al. 2010; Alley et al. 2010). Additionally, gerontology's growth as a distinctive discipline has been contested. In particular, scholars have debated the legitimacy of gerontology as a discipline and pointed to its limited research focus, disciplinary boundaries, accreditation issues in gerontology training and weak theoretical development (Achenbaum 1987; Bass and Ferraro 2000; Alkema & Alley 2006; Medeiros 2014). Given this intellectual background, the goal of this volume is straightforward: the chapters, drawing from different disciplines and theoretical orientations, blend theory and conceptual models with empirical clarity thereby acknowledging a common gerontological imagination (Bass and Ferraro 2000) and foregrounding social gerontology as an integrative discipline (Alkema & Alley 2006). A related effort has been to emphasize the importance of cultural diversity and plasticity in understanding the social process of age and aging. The authors in this volume have not only adopted interdisciplinary lens, but have also highlighted the unique historical, institutional and social structures that govern the subjective experience of aging in diverse contexts including India, Japan, Lebanon, Israel, China and Nigeria.

Studies on social gerontology and sociology of aging in the global south are not scarce. Scholars have made important contributions to the overall gerontological literature through their empirical insights on family structure, income, retirement, caregiving, intergenerational relationships, health and social relations of older adults (Hashimoto 1991; UN DESA 2005; Yount 2009; Wong and Palloni 2009; Lee and Mason 2012; Chen and Liu 2012; Samanta et al. 2015). Comparative studies are also gaining currency especially surrounding the questions of immigration, transnational identity and diaspora (Lan 2002; Dossa 1999; Cole and Durham 2007; Lamb 2009, 2010). Large-scale longitudinal surveys with internationally harmonized survey instruments have been conducted in the global south (e.g. China, India, Korea, Ghana and Mexico) to map the unprecedented rise in the older adult population in these countries and understand the challenges associated with this demographic shift (see Hermalin 2002; Kowal et al. 2012; *Global Aging Surveys* by the National Institute of Aging-RAND Corporation, United States). While a large body of empirical observations has been amassed, critics argue that inadequate attention to theory-driven research may lead to empirical generalizations with unsatisfactory explanations, subsequently constraining the cumulative knowledge building of the discipline (Bengtson et al. 1997; Hendricks et al. 2010).

Significantly, these concerns have been shared by leading research agencies globally (e.g. National Institute of Aging (NIA) and the National Science Foundation (NSF), international journals and the much cited handbooks on aging.[1]

Nonetheless, peer reviewed papers that have emerged from the multi-country, longitudinal datasets (referred earlier) have often remained atheoretical. A quick review of the of websites hosting these publicly available datasets reveals a large body of intellectual work devoted to the empirical questions of poverty, economic security, living arrangements, disease burdens, health care and disability[2] with no or very little acknowledgement to theory and/or conceptual models. While there is no denying that these studies profoundly improve our understanding of the economic and health conditions of older adults in the global south, but without a carefully designed theoretical framework it is impossible to (1) *explain* the *why* question under investigation and (2) *assess* if the outcome will have desired impacts on the lives of older persons (Alley et al. 2010). Since one of the central goals of gerontology from its inception has been to improve the lives of older persons (Achenbaum 1995; Putney et al. 2005), an atheoretical scholarship has limited ability to resolve the causal questions and bring transformative changes in the experience of aging. Therefore, the assertion of this volume is that theory and research are bound together; theorizing without empirical grounding is as unproductive and mute as sophisticated empirical analysis without theoretical propositions. However, this is not to suggest that applied research without a theoretical framework is not valuable. Hendricks and colleagues (2010) have reminded us of several scientific approaches such as randomized experiments and community based participatory research where an explicit theoretical framework might not be necessary. Nonetheless, where possibilities to harness the link between theory and applied research exist, maximizing such links have shown to enhance the likelihood of significant scientific discovery (Bengtson et al. 1997; Bengtson 2006).

This volume intends to re-establish gerontology as a discipline that has pragmatic links to policy and practice. I contend that collectively, the chapters will enrich public debates about the moral, cultural and economic questions surrounding aging thereby ameliorating the "problems" associated with aging societies. The chapters in this edited volume recognize the overarching sociological question of inequality (by age), which is crucial to the intellectual advancement of the field. Critical theorists have long argued that knowledge should be "emancipatory"

---

[1]See, for example, a peer report publication of the NSF, *The Role Of Theory In Advancing twenty-first century Biology: Catalyzing Transformative Research*, National Research Council 2008), leading journals in social gerontology and sociology of aging (e.g. *Journal of Gerontology: Social Sciences (JGSS), The Gerontologist, Ageing and Society, American Sociological Review (ASR)* and the *International Journal of Ageing and Later Life*) through their intentional inclusiveness for theory-heavy, humanistic and interpretative research methodologies (Bengtson 2006; Hendricks et al. 2010; Kivnick and Pruchno 2011) and with the publication of the second edition of the *Handbook of Theories of Aging* (Bengtson et al. 2009). In fact, in the *Handbook*, there is an explicit emphasis on integrating theoretical perspectives within and across disciplines through a featured section on "Theorizing aging across disciplines".

[2]See, for example, www.who.int/healthinfo/sage/articles_indepth.

(Habermas 1968). The current book has attempted to share this critical agenda by adopting a multi-paradigm approach (positivistic, interpretive and critical) to foster reflexive thinking and sharpen our analytical gaze. Finally, this volume reminds us that the experience of aging is not universal, rather social reality in constituted by meanings of everyday life and hence there is a need to integrate the familiar territory of macro data with micro level studies. This exercise in meaning-making will hopefully broaden the agenda of social gerontology in the global south.

1.2 The Social Gerontological Discourse

1.2.1 Epistemology

Over the years, gerontology has developed considerable intellectual capital as reflected through a growing number of funded large scale cross-national surveys, a rise in age-related scholarly journals and a steady proliferation of graduate education programs across the globe. However, the "data rich, theory poor" (Birren and Bengtson 1988) symptom permeates scholarship and policy. Recently, reflecting on the state of gerontology, Bass (2013) re-emphasizes the need to respond to the repeated demands of theory building and transcend discipline-specific guilds to successfully strengthen the discipline. Before reviewing the dominant theories of aging that have guided much of empirical research, I will first summarize some epistemological and methodological considerations that have shaped the social scientific enterprise of gerontology.

Theory and theorizing have been central to the evolution of "normal science" (Kuhn 1970). To cumulatively create knowledge, researchers engage themselves in *epistemology*, the analysis of origin, nature and limits of knowledge. One component of epistemology focuses on *methods*, the means by which empirical investigations are conducted to examine (or understand) phenomena in a way that is reliable and valid across the observations. Another important component of epistemology is *theory*. At its simplest, a theory is an attempt to explain *why* a phenomena occur (Alley et al. 2010). In Bengtson's (1997, 2006) sense, theories are "lenses" and theorizing "as putting together a puzzle" where "there is nothing so practical as a good theory" to advance cumulative knowledge development. Hence, an ad hoc, largely descriptive, model-based (rather than theory-based) approach to research is ineffectual overtime (Bengtson et al. 1997). Later, among others, Putney and Bengtson (2008) lend their voice to this debate and pointedly put it: theory-driven research helps to systematize what is known and explain the *how* and *why* behind the *what* of our data (p. 413). In his landmark *Structure of Scientific Revolutions*, Kuhn (1970) asserted that theory is a disciplinary matrix and an exemplar and that overtime revolutionary restructuring will render certain disciplines unable to adequately explain key questions using the existing paradigms. Kuhn called for "paradigm shift" where "one conceptual world view is replaced by another". Based on the current intellectual debates concerning the evolution of gerontology (Bass and Ferarro 2000; Alkema & Alley 2006; Medeiros 2014), it seems gerontology is uniquely poised to build conceptual linkages across biological, psychological and social processes of aging and emerge as an integrated, multi-paradigm discipline.

1.2.2 Challenges

Given its contested evolutionary roots as a discipline, it is no surprise that gerontology faces challenges in its intellectual development. Discipline specific methodological traditions (e.g. positivistic and interpretive) have contributed to a lack of unanimity in knowledge building. Gerontological research has been dominated by the *positivistic* paradigm where empirical studies drawing from population records and large surveys are common. The process of empirical examination involves observation and description of data, hypothesis testing and prediction. By contrast, *interpretive* researchers emphasize understanding the meaning and (subjective) experience of aging and are less involved with prediction and control. As articulated by Glaser and Strauss (1967), interpretive research begins with little or minimum a priori assumptions concerning the relationship/phenomena in question and meanings and interpretations emerge as the study progresses. Hence, interpretive researchers (those who focus on critical theory and social constructionist approaches) argue that positivistic paradigm is inherently value-laden which obscures understanding of non-anticipated empirical observations (Bengtson et al. 1997; Putney et al. 2005). On the other hand, since gerontology is applied in its scope and famed as a "problem-solving" discipline (Morrow-Howell et al. 2001), empirical researchers are less likely to concede the need for theory or a conceptual framework as their "investigations are problem-driven or bounded by pragmatic dimensions of policy and practice/monitoring/decision making in the real world and in real time" (Hendricks et al. 2010: p. 287). However, Hendricks and colleagues persuasively demonstrate the limits of this problem solving but atheoretical research. Borrowing Kuhn's classic work (1970), they argue that in absence of a theory, "all facts that could possibly pertain to the development of a given science are likely to seem equally relevant" (Kuhn 1970: p. 15, cited in Hendricks et al.). In other words, theoretical frameworks permits explanations to move *beyond* mere empirical generalizations and permits researchers to discover "why" events unfold as they do (Kuhn 1970: p. 15, cited in Hendricks et al.). Hence, it is important to note that notwithstanding these methodological chasms, the emphasis on theory is increasingly acknowledged by both positivistic and interpretive researchers in the study of aging (Bengtson et al. 1997). This volume builds on this acknowledgement and includes a combination of papers that adopt either or both the approaches.

1.2.2.1 Theories, Conceptual Models and Middle Range Theories

As noted earlier, the limited efforts in linking theory to applied research has been a long standing concern in social gerontology. In 1997, Bengtson and colleagues reviewed articles published between 1990 and 1994 in eight leading journals in social gerontology and sociology of aging and concluded that majority of articles (around 72 %) made no mention of any theoretical tradition as relevant to interpreting or understanding their findings. They summarized the dominant theoretical perspectives in the field and argued that lack of theory-based research is not due to paucity of relevant explanations within gerontology but a collective failure in the academic gerontological community to recognize the utility of theory as explanatory frameworks. Later, in 2010, two sets of studies—Hendricks and colleagues (2010) and Alley and colleagues (2010)—reviewed theory use in the 2000s. Both the studies reported a slight increase in theory use in gerontology especially the social constructionist, social exchange, life course, stratification and social exchange perspectives being the most frequently referenced. Specifically, Alley et al. (2010) calculated an overall 12 % increase in the use of theory over 10 years. Key drivers for this theoretical development include increased cross-fertilization between disciplines and changes in editorial policies in academia.

Nevertheless, critics contend that theory-driven research is still lacking in social gerontology, and while a shift towards theories and cross-disciplinary work is noted, in many cases conceptual models are often used as a supplement to or substitute for theory. Put simply, a model is a reproduction of the world, whereas a theory represents an attempt to explain the world (Marshall 1999, cited in Alley et al. 2010). Alley and colleagues (2010) reported that researchers in gerontology are using models (as opposed to conventional social science theories) as systems of explanation and contended that overtime such models may ascend to gain dominance in the field and emerge as important theoretical contributions to social gerontology. Echoing Merton's (1968) notion of "middle-range theories", Hendricks and colleagues (2010) asserted that these models may serve as a bridge between observable social problems and middle-range conceptualization. Paraphrasing Merton (1968), Hendricks et al. (2010) noted that middle range theories are a midpoint between raw empiricism and all-inclusive grand theory. The authors in this book build on this body of work and share the common goal that a well-grounded theory and a well-theorized applied model hold promise of cumulative knowledge development. Furthermore, part of the criticism of the current gerontological enterprise is that researchers incorporate micro-level analyses without connecting it with the larger political, cultural and social conditions that influence individual and group behaviours (Bass 2006). In other words, researchers are limited in their efforts to "think themselves away" (Mills 1959) and cultivate a shared gerontological imagination. Hagestad and Dannefer (2001) refer to this phenomenon as the "microfication of gerontology" where individual centred analysis prevails often ignoring the larger social context that influences these outcomes. The authors in this book are reflective of the socio-cultural context of their work while sharing their allegiance to theory-driven evidence based research.

Finally, it should be noted that the authors of this book do not necessarily *build* theory, but rather share the common recognition of the need for theory. To advance this overarching goal of the book, the authors have blended theories and conceptual models and carefully bridged disciplinary divides.

1.2.3 (Inter)Disciplinary Perspectives, Multiple Lenses

Interdisciplinary research has become the holy grail of scientific establishment (Bass 2009). Interdisciplinary inquiry refers to the contribution of two or more disciplines where the disciplinary boundaries are muted and the plurality of contributions creates a synergistic nature of original insight (Bass and Ferraro 2000; cited in Bass 2009). Critics have asserted that the social gerontological scientific establishment is at an interesting intellectual cusp. In particular, while acknowledging discipline-specific intellectual traditions, approaches and nomenclature, social gerontology scholars are slowly abandoning concerns over intellectual territory and integrating ideas and arguments from several disciplines including life sciences, medicine and the social sciences. This is evident in the works appearing in edited volumes (see Baars et al. (2006), *Aging, Globalization and Inequality*; Dannefer and Philipson (2010). *The SAGE Handbook of Social Gerontology*), journals, multi-disciplinary professional gerontological organizations and conference sessions. Since lives are experienced through a combination of "memories of a past and the anticipation of the future" (Fry 2009), the *life course perspective* (Elder 1974; Elder et al. 2003) emerge as a dominant lens through which age and aging can be examined. This is also an important perspective as it bridges both the micro and macro-social levels of analysis. A related concept, building on Elder and colleagues' pioneering conceptual life course based paradigm, is the *cumulative advantage–disadvantage theory* (Dannefer 2003). These perspectives have fostered interdisciplinary dialogues with their application and usefulness in explaining biological/psychological concepts such as stress (Finch and Seeman 1999), allostatic load (McEwen and Stellar 1993) and cognitive reserve (Stern 2002). Further, given its dynamic scope, life course based perspectives have also been utilized to explain how time, historical period and cohort shape the aging process for individuals as well as social groups (Elder and Johnson 2002). Longitudinal data and breakthroughs in specialized techniques (e.g. event history and hierarchical linear modelling) have further facilitated empirical investigations across the life cycle. It may be noted that although, the "theory" status of the life course based perspectives stand contested (Bengtson and Allen 1993), critics have suggested that in a field that is known for its paucity of a grand theory, the increased use of these perspectives and conceptual frameworks in the last decade suggest a new phase in social gerontological scholarship (Alley et al. 2010).

Critical perspectives in social gerontology have emerged as a popular micro-level perspective for understanding the social meanings and self-conceptions of age and aging across cultures. Critical gerontologists adopt phenomenological approaches (e.g. social constructivism and ethnomethodology) that have theoretical roots with early theorists including Theodor Adorno, Jurgen Habermas and Karl Marx. Phenomenology involves a systematic investigation of consciousness that is composed of thinking, imagining, constructing, perceiving and remembering (Schultz 1932; Longino and Powell 2009). Rather than contending that any aspect is a causal factor (as emphasized by positivists), "phenomenology views all dimensions as constitutive of others" (for a detailed discussion on the lineage of phenomenology and its application in gerontology, see Longino and Powell 2009). Theoretical perspectives within the phenomenological framework have interdisciplinary intellectual foundations including the political economy, feminist theory and the postmodern theory. Within gerontology, the social constructionist perspective has been employed to *understand* (as opposed to *explain*) the subjective experience of identity, sexuality, frailty and loneliness and often provides a stinging critique of the hegemonic presence of an overtly medicalized discourse of aging in the popular culture (Frank 1990; Powell and Wahidin 2007).

Clearly, these theoretical perspectives drawing from varied intellectual traditions "share a common orientation towards aging as continuity and change overtime within a physical, behavioural and social context, making gerontology a study of trajectories and transitions" (Alkema & Alley 2006: p. 577). Drawing from different disciplines and adopting multiple lenses, chapters in this book leverage on this intellectual tipping point of cross-fertilization, with a shared hope that this will pave the way to a more fertile ground for expansion of an integrative scientific enterprise.

1.2.4 Aging in Context

Understanding how people grow old in varied cultural contexts is a critical and a conceptual endeavour among social scientists. The ever growing and diverse older populations across the world require researchers to understand the differences in the cultural context of aging. Culture, as Goodenough (1999) asserts is a set of shared symbols, beliefs and customs that shape individual and group behaviour. In particular, anthropologists and researchers adopting phenomenological approaches remind us of the dangers of essentialism and the need to understand multiple modernities, identities, histories and memories that are associated with aging and personhood (Danely and Lynch 2013; Fry 2009). In fact, research has demonstrated that considerable disagreement exists about what constitutes "good old age" and "successful aging" (Rowe and Kahn 1997) across varying cultural contexts (Fry et al. 1997; Lamb 2010, 2014). For example, Christine Fry and colleagues (1997) in their cross-cultural (included study sites from Africa, United States, Hong Kong and Ireland) research project examining the meaning and culture of "good old age" found no monolithic, universal interpretation of "goodness"; old age was experienced differently in different cultural contexts with disparate levels of importance attached to good health, social security frameworks, family support and kinship, independence and social connectedness. Similarly, in a poignant piece based on her ethnographic work in Bengal (India) and the diasporic community in North America, Lamb (2010, 2014) problematizes the (western) notion of "successful aging". She contends that the successful aging movement with its emphasis on "independence, productivity, self-maintenance and the individual self as a project" is not only a biopolitical and ageist model but also has limited ability in explaining the subjective experience of aging in contexts where human transience is accepted and a "meaningful decline" is constructed as an idealized form of aging.

Dilworth-Anderson and Cohen (2009) put it pointedly: when we ask these type of questions, particularly about their cultural metaphors and values and their historical and socio-political realities, we are helping to create a "firewall" of culturally competent boundaries of our theoretical thinking (p. 490). The chapters under the section on "Aging and Context" draw on these cultural metaphors, meanings and values thereby retooling and informing the links between theory and methods.

1.3 Bridging the Gap: This Volume

This volume is uniquely cross-cultural, theory-driven and cross-disciplinary. It fills a gap in the gerontological scholarship of the global south that is predominantly descriptive and empirical. In fact, this volume echoes Sokolovsky's (2009: *The Cultural Context of Aging*) assertion of the importance of cross-cultural research on aging. Sokolovsky argues how inter cultural analysis helps us understand the policy response to aging in varying contexts at varying levels of the demographic transition. As noted earlier, this contention coincides with the shift that aging studies have encountered in late modernity. That is, a shift away from structure "toward agency in which themes of identity, reflexivity and individuation are prominent" (Giddens 1991, cited in Twigg and Martin 2015). As discussed in the next few sections, the varied temporal and political contexts of India, China, Lebanon, Israel, Nigeria, Japan and Europe underscore how cultural change and continuity are preserved, destabilized or ruptured. The attempt, however is not in drawing comparisons and commonalities across these contexts, but rather in highlighting how the experience of "being in society" (Rojek and Turner 2000) increasingly shapes subjectivity, identity and social reality of older adults. In this way, this volume advances gerontology's agenda by embodying the two impulses that seem to influence aging studies in the past decades—the epistemological and the historico-social (Twigg and Martin 2015). In sum, a global perspective achieved through cross cultural scholarship creates an interactive laboratory where innovations are absorbed, transformed and sometimes improved as they pass into different contextual settings (Weaver 2008).

The chapters in this volume are divided into three sections. (I) *Frameworks and Integrative Approaches* (this Introduction and Chaps. 2–3), (II) *Culture, Contexts and Aging* (Chaps. 4–8) and (III) *Health and Wellbeing* (Chaps. 9–13). Though each of the chapters is unique in its inquiry, cultural context and methodological approach, the intellectual binding knot lies in the shared recognition of theory and theorizing as powerful analytical tools to evidence based research.

Frameworks and Integrative Approaches

Ghosh and colleagues (Chap. 2) expands on this Introduction and takes the intellectual agenda of advancing social gerontology's mission of establishing itself as a cross-disciplinary and culturally relevant discipline. In particular, this chapter addresses the question whether contemporary theories of caregiving, majority of which evolved in the West, can explain socio-cultural realities surrounding caregiving in non-western settings, primarily India. Drawing from the gerontological scholarship on India that highlights the importance of multigenerational living as a site for aging, security and wellbeing, Ghosh and colleagues contrast this with the western model of aging where co-residence with adult children is rare. Reminding us of the atheoretical nature of gerontology when it comes to explaining the social context of care, this chapter critically reviews dominant theoretical frameworks and models including Pearlins' stress system and coping models, Kahn & Antonucci's (1980) convoy model, role theory (Marks 1977), the life course perspective and feminist economic theories of care. Ghosh and colleagues raise critical questions that not only push the boundaries but also reveal the cultural limits of these existing theories. For example, empirical literature on caregiving is inundated with the notion of a "dependent" elder who is in need of caregiving; this chapter argues that in settings where filial piety is common and the intergenerational social contract is carefully maintained, older adults might consider themselves as legitimate "recipients" of care from younger family members as an unintended consequence of a long cultural tradition of intergenerational reciprocity. Addressing such conceptual gaps, Ghosh and colleagues conclude that contemporary (western) theories, while important, are inadequate to understand the generational complexity and the shifting cultural context of caregiving in the global south. They stress on the importance of developing more comprehensive and culturally nuanced theories that pay attention not only to the recipients of care but also to the caregiver.

Echoing the overarching assertion of this Introduction, Marques and colleagues (Chap. 3) demonstrate how social psychological theories substantively shape gerontological questions on age and aging and subsequently examine the effect of age discrimination on older adults' wellbeing in Europe. A related goal of this chapter is to bridge the gap between research and practice in terms of policy interventions to reduce ageism and age-based negative stereotyping. Significantly, this chapter discusses the relevance of a social psychological approach in examining gerontological questions of ageism as social interactions and experiences operate both at micro and meso levels. For example, borrowing the concept of methodological collectivism, Sibila and colleagues demonstrate that when a society as a whole formulates a collective reality of prejudices and stereotypes, it has impacts on individual members' (here, older adults) self-concept and identification. In sum, the authors argue that "both objective macro-environments as well as cultural and collective phenomena relate to human mind and behaviour". The study draws data from a unique module-*Experiences and Expressions of Ageism in Europe* (2008–2009)—of the European Social Survey. The authors' empirical investigations reveal a disturbing social reality: identifying oneself as part of a devalued social group (here, old age) has negative effects on self-concept and subjective health. As such, this finding holds even after country-level differences on wealth and income are controlled for. The strength of this study lies not only in bonding the fertile links between social psychology and gerontology but also in expanding the gerontological imagination, as noted earlier in this Introduction. The chapter concludes with a helpful discussion on the alternative policy interventions and initiatives that are either already undertaken or are recommended. In a way, this chapter accomplishes the coveted goal of turning a scientific gerontological inquiry into action.

Culture, Contexts and Aging

As noted earlier, the chapters in this section examine the process of aging in varied cultural systems: Japan (Chaps. 4 and 7), Nigeria (Chap. 5) and India (Chap. 6). One of the promises of cross-cultural gerontology is "to gain an understanding of aging, divorced from the narrow boundary of a single case" (Sokolovsky 2009: p. 4). This goal fits well within the overarching assertion of this book: a demonstrated lack of phenomenological lens to the study of aging in the global south. Each chapter shows how social, cultural and historical narratives construct the subjective and everyday experience of aging and its associated meanings, expectations, aspirations, challenges and struggles.

Drawing from extensive ethnographic fieldwork in Japan, Danely (Chap. 4) and Ricart (Chap. 7) focus on the continuities and discontinuities in the experience of aging. In the process, both reflect and theorize on the anthropological thinking on aging and life-course and effectively illustrate convergence and complementarity with other disciplines and scientific schools of thought.

Danely (Chap. 4) builds on the anthropological scholarship on subjectivity (often defined as "the emotional world of the political subject") and demonstrates how subjectivity helps in understanding the new politics of old age in Japan and how that affects notions of loss, dependence, among older Japanese. Danely notes, "Exploring subjectivities of older adults in this way, could help us understand what kinds of lives older Japanese adults felt possible for themselves, and what kinds of futures become disrupted, blocked, or foreclosed due to their cultural and political circumstances". Not surprisingly, both Danely and Ricart (Chap. 7 in this volume) engage in the anthropological quest of understanding the *future* of an aging society in crisis. In Danely's sense, it is older adults managing their "future moral selves" while Ricart directly theorizes future as an ethnographic object. In the process, both reflect on the cultural definitions of "successful" aging and examine alternative models of care in the Japanese society. Danely's focus, however, is on understanding the cultural model of the Japanese self. Drawing from his ethnographic fieldwork on aging and grief in Kyoto, Danely shows how the Japanese older adults are constructing alternative narratives of the self-one that is soaked in cultural beliefs and practices of caring for the spirits of the departed ancestors. This, Danely, argues is a unique cultural form of "dependence" while sharply rejecting the more traditional form of dependence on adult children and family members. A remark from one of Danely's study participants poignantly exemplifies this rejection of affective bonds and the burden anxiety among Japanese older adults. Danely notes: "In his (a 70-year-old widower) words, he wanted to "die alone" without a funeral, since that would be costly and an unnecessary burden".

Ricart (Chap. 7) theorizes future as an ethnographic object by adopting the theory of ontogenesis—a study, she explains, of becoming and ordering. Her focus on Japan as a critical site for this ethnographic inquiry is significant. She asserts that as the aging society crisis of Japan permeates the scientific discourse and policy anxiety, understanding "futurity as a force of change" is crucial in shedding light on the emerging ontologies of the self, the life world and the formation of institutions of care. She is critical of the binary discourse of care (*kaigo*) and care-prevention (*kaigo yobo*) in the gerontological discussion of the Japanese aging society crisis. She demonstrates that "care and care-prevention are discrete, occupying different configurations of institutions, knowledge, and practice. However, they share an ontology of the aging process, and, temporally, both are oriented around and motivated by the future aging society crisis". Further, Ricart discusses the conceptual and thematic overlaps of ontogenesis with the cybernetic theory (Simondon 1980) to analyse the social imaginaries of the future. Consistent with her intellectual agenda of theorizing the future, she argues that a cybernetic approach empowers the researcher to study the future both as a force of change but yet transformable: "The aging society crisis, then, recursively feeds back into itself, inciting in its parts that in turn can bring about change in the whole".

Later, in her discussion on continuities, healthy aging, mobility and decline, Ricart draws from her interviews with Japanese gerontologists, care-prevention staff, local and national government employees. We learn that the Japanese emphasize on "active aging" marked by mobility, social activity and vitality to prevent the onset of age-related physical, mental decline and enrolment in the Long Term Care Insurance (LTCI) system. Her study participants referred to the most desired and idealized form of death as-*Pin Pin Kori*-which implies "lively, active and a sudden" death instead of a slow and gradual decline. Culturally speaking, this idea closely resembles the self-reliant "western" way of "successful" aging and stands in contrast to an idea of "meaningful decline" marked by disengagement and social withdrawal, often aspired by older Indians (Samanta and Gangopadhyay, in Chap. 6 discuss this phenomena in their cultural critique of the link between aging and social capital). This comparison illustrates cultural heterogeneity and challenges the assumption of a singular understanding of piety and intergenerational reciprocity, often associated with Asian societies Finally, Ricart's chapter successfully demonstrates the potential of ethnographic research to speak to policy action.

Agunbiade and Akinyemi (Chap. 5) take us to Nigeria where we learn about the challenges faced by older adults in two post-conflict Yoruba communities of Nigeria against the backdrop of a shifting cultural and political regime. Adopting both a micro-level interactionist approach and a macro-level political economy approaches, Agunbiade and Akinyemi examine how neo-liberal social policies have eroded the economic and social safety net of older adults in Nigeria. The chapter asserts the importance of resilience as a mitigating and empowering tool that is often adopted by older adults to navigate through the dismal politics of poverty, neglect and abuse in a post-conflict regime. Consistent with other authors in this section, Agunbiade and Akinyemi's qualitative work is a poignant call for the need of recognizing a cultural model in guiding economic policies.

Using a social constructionist perspective, Samanta and Gangopadhyay (Chap. 6) provide a critique of the mainstream understanding of social capital, often considered as an unambiguous boon in the health literature. The authors draw from their in depth interviews of older adults residing in elder care homes (commonly known as "old age homes") in Ahmedabad, a prosperous city in the western Indian state of Gujarat. Contrary to the much celebrated bonding, bridging and empowering potential of social capital, Samanta and Gangopadhyay demonstrate how moral-cultural scripts rupture the motivation among older Indians in these homes to actively unlock the benefits of social capital, despite intimate living spaces, shared histories and similar life circumstances. As with other authors, while reflecting on the notion of "successful aging", this chapter reminds us of the varied cultural meanings of decline, death and transience. By focusing on the "dark side" of social capital, this chapter provides a critical rethinking on the policies that aim at providing alternative non-familial support to the elderly. Notably, all chapters in this section reflect both on the continuity and temporality of the life course and examine how older adults in different cultural systems are making sense of shifting cultural scripts and social transformations.

Health and Wellbeing

Overall, inequality has been a major theoretical and conceptual focus in the social science scholarship on health and wellbeing. Husain and Ghosh (Chap. 10), Guo (Chap. 12) and Joe and colleagues (Chap. 13) demonstrate a convergence of thought on the substantive significance of inequality. Rajan and colleagues (Chap. 8) provide a descriptive analysis of the determinants of subjective wellbeing among older adults in the southern state of Kerala, India. Spatial and historical contexts are powerful forces that influence later life outcomes. Using a life course lens, Aydin and colleagues (Chap. 9) demonstrate how historical circumstances and individual choices shape trajectories of wellbeing later in life among community-residing older Lebanese. Building on the activity theory of aging, Isaacson and colleagues (Chap. 11) draw from their GPS based study on Israel and Germany to examine how wellbeing of older adults is spatially governed. Notably, the studies by Isaacson et al. and Aydin et al. are illustrative examples of gerontology's quest for an integrative discipline: transcending disciplinary guilds, blending methodologies and a shared recognition of the importance of multidisciplinary teamwork.

In Chap. 8, Rajan and colleagues adopt a social-psychological lens, particularly building on the convoy (Antonucci 1980) and the cognitivist (Peplau et al. 1982) models to examine the determinants of subjective wellbeing among older persons in

Kerala, India. This examination is noteworthy as the state of Kerala is often discussed as a successful model of human development with significantly higher than national average rates of education, nutrition, life expectancy and several other health indicators. The state of Kerala uniquely receives massive amounts of remittances as a result of a long standing tradition of younger Keralites migrating to the Gulf nations in search of jobs, education and related opportunities. What has been missed in this celebratory development discourse is a systematic investigation of whether such health gains are enjoyed by the left-behind older people in terms of their subjective health (often measured by questions on happiness, life satisfaction and allegiance to cultural expectations). Consistent with the gerontological scholarship elsewhere in Asia, the authors find a complex interplay of age, gender, marital status, cultural expectations and role continuity as determinants of subjective wellbeing among older adults. These findings remind us that psychological health gains are not automatic by-products of development; rather they are culturally and morally governed.

Husain and Ghosh (Chap. 10) revisit the conceptual distinction between perceived and actual health and how it relates to gender. In particular, they argue that the experience of health and aging is not an isolated event, rather a reflection of the existing social and gender hierarchies. The authors remind us that this distinction is crucial in India where gender is an organizing principle of social structures, identity and cultural scripts. In a culture where there is a pervasive social tendency to normalize women's deprivation, expectations about health will be different for men and women. Husain and Ghosh's study is an empirical examination of this "positional objectivity" where women are conditioned to accept their survival disadvantages and poor health status as the "normal" way of living. Drawing data from two rounds of the National Sample Survey Organization (NSSO, India), the authors demonstrate that while gender differences in perceived health might be insignificant, there does exist differences in symptomatic measures of health. The authors conclude raising pertinent questions on the limits of "objective" measures of health, and calls for a construct that is both culturally relevant and gender sensitive. Similar to other contributors in this volume, the importance of employing a cultural lens in the pursuit of empirical questions, remain a germane assertion.

Joe and colleagues' study (Chap. 13) focuses directly on the issue of gender inequality in the context of health care utilization and financing. Theoretically, the study is located within the dominant frameworks of household allocation-unitary, collective and intersectionality where bargaining, information and relational identities determine allocation of resources among household members. Arguably, in a context where household resource allocation and sharing are embedded within a larger social structure of gender based hierarchies, (elderly) women are caught in the interlocking disadvantages of gender, wealth and age. Joe and colleagues' study provide compelling evidence to the fragile nature of financing for older persons in India and the sharp gender difference in health care utilization and financing. The authors assert the importance of a universal health care coverage and community based health care as national policy imperatives to successfully address the ongoing demographic and epidemiological transitions. Extending the conceptual and empirical debate on inequality, Guo and Tu (Chap. 12) provide an empirical estimation of inequalities that are rooted in age, gender and geographical location. Drawing data from China and employing the widely used Sullivan method the authors demonstrate growing disparity in life expectancy (LE) and disability free life expectancy (DFLE) among rural and urban Chinese older adults as well as among older men and women. The authors caution us about the need for community support and revised pension models to address China's impending demographic and social forces-a gradual erosion of intergenerational support systems, a shrinking younger population and increasing disease and disability burdens among older Chinese. It is interesting to note that under shifting social, cultural and economic realities, both the studies from China (Guo and Tu) and India (Joe and colleagues) suggest similar policy initiatives that emphasize on a movement away from the traditional family system to market and community based models as emerging alternatives of support for the elderly. In a sense, this confirms to the larger promise of gerontology as a discipline in action, one that is "in continuous conversation with its publics" (Putney et al. 2005) and persistently addressing the "so what" question to ameliorate the challenges of aging.

As noted earlier, from a life course perspective, time and space construct meanings, memories, histories, struggles and imaginations which actively influence the path one takes as one grows older. This notion is exemplified in Aydin and colleagues (Chap. 9) study on Lebanese older adults who have lived a shared history of the Lebanese civil war which extended for over 16 years (1975–1991). Drawing from a sample of older adults in Lebanon, who were young adults during the civil war, the authors investigate how occurrence, timing and life stage of critical transitions, such as marriage, divorce, work and health affect current psychological wellbeing. The authors find evidence that older Lebanese who experienced health decline during the wartime as adolescents are also the ones who have higher likelihood of depressive symptoms and vulnerability to stressful situations later in the life course. Notably, these findings are consistent with studies conducted on older adults who were confronted with turbulent social, political and economic histories such as the World War II or the Great Depression during their lifetime (see Elder 1974). As with the other authors throughout this book, Aydin and colleagues' study not only bridges the micro-macro social levels, but also unpacks how history, temporal context and the norms of "on-time" and "off-time" events dynamically shape the social experience of aging. Isaacson and colleagues (Chap. 11) both theorize and model space as a determinant of wellbeing among older adults. Specifically, they ask: does the widely used and much cited activity theory (Lemon et al. 1972) have a spatial dimension; if so, what does it look like? Adopting a gero-geographical perspective and employing a non-standard methodological approach (GPS based measures), they focus on *volume* and *quality* of out-of-home mobility among older adults in Israel and Germany. While the study reported association between volume of out of home mobility and life satisfaction among older adults, it also asserted the importance of non-spatial communication technologies in the lives of older adults. Isaacson and colleagues' study reminded us of the importance of pushing the disciplinary horizons to heighten the possibility of an unexpected discovery.

1.4 Moving Forward: On Gerontology's Future

Compared with other scientific disciplines, gerontology is a late starter. But despite its late start, the pace of growth and the breadth of disciplinary interests is impressive (Bass and Ferrrao 2000). Reflecting on gerontology's future success as a distinctive and integrated discipline, critics have recognized the multidisciplinary scope of the field with a generosity to foster synoptic correlation of diverse viewpoints and findings (Hendricks and Achenbaum 1999; Bass 2013). Further, critics have asserted that gerontology is being increasingly "owned" by both social sciences and humanistic disciplines which have substantively enhanced gerontological thinking on specific aspects of caregiving, retirement, health and disability, resilience and friendships (Cole et al. 2010; Medeiros 2014). Needless to say, this integration and cross-fertilization are the strengths of the discipline. However, Gans and colleagues (2009) warn us that it is not enough to understand specific aspects of aging, rather the interconnectedness is the key to scientific progress and cumulative development of the discipline—a shared recognition that the "whole is greater than just the sum of its parts" (p. 728). Notwithstanding these successes, criticisms abound concerning the inadequate theoretical foundations in gerontology. As emphasized earlier, there have been repeated calls for theory building as a route to cumulative development of the discipline. As such, the discipline of gerontology has witnessed an emerging movement toward cross-disciplinary theorizing. Review studies comparing theory-based research in gerontology have concluded increase in some theory use with researchers increasingly adopting conceptual models and paradigms. Gans and colleagues refer this shift as "cross-disciplinarity as a continuum" where at the lower end theorists may still operate from a paradigm that is consistent with a unitary discipline whereas at the higher end there is full interdisciplinary theorizing and collaboration. The chapters in this book exemplify this intellectual phenomenon by leveraging on the advantages of both theory and applied research. While we celebrate this emerging intellectual movement in gerontology, Bass (2013) reminds that cross-fertilization and interdisciplinary collaboration necessitate precision in nomenclature, concepts, approaches and theories. He urges the gerontological community to capitalize on its own strengths but at the same time address its weaknesses—lack of clarity in basic concepts, terms, methods and theoretical richness. To achieve this, a multi-level, synergistic effort is crucial among members of the scientific community, policy, government agencies and gerontology research associations.

This book has brought together theoretical and empirical insights from varied cultural contexts and across disciplines. And though there might be theoretical and methodological contrasts and tensions, I have, in this Introduction, highlighted some of the striking commonalities—theorizing the future, interpreting varying models of "successful" aging, the complex interplay of age, gender and health and finally, examining aging as a lifelong process. In their concluding chapter of the *Handbook of Theories of Aging* (2009), Gans and colleagues underscore the importance of identifying common themes as a tool for theory building, future research initiatives and advancing critical inquiry. I hope, we have collectively addressed that goal and have chartered an intellectual journey of new questions, hopes, explanations and directions.

References

Achenbaum, W. A. (1987). Can gerontology be a science? *Journal of Aging Studies, 1*(1), 3–18.

Achenbaum, W. A. (1995). *Crossing Frontiers: Gerontology emerges as a science*. New York: Columbia University Press.

Alkema, G. E., & Alley, D. E. (2006). Gerontology's future: An integrative model for disciplinary advancement. *The Gerontologist, 46*(5), 574–582.

Alley, D., Putney, N. M., Rice, M., & Bengtson, V. (2010). The increasing use of theory in social gerontology: 1990–2004. *Journal of Gerontology: Social Sciences, 65*(B), 583–590.

Baars, J., Dannefer, D., Philipson, C., & Walker, A. (2006). *Aging, globalization and inequality: The new critical gerontology*. Amityville, NY: Baywood Publishing.

Bass, Scott. (2006). Gerontological theory: The search for the Holy Grail. *The Gerontologist, 46* (1), 139–144.

Bass, S. A. (2009). Toward an intergrative theory of social gerontology. In Bengtson, et al. (Eds.), *Handbook of theories of aging* (2nd ed.) (pp. 347–374). New York: Springer.

Bass, S. (2013). The state of gerontology-Opus one. *The Gerontologist, 53*(4), 534–542.

Bass, S., & Ferraro, K. (2000). Gerontology education in transition: Considering disciplinary and paradigmatic evolution. *The Gerontologist, 40*(1), 97–106.

Bengtson, V. L. (2006). Theorizing and social gerontology. *International Journal of Ageing and Later Life, 1*(1), 5–9.

Bengtson, V. L., & Allen, K. R. (1993). The life course perspective applied to families overtime. In P. G. Boss et al (Eds.), *Sourcebook offamily theories and methods: A contextual approach*. New York: Plenum Press.

Bengtson, V. L., Gans, D., Putney, N. M., & Silverstein, M. (2009). *Handbook of Theories of Ageing*. New York: Springer.

Bengtson, V. L., Burgess, E., & Parrott, T. (1997). Theory, explanation and a third generation of theoretical development in social gerontology. *Journal of Gerontology: Social Sciences, 52B* (2), S72–S88.

Birren, J. E., & Bengtson, V. L. (Eds.). (1988). *Emergent Theories of Aging*. New York: Springer.

Chen, Feinian, & Liu, G. (2012). The health implications of grandparents caring for grandchildren in China. *The Journals of Gerontology, Series B: Psychological and Social Sciences, 67*(1), 99–112.

Cole, J., & Durham, D. (2007). *Generations and globalization: Youth, the and family in the new world economy*. Bloomington, IN: Indiana University Press.

Cole, T., Kastenbaum, R., & Ray, R. (Eds.). (2010). *Handbook of humanities and aging* (2nd ed.). New York: Spriênger.

Danely, J., & Lynch, C. (2013). Transitions and transformations: Paradigms, perspectives and possibilities. In C. Lynch & J. Danely (Eds.), *Transitions and transformations: Cultural perspectives on aging and the life course* (pp. 3–20). New York: Berghahn.

Dannefer, D. (2003). Cumulative advantage/disadvantage and the life course: Cross fertilizing age and social science theory. *Journal of Gerontology: Social Sciences, 58B*, S327–S337.

Dannefer, D., & Philipson, C. (2010). *The SAGE handbook of social gerontology*. SAGE.
Dilworth-Anderson, P., & Cohen, M. D. (2009). Theorizing across cultures. In Bengtson, Gans, Putney, & Silverstein (Eds.), *Handbook of theories of aging* (pp. 487–498). New York: Springer Publishing.
Dossa, A. P. (1999). (Re)imagining aging lives: Ethnographic narratives of Muslim women in diaspora. *Journal of Cross-Cultural Gerontology, 14*(3), 245–272.
Elder, G. H. (1974). *Children of the great depression: Social change in life experience*. Chicago: Chicago University Press.
Elder, G. H., & Johnson, M. K. (2002). The life course and aging: Challenges, lessons and new directions. In R. A. Setterson (Ed.), *Invitation to the life course: Toward new understandings of later life* (pp. 49–81). Amityville, New York: Baywood.
Elder, G. H., Johnson, M. K., & Crosnoe, R. (2003). The emergence and development of life course theory. In J. T. Mortimer & M. J. Shanahan (Eds.), *Handbook of the life course* (pp. 3–22). New York: Plenum Press.
Finch, C. F., & Seeman, T. F. (1999). Stress theories of aging. In V. l. Bengtson & K. W. Schaie (Eds.), *Handbook on theories of aging*. New York: Springer.
Frank, A. (1990). Review article: Bringing bodies back in a decade. *Theory, Culture and Society, 7*, 137–162.
Fry, C. (2009). Out of the armchair and off the veranda: Anthropological theories and the experiences of aging. In Bengtson, Gans, Putney, & Silverstein (Eds.), *Handbook of theories of aging* (pp. 499–515). New York: Springer Publishing.
Fry, C., Dickerson-Putman, J., Draper, P., Ikels, C., Keith, J., Glascock, A. P., et al. (1997). Culture and the meaning of a good old age. In J. Sokolovsky (Ed.), *The cultural context of aging-worldwide perspectives* (2nd ed.). Westport, CT: Bergin & Garvey.
Giddens, A. (1991). *Modernity and Self Indentity: Self and society in late modern age*. Standford University Press.
Glaser, B., & Straus, A. (1967). *The discovery of grounded theory*. Chicago: Aldine
Goodenough, W. H. (1999). Outline of a framework for a theory of cultural evolution. *Cross Cultural Research, 33*, 84–107.
Habermas, J. (1968). *Knowledge and human interests*. Beacon Press.
Hagestad, G. O., & Dannefer, D. (2001). Concepts and theories of aging: Beyond microfication in social science approaches. In R. H. Binstock & L. K. George (Eds.), *Handbook of aging and social sciences* (5th ed., pp. 3–21). San Diego: Academic Press.
Hashimoto, A. (1991). Living arrangements of the aged in seven developing countries: A preliminary analysis. *Journal of Cross-Cultural Gerontology, 6*, 359–381.
Hendricks, J., & Achenbaum, A. (1999). Historical development of theories of aging. In V. L. Bengtson & K. W. Schaie (Eds.), *Handbook of theories of aging* (pp. 3–20). New York: Springer.
Hendricks, J., Applebaum, R., & Kunkel, S. (2010). A world apart? Bridging the gap between theory and applied social gerontology. *The Gerontologist, 50*(3), 284–293.
Hermalin, A. I. (2002). *The wellbeing of the elderly in Asia: A Four-country comparative study*. Ann Arbor: University of Michigan Press.
Kahn, R. L., & Antonucci, T. C. (1980). Convoys over the lifecourse: Attachment, roles and social support. In P. B. Baltes & O. Brim (Eds.), *Lifespan Development and Behavior*, (Vol. 3, pp. 254–283). New York: Academic Press
Kivnick, H. Q., & Pruchno, R. (2011). Bridges and boundaries: Humanities and arts enhance gerontology. *The Gerontologist, 51*(2), 142–144.
Kowal, P., Naidoo, N., Chatterji, S. (2012). Health status, chronic conditions and healthcare utilization is respondents aged 50 and older from 70 countries in the World Health Survey. SAGE Working Paper, 1.
Kuhn, T S (1970). The structure of scientific revolutions. Chicago: University of Chicago Press.
Lamb, S. (2009). *Aging and the Indian diaspora: Cosmopolitan families in India and abroad*. Bloomington, IN: Indiana University Press.

Lamb, S. (2010). Love and aging in Bengali families. In D. P. Mines & S. E. Lamb (Eds.), *Everyday life in South Asia*. Bloomington, IN: Indiana University Press.
Lamb, S. (2014). Permanent personhood or meaningful decline? Toward a critical anthropology of successful aging. *Journal of Aging Studies, 29*, 41–52.
Lan, P. (2002). Subcontracting filial piety: Elder care in ethnic Chinese immigrant families in California. *Journal of Family Issues, 23*(7), 812–835.
Lee, R., & Mason, A. (2012). Population aging, intergenerational transfers and economic growth: Asia in a global context. In J. P. Smith & M. Majumdar (Eds.), *Aging in Asia: Findings from new and emerging data initiatives*. Washington DC: National Academies Press.
Lemon, B. W., Bengtson, V. L., & Peterson, J. A. (1972). Exploration of activity theory of aging —activity types and life satisfaction among in-movers to a retirement community. *Journals of Gerontology, 27*, 511.
Longino, C. F., & Powell, J. L. (2009). Toward a phenomenology of aging. In Bengtson, Gans, Putney, & Silverstein (Eds.), *Handbook of theories of aging* (pp. 375–387). New York: Springer Publishing.
Marks, S. R. (1977). Multiple roles and role strain: Some notes on human energy, time and commitment. *American Sociological Review, 42*(6), 921–936.
Marshall, V. (1999). Analyzing social theories. In V. Bengtson & K. W. Schaie (Eds.), *Handbook of theories of aging* (pp. 434–455). New York: Springer.
McEwen, B. S., & Stellar, E. (1993). Stress and the individual: Mechanisms leading to disease. *Arch Internal Medicine*, 153, 2093–2101.
Medeiros, K. (2014). Who "owns" gerontology? The importance of thinking beyond the sciences. *The Gerontologist, 54*(4).
Merton, R. (1968). *Social theory and social structure*. New York: Free Press.
Mills, C. W. (1959; 2000). *The sociological imagination*. Oxford University Press. 40th Anniversary Edition.
Morrow-Howell, N., Hinterlong, J., & Sherraden, M. (2001). *Productive aging: Concepts and challenges*. Baltimore: Johns Hopkins University Press.
Peplau, L. A., Miceli, M., & Morash, B. (1982). Loneliness and selfevaluation. In L. A. Peplau & D. Perlman (Eds.),*Loneliness: A sourcebook of current theory, research, and therapy* (pp. 131–151). New York: John Wiley.
Powell, J., & Wahidin, A. (2007). Understanding aging bodies: A postmodern dialogue. In J. Powell & T. Owen (Eds.), *Reconstructing postmodernism* (pp. 27–32). New York: Nova Science.
Putney, N. M., Alley, D., & Bengtson, V. L. (2005). Social gerontology as public sociology in action. *The American Sociologist, 36*, 88–104 (Fall/Winter).
Putney, N. M., & Bengtson, V. L. (2008). Theories of aging. In Carr, Crosnoe, Hughes, & Pienta (Eds.), *Encyclopaedia of the life course and human development* (pp. 413–423), Farmington Hills, MI: Gale Group.
Rojek, C., & Turner, B. (2000). Decorative sociology: Towards a critique of the cultural turn. *Sociological Review, 48*, 629–648.
Rowe, J. W., & Kahn, R. L. (1997). Successful aging. *The Gerontologist, 37*(4), 433–440.
Samanta, T., Chen, F., & Vanneman, R. (2015). Living arrangements and health of older adults in India. *Journals of Gerontology Series B: Psychological and Social Sciences, 70*(6), 937–947.
Schultz, A. (1932). *The phenomenology of the social world*. London: Heinemann.
Marks, SR (1977). Multiple roles and role strain: Some notes on human energy, time and commitment. American Sociological Review, 42 (6), 921-936.
Simondon, G. (1980 [1958]). *On the mode of existence of technical objects: Part 1* (M. Ninian, Trans.). Ontario, Canada: University of Ontario.
Sokolovsky, J. (2009). Introduction: Human maturity and global aging in cultural context. In J. Sokolovsky (Ed.), *The cultural context of aging: Worldwide perspectives* (3rd ed.). Praeger.
Stern, Y. (2002). What is cognitive reserve? Theory and research application of the reserve concept. *Journal of the Neuropsychological Society, 8*, 448–460.

Twigg, J., & Martin, W. (2015). The challenge of cultural gerontology. *The Gerontologist, 55*(3), 353–359.
United Nations. (2005). *Living arrangements of older persons around the world.* New York: Department of Economic and Social Affairs (DESA), Population Division.
Weaver, R. (2008). *Bridging the social security divide: Lessons from abroad.* Policy Brief Series, No. 166. Washington DC: Brookings Institution.
Wong, R., & Palloni, A. (2009). Aging in Mexico and Latin America. In *International handbooks of population* (Vol. 1, pp. 231–252). Springer.
Yount, K. M. (2009). Gender and intergenerational co-residence in Egypt and Tunisia. *Population Research and Policy Review, 28*, 615–640.

Chapter 2
Who Will Care for the Elder Caregiver? Outlining Theoretical Approaches and Future Research Questions

Subharati Ghosh, Benjamin Capistrant and Greta Friedemann-Sánchez

Abstract The goal of the current chapter is to explore some of the popular theoretical frameworks used to study the experiences of caregivers. The chapter discusses the relevance of these theoretical frameworks in the context of India, and compare and contrasts them to research in the West. It critically examines household living arrangements and complexities in identifying the primary caregiver in India, especially in the context of elder care in multigenerational households. The chapter concludes by discussing the knowledge gaps and the way forward to study caregiving issues around the elderly in India.

Keywords India · Caregiving · Theories · Family structure · Living arrangement

2.1 Introduction

The world's population is growing and aging. Globally, the population is projected to reach 7 billion by the end of 2011. The percent of the world's population who are aged 65+ has been rising slowly and are about to spike in growth. In the United States, the first wave of "Baby Boomers" is beginning to reach retirement age. Between 2000 and 2010, the percent of the population aged 65+ grew from 12.3 to 13 %; by 2020, this is projected to increase to 16.1 % (see Fig. 2.1). These trends are even more pronounced in less developed settings where, historically, life expectancy and the old age population have been small. For example, in Mexico,

S. Ghosh
Tata Institute of Social Sciences, Mumbai, India
e-mail: subharati@gmail.com

B. Capistrant (✉) · G. Friedemann-Sánchez
University of Minnesota, Twin-Cities, Minneapolis, USA
e-mail: bcapistr@umn.edu

G. Friedemann-Sánchez
e-mail: gfs@umn.edu

T. Samanta (ed.), *Cross-Cultural and Cross-Disciplinary Perspectives in Social Gerontology*, DOI 10.1007/978-981-10-1654-7_2

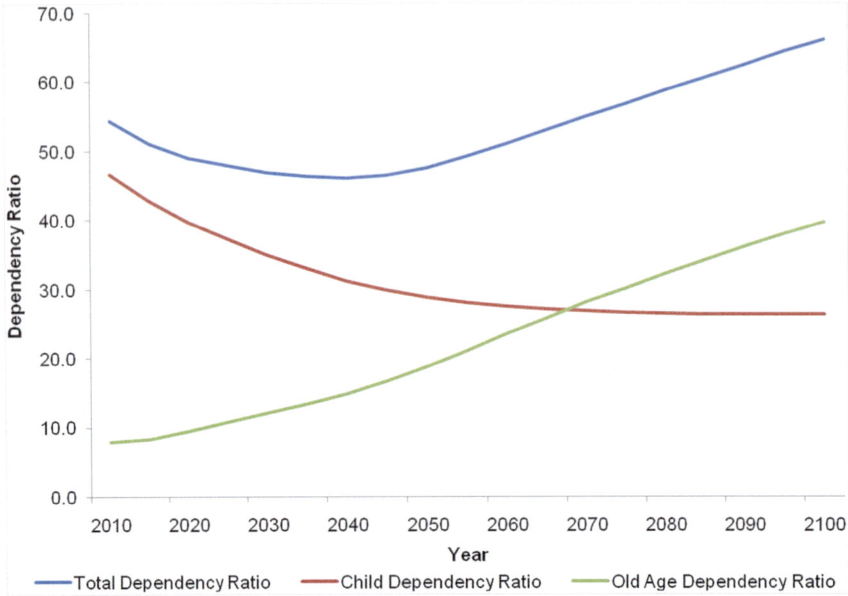

Fig. 2.1 Dependency ratio. *Source* United Nations (2014). Probabilistic population projects based on the world population prospects: the 2012 revision. Population Division, DESA/ST?/ESA/SER. A/353. http://esa.unc.edu/unpd/ppp

the proportion of the population who are aged 65+ is projected to almost quadruple between 2000 and 2050, from only 5.2 to 22.1 %.

Less developed regions will also be home to the vast majority of the world's aged by 2050. Though the distribution of world's aged population was roughly evenly divided between more and less developed regions in 1950, less developed areas represented 62 % of the world's aged as of 2010, which is projected to grow even further to 78 % by 2050.

The implications of population aging for anticipating the informal or at-home care that older adults need for their health are extremely relevant in lower and middle-income countries (LMICs). Disability from factors like depression, accidents, and chronic (particularly, cardiovascular) and communicable (notably, HIV/AIDS) diseases are extremely common among elderly in LMICs (Murray and Lopez 1997a, b). Recent evidence suggests that morbidity is expanding worldwide, including in LMICs, which also translates to more years with care needs (Murray et al. 2012; Salomon et al. 2012; Vos et al. 2012). Moreover, older people in developing settings are socially and economically vulnerable. A significant proportion of the elderly live below the poverty line (Barrientos et al. 2003; Barrientos and Hulme 2008; Deaton and Paxson 1997; Harper 2006) and generally lack access to pension and other benefits that are common support structures in developed settings (Bloom et al. 2010; Harper 2010). Health care access is a further challenge for the elderly in developing settings, exacerbated because older people tend to live in rural areas where health facilities, in particular, facilities specializing in geriatrics, are scarce (Lloyd-Sherlock 2000, 2002, 2010; Shrestha 2000). Given the symbiotic relationship between care recipient and caregiver, a corollary implication of the aging trends is the increasing numbers of informal caregivers who will face health and financial consequences as they take on the role of caregivers of elderly with health needs.

In this chapter, we question whether contemporary theories of elder caregiving, developed in the Unites States and Western Europe, apply to LMICs and specifically to the Indian context. A corollary area of inquiry regards *emic* understandings of caregiving and the *etic* and theoretical implications. To do so, we explore the history of family structure and norms of elder care. Since theories are in part a representation of social reality, this discussion will lay the ground to understand differences in normative beliefs around elder care and living arrangements between Euro-Americans and South Asians, and the extent to which the norms may have implicitly or explicitly impacted the theories of caregiving, and their relevance in the current Indian context. The chapter would then describe the different theoretical models of the health and social sciences. We close the chapter by identifying theoretical areas that are problematic when used in the Indian context.

2.2 Western and Indian Family Structures and Norms of Elder Care

Much of the early discussion and research around elder care grew out of concern for the changing family structure: a move from a traditional multigenerational set-up to a nuclear structure, which arguably gave rise to questions about who would care for the elderly, and how and where would care be located.

There is adequate evidence to show that the living arrangements of the elderly have shifted significantly over the decades in almost every Western industrialized country (Alter et al. 1996; Andorka 1995; Ruggles 2007; United Nations 2005). In the USA, intergenerational living by the elderly declined steadily from 70 % in the mid-1800s, to less than 15 % in the 1990s (cited in Ruggles 2011). Living with or adjacent to an adult child's home was replaced by living only with one's spouse, alone or in institutions. The societal changes stemming from economic development help to explain the decline of intergenerational co-residence. These include industrialization, migration to cities, urbanization, reduced ties to the land and the rise of wage labour.

Although daughters and spouses continued to remain the primary caregivers of older adults, the division of labour by gender along with different family and kinship structures brought about by divorce, remarriage and stepfamily formation increased the complexity of who would take on the caregiving role (Silverstein and Giarrusso 2010). As such there is no clear pathway to who is to care for the elderly and whether this is a private or a public responsibility.

In contrast, intergeneration family continues to be the site for elder care in India. However, it needs to be cautioned that intergenerational family in the context of eighteenth century US/Western Europe was no way comparable to an intergenerational family in the Indian context. According to Le Play (cited in Ruggles 2010), the most dominant family structure of peasant households in mid-eighteenth century Europe was the "stem family" structure. The arrangement involved one child usually selected by the father to live near parental homestead and work on the farm while the other's left home on reaching adulthood to form their nuclear families. This is in contrast to family structures seen historically in several Asian countries, including India and China, where three or more generations live under the same roof, referred to as the joint family. Therefore historically, the joint family structure was never part of the Western European family structure.

In India, the family has long been viewed as the primary site of age (Rajan and Kumar 2003; Jamuna 2003). The ideal Indian joint family (may also be referred to as extended family) includes three or more generations living together, with the patriarch being the head of the household. A joint family household could have grandparents, aunts, uncles, nephews and nieces, all living under the same roof. This family structure has historically served as a safety net for the elderly (Jamuna et al. 2003) irrespective of their health status. Although the original form of joint family structure has changed, an intergenerational household arrangement that is living with only adult children, preferably the son, continues to be the dominant living arrangement for majority elderly Indians above the age of 60. In a study of 9852 elderly above the age of 60 in 2011, 70.03 % reported living in intergenerational households (UNFPA 2011).

Historically, the norms of eldercare revolve around the concept of filial piety, which underpins elderly caregiving in India. Caregiving, also termed as "*seva*" (incorporates components of respect, care, love, moral and spiritual indebtedness and obligation), is a family responsibility, passed on from one generation to the other, with usually the elder son (although there are regional variations) taking upon the primary responsibility of aging parents. In the context of India, sons represent old age security (Jamuna 2003). However, it needs to be cautioned that while sons take upon the caring role, a significant amount of the assistance with daily living activities for the elderly are provided by the spouse, followed by daughters-in-law who usually move in with their parent-in-law upon marriage. Therefore, *emic* understandings of intergenerational caregiving consider it natural, obligatory, guided by a sense of filial piety, and simultaneously sustaining and challenging.

From an *etic* perspective, the ideal arrangement of elder care is possible if families continue to live jointly. However, research from Asian countries, such as China, Korea, Taiwan, Japan, which traditionally had joint family structures (see Ruggles 2010 for review), show patterns of declining intergenerational co-residential structures (Martin 1990; Hirosima 1997; Hermalin et al. 1992; De Vos and Lee 1993; Chattopadhyay and Marsh 1999; United Nations 2005). Forces of development have been cited as potential determinants of the change. Although there is no empirical evidence to support the same, there is a consensus that similar patterns are emerging in India too. Under such circumstances, much is left to understand how the change in family structure, would eventually impact care and location of care of the elderly. Second, the *etic* perspective, focusing on "the family" as the site of care rather on the specifics of who is performing the care, fails to highlight the pivotal role of daughters-in-law, daughters, and spouses who provide the majority of the care. Unlike the West, where daughters tend to be the primary providers of care to an aging parent, in the Indian context daughters usually relinquish their family of origin after marriage and are not dependent for care (Lamb 2005). Instead, spouse and daughters-in-law are expected to take on the role of caregiving (Jamuna 1997).

Given that social theories are a replication of social reality, it is but natural to ask whether the social science theories widely applied in the Western literature on caregiving can be validly applied in the context of India. To do so, we first review the theories of care from the health and social sciences to be able to ground our analysis of whether the theories apply to the Indian context and signal areas of fruitful theoretical development in the future.

2.3 Theories of Caregiving

It needs to be cautioned that there are no specific theories of caregiving per se. Instead, the field of caregiving research has borrowed theories and models extensively from disciplines such as sociology and psychology to explore the experiences of caregivers and their elderly care recipients.

Some of the core sociological theories discussed are the role theory and the cumulative advantage/disadvantage theory. In addition to the theories, are models and perspectives, such as the life course perspective, Pearlin's stress and coping model and the convoy models, used extensively to understand caregiver experience and the impact of caregiving on caregiver's physical, social and psychological wellbeing. It also needs to be emphasized that life course perspective, the stress process model and the role theories, are not stand-alone models and theories, but share substantial overlap among themselves. Recent work by Pearlin (2010) has attempted to find a "conceptual overlap and cross-fertilization" between the models of stress and coping and the life course perspective. These models are continually under integration and development and frame the way for future theories of caregiving. As such there is much scope for its use to understand better caregiving experiences both in the developed north as well as in the global south.

2.3.1 Pearlin's Stress and Coping Model

The most widely used model in caregiving research, including in gerontological caregiving, is Pearlin's stress and coping model, developed by Pearlin et al. (1990). This model that comes from the health sciences postulates that stressors have the ability to affect wellbeing, but the extent of their impact on wellbeing is determined by the availability of resources to counter the negative effects or burden as these are technically known. Included in the caregiving and stress process model is moderating factors, such as coping strategies and social support. Although developed in the context Alzheimer's care, this framework is extensively used across different ethnic groups in the U.S. because it provides a path towards caregiver interventions that seek to reduce caregiver burden. The model is used in the context of health-related caregiving. The model has been critiqued for its emphasis on burden which may overlook positive aspects of caregiving. However, the model can be adapted for use in the Indian context. Outstanding questions to adapt the model to India are the discovery of specific strategies of coping, the detailing of social support, factors that constitute burden and whether the pathways or processes of burden and wellbeing hold true in the Indian context. All three areas may benefit from culturally appropriate measures that can capture coping, social support and burden. For example, greater attention needs to be paid to the constructs of filial piety as a potential moderating factor and its intersection with gender and kinship relationship. A study by Gupta et al. (2009) found a distinct gender difference in determinants of burden among caregivers to the elderly. For the men, greater role overload predicted significantly higher burden, but the impact of role overload was higher for women who had higher adherence to Asian cultural norms, compared to women who had lower adherence to Asian cultural norms, specifically filial piety [measured using a 22-item scale by Ho and Lee (1974)]. Much work is needed to deconstruct the components of filial piety in the Indian context and theorizes how it relates to care in later life.

Because the model is used in the context of health-related caregiving, it remains to be seen how it can be adapted to the study of elder caregiving in the absence of a disabling condition in India, where the definition of a "dependent" elder is even more ambiguous than it already is in the West. Elderly recipients of care may not necessarily consider themselves as dependents, but "receivers", who hold legitimate right over the received care because of their prior contribution towards the child and the family (Lamb 2005). On the flip side, caregivers may not consider themselves "caregivers", but simply as fulfilling a kinship role of reciprocity by providing material, financial, physical and emotional support (Lamb 2005), to an extent that it is less common in the Western context. In summary, one additional gap in the application of the model and fact in general when thinking about elder caregiving for the Indian context is who is a caregiver and who considers themselves to be caregivers. An understanding of when it is that "receivers" cross the threshold, if at all, to being dependents and who is a caregiver rather than simply a spouse or a daughter-in-law will be key in defining sampling populations who may have unmet needs.

2.3.2 Life Course Perspective—Aging with Disability, Life Course Perspective

The life course perspective is a framework extensively used to understand various facets of caregiving. It is not a theory of caregiving per se, but its concepts have been used extensively to understand caregiving experience. Although primarily used in the context of health-related care research, the perspective offers much scope to understand the non-health-related care, and thus is useful in the Indian context, in conjunction with the stress and coping framework. The framework is inherently dynamic (Dannefer and Sell 1988) and multidisciplinary, and has borrowed heavily from disciplines such as sociology, psychology, history, anthropology (Bengtson and Allen 1993) and demography, to explain individual phenomena via a life course lens. Using the life course perspective caregiving is conceptualized as a "career" or trajectory, which involves entering or exiting the role several times over the life span, for instance becoming a parent or a grandparent, suddenly having a spouse with a spinal cord injury requiring much care. Unlike Pearlin's stress and caregiving model, the life course perspective by incorporating historical and demographic processes, does not necessarily pre-suppose caregiving as stressful and having negative impact on caregiver adaptation, but a normative part of life course (Elder et al. 1996) shaped by a range of macro- and micro-level socio-economic-historical, familial and individual structures. These micro- and macro-level factors render advantage or disadvantage accumulated over the life course, which leads to systematic divergence or heterogeneity in experience (Dannefer 2003; Dannefer and Sell 1988). Several concepts, such as (1) linked lives (2) social pathways, trajectories and transitions, and (3) timing of an event, off-timedness and duration, concepts core to the life course perspective has been extensively used to understand the heterogeneity in experiences of both aging care recipients as well as their caregivers. According to the authors, these concepts are universal and could be applied to any caregiving situation in any cultural context.

The life course perspective lays significant importance to historical time and space. This allows an understanding of how the various demographic factors interact with sociocultural and developmental changes over time, and the effect of these on family size, patterns, generational relationships, all of which have consequences for old age (Elder 1994). The life course perspective helps us understand the experiences of the different cohorts of aging adults, and how their early life or historical experiences have shaped their understanding and expectations of kin support in old age and their interaction with formal systems of care, welfare agencies and institutions (Hareven 1994). The concept of historical time and space is significantly important in understanding the intergenerational differences in filial piety, morality, and values attached to elder care; which may significantly impact both caregiver and care recipient wellbeing.

One of the significant contributions of the life course perspective is that it could be applied in any caregiving context to understand an individual's social pathway. Social pathways are composed of multiple interdependent pathways, which includes both family and non-family roles (Macmillan and Eliason 2003). It is the timing, quality and structure of these roles that determines wellbeing (Settersten 2003). The life course perspective thereby underscores the importance of multiple role trajectories in shaping an individual's experience.

Another important concept widely used and could be applicable in the Indian context are transitions, perhaps, a construct that has received more interest than any other, in the caregiving literature. According to the construct, caregiving is a role that likely involves several transitions marked by "entry" or "exit" one or more times during adulthood (Aneshensel et al. 1995; Moen et al. 1994, 1995). Transitions, especially entry and exit out of the caregiving role could be stressful. For example, research shows that transitioning into a caregiving role is associated with higher levels of depression, poorer physical health, greater financial burden and lower psychological wellbeing among caregivers (Burton et al. 2003; Ghosh et al. 2012; Kramer and Lambert 1999; Marks et al. 2002; Seltzer and Li 2000), and often gives rise to "competing commitments" (Brody 1981). It is important to explore transitions in the context of India and for countries where caregiving is bound by ties of filial piety. It is not known if the transition to a caregiving role is natural and expected and, therefore, could be less stressful, or could be equally stressful for caregivers as their Western counterparts, and whether there are gender and kinship differences. Given the popular discussion on the erosion of filial piety among the younger cohorts of adults, or the potential caregivers to adults currently in their sixties and seventies, it is also important to investigate whether the impact of transitioning to elder caregiving role is subject to cohort effects. Finally, it is also important to investigate whether the transition to taking on the responsibility of an aging parent who is physically fit, is different from caring for an aging parent who needs active assistance within the Indian context.

Two other concepts, timing and off-timedness, and duration are integral to describing the impact of transitions on an individual's life course. Timing refers to both the chronological age and life stage when transitions occur. Duration refers to the length of time an individual spends in the state, for example caregiving. Therefore, similar events could evoke different experiences for different people, depending on when they occur in the life course (George 1993). A related concept off-timedness refers to life events that are not age appropriate or life stage appropriate, as these events do not conform to age-appropriate norms, thereby exposing individuals to a range of vulnerabilities (Neugarten et al. 1965). For instance, a daughter-in-law who has young children and relatively young father-in-law in need of care due to ALS, or for an elderly mother or mother-in-law to provide care to an adult son or a daughter-in-law. In the context of India, where sons and daughters-in-law are expected to care for an older adult, the role reversal could significantly violate the age-appropriate norms, and could be a significant source of stress for the caregiver. Therefore, the timing of an event is shown to play an important role in determining caregivers' unique source of burden, their coping strategies, and the resources available to buffer the effects of stress. The concept of transitions probably plays out in different ways in the Indian context according to the life course of a caregiver. A significant amount of work is needed to understand how the concept of off-timedness plays in the context of India, and investigate some of the mediating and moderating factors that could mitigate the effect of off-timedness on caregiver's wellbeing. A gap in theory and research is the role of transitions with regard to elder caregiving in India.

2.3.3 Role Theory

The role theory has been extensively used in the gerontological research to understand the experiences of middle-aged women, engaged in a parental caregiving role, while at the same time, having to manage additional roles of that of spouse, mother, working woman, friend and so on. Two major perspectives within the role theory that have guided research studies on how women's multiple roles affect their psychological wellbeing are role occupancy theory and role quality theory. Role occupancy theory focuses on the number of roles an individual occupies. Within this theory, there are two competing hypotheses, known as (1) the scarcity/depletion hypothesis and (2) the expansion hypothesis. The scarcity/depletion hypothesis, as referred to by Marks (1977), contends that multiple roles have negative consequences resulting in psychological distress (Gersen 1976; Goode 1960), such that the demands from one role can often result in role strain in another role (Repetti et al. 1989). The hypothesis also asserts that "individuals have a fixed amount of energy and personal resources to allocate across roles" (Bainbridge et al. 2006, p. 490). Therefore, when demands exceed resources, an individual is likely to experience maladaptation, and the onset of a new role could lead to "competing commitments" (Brody 1981).

In contrast, the expansion hypothesis, otherwise known as the role-enrichment hypothesis (Marks 1977; Sieber 1974), contends that multiple roles need not be stressful. Indeed, multiple roles may provide opportunities for personal growth and satisfaction. "The enrichment argument assumes that the benefits of multiple roles outweigh the costs, leading to net gratification" (Rothbard 2001, p. 656). Thus, the loss of resources that might contribute to stress (Hobfoll 1988, 1989), is offset by benefits through participation in other meaningful roles, such as that of a wife, an employee, a parent and a neighbour (Parker 1993; Hong and Seltzer 1995). Further, researchers contend that it is not the number of roles but the quality of the roles that have important impact wellbeing (Aneshensel and Pearlin 1987; Barnett and Baruch 1985).

In addition to facilitating the systematic study of the importance of multiple roles on caregiver wellbeing, the role theory substantively addresses the effects of various role combinations on caregiver adaptation. More specifically, role theory contends that different roles have different effects (Barnett and Baruch 1985; Menaghan 1989). Further, according to Menaghan (1989), the effects of roles and role combinations on psychological wellbeing are dependent on the normative appropriateness of the roles. An important determinant of roles on caregiver wellbeing is the volition of the roles. Take for instance, the experience reported from a qualitative study of caregivers in India. A primary female caregiver was hospitalized for back problems after being assaulted by the care recipient who was her father-in-law. "[The father] moved to live with another son who accepted the responsibility of care, though reluctantly. However, he insisted on the father returning, as soon as his sister-in-law returned from the hospital. This led to dispute, ending in a fight between the two sons requiring intervention by the police. Finally, the person with dementia was brought back to the original caregiver and was looked after by her without support from any other relatives" (Shaji et al. 2003). Here, care is unquestioningly provided by women, though the decisions about care are made by men. This theme of care roles and choice suggests that the extent of women's provision of care may be in fact out of their control. Significant work is needed to understand the effect that choice or volition has on the wellbeing of caregivers and whether the relationship is moderated by filial piety. Other determinants might include caste, class, educational level and economic independence of the caregiver, which need to be empirically validated.

2.3.4 Convoy Model

The convoy model was first introduced by Kahn and Antonucci (1980). This model is primarily used to understand the social relationships of caregivers and care recipients, because of their far-reaching effects on their health and wellbeing. According to Kahn and Antonucci (1980), individuals move through the life course, surrounded by a network of individuals arranged in a concentric circle, also referred to as convoys. At the centre of the circle is the target individual, and those immediate in the concentric circles are individuals who are closest to the central figure. Although convoys change over the life course, members closest to the centre of the circle are considered to be relatively stable and provide several forms of social, emotional, psychological and instrumental assistance, and also, help mobilize resources to meet the needs. The convoy provides a protective base to the central individual, and has both objective and subjective components and is shaped by personal (gender, education, personality, race) and situational factors (norms, social roles, normative expectations and demands). In general, older adults tend to have a restricted network, where close family members take on a pivotal role, and the structure and the quality of the convoy network are determined by a family's history, gains and losses due to widowhood, historical contexts, within network dynamics, its structure, function and quality (Kahn and Antonucci 1980). Thus, the convoy model normalizes intergenerational caregiving (Walker and Pratt 1991), as a continuum of aid provided to the elderly.

The convoy model has been used extensively in gerontological literature in a variety of contexts. According to several studies conducted (Antonucci and Akiyama 1987; Depner and Ingersoll-Dayton 1988; Walker and Pratt 1991), adult children have been shown to be the ones who are the closest to the inner social circle for middle-aged and older adults, and provide them with significant social, emotional, psychological, and instrumental and health support. The daughters are important supportive convoys of care for the aging mother, which was evident from a study by Walker and Pratt (1991), who found daughter caregivers to provide aid to mothers, irrespective of whether the aging mothers were "self-sufficient" or "dependent", reaffirming intergenerational functional solidarity (Antonucci and Akiyama 1991). Is this theory applicable in the Indian context, where daughters-in-law are usually considered the primary caregivers yet their entry into the family cycle is usually fairly late? Do these findings also explain the popular perception of significant family strife between parents-in-law and daughters-in-law? Could there be some universal truth to the fact that adult children are the ones to remain a continuous source of emotional support through life, and while other individuals get added on, they never come close to the inner circle regardless of the amount of material or instrumental support they provide? Findings from a study by Datta et al. (2003) provide evidence for the same. In the study, the authors compared Indian and Belgian daughters and daughters-in-law caring for their aging mothers and mothers-in-law. The results of the study showed that Indian daughters-in-law were less close to their mothers-in-law compared to their Belgian counterparts, yet provided more care. The authors concluded that actual behaviours associated with care by daughters-in-law are more likely to follow cultural norms, at least in the Indian setting, but are less likely to influence care provided by daughters to their mothers.

2.3.5 Feminist Economic Theories of Care

Currently, there is no consensus on a theoretical framework for the study of care in developing contexts. The recent focus on care defines the care work as any activity that supports direct (feeding, bathing) or indirect (preparing food, collecting water) the reproduction of people economy (Esplen 2009). Care work can be performed within the paid economy or outside of it. Debates in the field hinge on whether caregiving should incorporate all forms of unpaid labour (subsistence agriculture, domestic work and direct care of persons) or not. Some have argued that a narrow definition of unpaid caregiving based upon the dependence of individuals allows researchers to distinguish between the developmental stages of various care recipients (baby, child, adolescent, adult, elderly) and their conditions (healthy, ill, chronically ill, disabled) (Friedemann-Sanchez and Griffin 2011). The authors argue that this definition offers the possibility of isolating the benefits to care recipients in developing economies, where home-based work and subsistence work are prevalent, allowing for the measurement of both direct and indirect forms of support for dependents and non-dependents. In this sense, they understand unpaid caregiving of dependents to be a subset of the larger concept of "care work", which includes both dependents and non-dependents, and of "unpaid labour", which in turn incorporates unpaid labour force work, unpaid housework and care work. Other scholars may find this definition too narrow arguing that such distinctions are not possible in developing contexts such as India where an individual may be simultaneously performing paid and unpaid activities that support the care of persons who may or may not be dependents.

Despite the absence of consensus on the definition and operationalization of the caregiving, there is much consensus that most caregiving work is done by women who face financial penalties. Unpaid care has negative effects on the participation of women in the paid labour force (Lilly et al. 2007). In fact, studies have shown that income decreases as caregiving time increases, and increases in caring responsibilities have a larger and more immediate effect on the earnings of women than men on the financial penalties faced by caregivers (Bittman et al. 2007). Little research has explored the economic effects of elder caregiving in developing countries. One of the few studies on financial penalties in the developing countries explored the ability of working parents in Vietnam to care for their children and found that 63 % of parents, most of whom were women, lost income or promotions or had difficulty retaining jobs due to caregiving (Vo et al. 2007).

2.4 Gaps in Knowledge

2.4.1 The Definition of the Caregiver

Given the discussion on some of the most widely applicable theories used in the study of caregiving in the West, one of the most prominent gaps arises from the fact that the theories are used exclusively to study caregiver stress and burden in the context of adult disability or health-related condition. The core issue that centres on the applicability of Western models of caregiving on non-Western population lies in the definition of caregiver and care recipient. When does a functional adult cross the threshold to become a dependent? Should researchers apply *emic* parameters or *etic* parameters when bounding the concepts? Should the study of caregiving limit itself with the physical and mental wellbeing of care recipient? Should it include instrumental support assuring social and financial wellbeing? Should caregiving research consider only the health effects of caregiving on caregivers or incorporate financial and social outcomes?

Care is often conceived more broadly in LMICs countries than just assistance with IADL/ADL limitations alone or other health-specific tasks. Qualitative evidence suggests that many consider keeping elder family members company and "watching out" for them as part of the range of duties of care (Abdulraheem 2005; Mendez-Luck et al. 2008). Thus, prevalence estimates based solely on physical and health-related care may underestimate what is considered to be the full scope of a caregiver's duties. By the same token, it is possible that people's report of time spent caring may reflect these other activities that are not otherwise captured by a structured, closed-ended quantitative survey of the types of health care provided. To have a full discussion of what it means to care for an elder family member, we consider some of the data available on care recipients, which imply different definitions of care.

2.4.2 Time-Use Data

Since there is no standard definition of what a caregiver is or who is a care recipient, research to date has used a variety of proxy measures to determine the prevalence of caregiving. One of these comes from time-use studies. One cross-national study investigated the prevalence of care work in six countries. They used existing time-use surveys from Argentina, Nicaragua, South Africa, Tanzania, India and Korea to analyse the extent of time spent on caring for other persons. Although this study's use of time-use data was limited in its ability to differentiate old age care from childcare, it is notable that in India, 44 % of women and 17 % of men participated in person care (Budlender 2008). Clearly, any child under the age of seven is a dependent, at least in as far as physical safety and security are concerned. A limitation of time-use surveys is that it is not able to discern if the care performed is on behalf of a dependent or not (Friedemann-Sanchez and Griffin 2011). For instance, caring for individuals with chronic or acute health conditions is different than doing domestic labour on behalf of an adult in good health. Time-use surveys replicate the problematic lack of distinction with regard to unpaid labour in the dominant economic development literature ((Friedemann-Sanchez and Griffin 2011). Furthermore, time-use surveys probably underestimate the time demands on caregivers when the care recipients' health status critically depends on the availability of caregivers such that the caregiving activities themselves do not take long, but may conjoin caregiver and care recipient on an intermittent yet hourly basis limiting caregivers from engaging in paid labour or other activities (Folbre 2014).

2.4.3 Dependency Ratios

The second proxy measure to assess the prevalence of care recipients is dependency ratio. The old age dependency ratio is the proportion of older persons (aged 65+) to those who are economically productive (15–64). Neither the ages in the numerator nor the denominator of this ratio are fixed; the ratio could be of the population aged 65+ or 70+ about the number of the population aged 20–64 (or 20–69, in the case of the 70+ dependency ratio). The old age dependency ratio is sometimes combined with child dependency ratios (# aged 0–5 per 100 persons aged 15–64) for a "total" dependency.

Many critique the usefulness of the dependency ratio. Of the total dependency ratio, others note that the kind of care required by a child under 5 is very different than the care required by an older adult (Folbre and Nelson 2000). Moreover, the total ratio does not offer insight into the distribution of the "dependents". Since one driver of total population aging is declining fertility, we would expect old age care to increase.

However, the total dependency ratio would not reflect this change in the distribution of the numerator (i.e. dependents) explicitly; instead, if the number of

Table 2.1 Old age dependency ratio (# aged 65+ per 100 aged 15–64): 2000–2050

Column 1	2000	2010	2020	2030	2040	2050
World	11	12	14	18	22	25
Less developed regions	8	9	11	15	19	23
More developed regions	21	24	29	36	41	45
U.S.	19	19	25	32	34	35
China	10	11	17	24	35	38
Ghana	6	6	7	8	10	12
India	7	8	9	12	15	20
Mexico	9	10	13	18	28	36
Russia	18	18	23	30	32	39
South Africa	6	7	10	12	13	15

Source Population Division of the Department of Economic and Social Affairs of the United Nations Secretariat, World Population Prospects: the 2008 revision, http://esa.un.org/wpp/, Medium Variant Projections
United Nations (2008)

dependents stayed the same, we would not know from the total dependency ratio alone the extent to which the burden was from older adults or from children. For instance, although the total dependency ratio in India is projected to decline until 2040, this decline masks a steady increase in the old age dependents (Fig. 2.1). Nonetheless, dependency ratios, either old age or total, are commonly used as a broad measure of the burden of population aging.

Outside of India, the old age dependency ratio is projected almost to triple for less developed regions between 2000 and 2050 and double in more developed regions. The old age dependency ratio is projected to spike between 2010 and 2020 during which time the burden of caring for the elderly will begin to grow a great deal, and it will maintain or continue to grow through 2050 (Table 2.1).

2.4.4 Intergenerational Co-residence

Family demographers also commonly focus on how families and households are arranged and composed. With whom older adults live offers a useful impression of the extent to which an older adult has care available to her. Information on intergenerational co-residence is perhaps the proxy indicator that has had the most influence in caregiving studies.

In lower and middle-income countries, the care needs of older persons nearly always fall to family members alone (Apt and Gricco 1994; Cattell 1990; Wachter 1997). In many areas of the world, there is a cultural expectation that older people will live with and receive their care from relatives than in developed settings (Bongaarts and Zimmer 2002; De Vos 2003; Ruggles and Heggeness 2008). Among households with older people in China, the 2000 Census found that about

11 % were older person living alone or an older couple living alone, respectively (Kincannon et al. 2005). Others cite that institutionalized care in China is primarily held only for the "three-no" elders: no living children/relatives, little or no income, and no physical ability to work (Gu et al. 2007). Data from the Chinese Longitudinal Healthy Longevity Survey and elsewhere suggest that nearly 60 % of the institutionalized elders do not have children to care for them (Zhan et al. 2005; Zhan 2002, 2005).

In India, over 70 % of the older population lives with family members (Bloom et al. 2010). One study used Demographic and Health Survey and census data from over 60 countries, including India. They found that Indian women aged 60 had an almost 2.5-fold greater odds of living alone than men ($p < 0.001$) (United Nations 2005). Intergenerational co-residency is an important indicator of availability of care, as much of the informal care comes from family members, typically daughters and/or daughters-in-law. Though economic development has changed the family structure, traditional patterns still persist like the "joint family system" in India (Gupta 2009). Patrilineal family patterns where a woman would live with her husband's family even after his death. (Lloyd-Sherlock 2010) Older women, particularly widows (Patel and Prince 2001; Shaji et al. 2002, 2003) are often disenfranchised, yet we know little about the care demands placed on them.

Because informal care from family members is culturally normative in many lower and middle-income settings, there is typically a weak formal care infrastructure, such as long-term or residential care facilities (Lloyd-Sherlock 2010), to support older adults if they have more substantial needs than what the family can provide.

2.5 Caring for the Caregiver: Implications

Research in the developed West shows the significant burden of care borne by caregivers of older adults with disabilities, in the form of poor physical and psychological wellbeing, lower social support and financial constraints. In contrast, literature of caregivers is scant and much of the literature on aging in India (Jamuna 1995) have risen out of concern for the elderly, focusing on their poor wellbeing and unmet needs, yet the unmet needs of caregivers or the effects of caregiving has little to no attention.

India, a society steeped in filial piety, family is still considered to be the ideal setting for elder care (see Lamb 2013). However, traditional structures, such as the joint family that had been responsible for elder care have changed or are fast changing. Further, laws and policies often get framed to support the elderly and not much to protect their caregivers, who are primarily women. These raise questions and speak to the extent to which cultural and moral visions around elder care dominate public and academic discourse, and raises pertinent questions on where the future of elder care is to be located, and who would be responsible for their care: the family or the state. For example, the Maintenance and Welfare of Parents and

Senior Citizens Act in India, implemented by the Ministry of Social Justice and Empowerment, India, in 2007 made it legally obligatory for children and heirs to provide maintenance to senior citizens and parents, by providing them with monthly allowance. The primary goal of this Act was to protect the lives and ensure financial security of the elderly in India by making it legally binding for families to care for them. Arguments provided by activists, bureaucrats, academics and demographers for the creation of the Act, centre on the breakdown of family, brought about by forces of development, such as modernization, migration, urbanization and the like. Thus, there was an implicit belief that the breakdown of the intergenerational family was the root of the old age depravity and a need for state intervention. Similar arguments were also provided by the early framers of the US Social Security Act in 1936 stated: "in the old days, the old-age assistance problem was not so great so long as most people lived on farms, had big families, and at least some of the children stayed on the farm". (Eliot 1961). However, although similar arguments were provided, the solutions sought in the two different contexts reflect a fundamental difference in the approach in how, where and who is responsible for the long-term care of the elderly. Unlike the U.S. Social Security Act, which meant "state" and only recognizes as beneficiaries those employed in the paid labour force and who have contributed financially to the system, the Maintenance and Welfare of Parents and Senior Citizens Act in India through state intervention made it legally obligatory for elderly to be cared for by their children/caregivers. The solutions sought were therefore to an extent a reflection of culture and what it values as the place and person to be involved in the care for an elderly.

2.6 The Way Forward

Given the status of caregivers of the elderly and their invisibility from popular discourse as well as policy formulation how can we move forward the agenda that makes caregivers of the elderly visible and that responds to the needs of caregivers? Who will care for the caregiver? How can we care for caregivers? In order to advance this research agenda, we have identified some theoretical and empirical questions, which can be summarized as follows:

- Identify what is meant by caregiving in the context of India.
- Identify factors associated with becoming a caregiver.
- Identify the unique sources of burden and stress among caregivers and also the various factors that might mitigate the relationship between caregiver stress and wellbeing.
- Empirically assess whether the existing theoretical frameworks could be applied in the study of caregiving.

- Study how changes in household composition are affecting caregivers. Such changes are not reflected in current theoretical approaches to the study of elder care.
- Time frames used in studies of elder care in Western contexts make invisible intermittent long-distance care. *Emic* definitions of caregiving can incorporate such type of care to better reflect caregiving realities.

A next step will be to integrate current theories and gaps in order to develop a comprehensive framework for the study of elder caregiving in India.

References

Abdulraheem, I. S. (2005). An opinion survey of caregivers concerning caring for the elderly in Ilorin metropolis, Nigeria. *Public Health, 119*(12), 1138–1144.

Alter, G., Cliggett, L., & Urbiel, A. (1996). *Household patterns of the elderly and the proximity of children in a nineteenth century city: Verviers, Belgium, 1831–1846* (pp. 30–42). Berlin: de Gruyter.

Aneshensel, C. S., Pearlin, L. I., Mullan, J. T., Zarit, S. H., & Whitlatch, C. J. (1995). *Profiles in caregiving: The unexpected career*. San Diego, CA: Academic Press.

Aneshensel, C. S., & Pearlin, R. F. (1987). Structural contexts of sex differences in stress. In R. C. Barnett, L. Biener, & G. K. Baruch (Eds.), *Gender and stress* (pp. 75–95). New York, NY: Free Press.

Andorka, R. (1995). Household systems and the lives of the old in eighteenth-and nineteenth-century Hungary. In *Aging in the past: Demography, society and old age* (pp. 129–155).

Antonucci, T. C., & Akiyama, H. (1987). An examination of sex differences in social support among older men and women. *Sex Roles, 17*(11–12), 737–749.

Antonucci, T. C., & Akiyama, H. (1991). Convoys of social support: Generational issues. *Marriage & Family Review, 16*(1–2), 103–124.

Apt, N. A., & Gricco, M. (1994). Urbanization, caring for elderly people and the changing African family: The challenge to social policy. *International Social Security Review, 47*(3–4), 111–122.

Bainbridge, H., Cregan, C., & Kulik, C. (2006). The effect of multiple roles on caregiver stress outcomes. *Journal of Applied Psychology, 91*(2), 490–497.

Barnett, G. K., & Baruch, R. C. (1985). Role quality, multiple role involvement, and psychological wellbeing in midlife women. *Journal of Personality and Social Psychology, 51*(3), 578–585.

Barrientos, A., Gorman, M., & Heslop, A. (2003). Old age poverty in developing countries: Contributions and dependence in later life. *World Development, 31*(3), 555–570.

Barrientos, A., Hulme, D., University of Manchester Brooks World, P. I., & Chronic Poverty, R. C. (2008). *Social protection for the poor and poorest: Concepts, policies and politics*. Basingstoke England; New York: Palgrave Macmillan.

Bengtson, V. L., & Allen, K. R. (1993). The life course perspective applied to families over time. In *Sourcebook of family theories and methods* (pp. 469–504). New York, USA: Springer US.

Bittman, M., Hill, T., & Thomson, C. (2007). The impact of caring on informal careers' employment, income and earnings: A longitudinal approach. *Austrial Journal of Social Issues, 42*(2), 255–273.

Bloom, D. E., Mahal, A., Rosenberg, L., & Sevilla, J. (2010). Economic security arrangements in the context of population ageing in India. *International Social Security Review, 63*(3–4), 59–89.

Bongaarts, J., & Zimmer, Z. (2002). Living arrangements of older adults in the developing world: An analysis of demographic and health survey household surveys. *Journals of Gerontology. Series B, Psychological Sciences and Social Sciences, 57*(3), S145–S157.
Brody, E. M. (1981). Women in the middle and family help to older people. *The Gerontologist, 21*, 471–480.
Budlender, D. (2008). *The statistical evidence on care and non-care work across six countries* (Vol. 4, pp. 1–61). Geneva: United National Research Institute for Social Development.
Burton, L. C., Zdaniuk, B., Schulz, R., Jackson, S., & Hirsch, C. (2003). Transitions in spousal caregiving. *The Gerontologist, 43*(2), 230–241.
Cattell, M. G. (1990). *Models of old age among the Samia of Kenya: Family support of the elderly*. Dordrecht: Springer Netherlands.
Chattopadhyay, A., & Marsh, R. (1999). Changes in living arrangement and familial support for the elderly in Taiwan: 1963–1991. *Journal of Comparative Family Studies, 30*, 523–537.
Dannefer, D. (2003). Cumulative advantage/disadvantage and the life course: Cross-fertilizing age and social science theory. *The Journals of Gerontology Series B: Psychological Sciences and Social Sciences, 58*(6), S327–S337.
Dannefer, D., & Sell, R. R. (1988). Age structure, the life course and "aged heterogeneity": Prospects for research and theory. *Comprehensive gerontology. Section B, Behavioural, Social, and Applied Sciences, 2*(1), 1–10.
Datta, P., Poortinga, Y. H., & Marcoen, A. (2003). Parent care by Indian and Belgian caregivers in their roles of daughter/daughter-in-law. *Journal of Cross-Cultural Psychology, 34*(6), 736–749.
De Vos, S. (2003). *Research note: Revisiting the classification of household composition among elderly people*. Dordrecht: Springer Netherlands.
De Vos, S., & Lee, Y. J. (1993). Change in extended family living among elderly people in South Korea, 1970–1980. *Economic Development and Cultural Change, 41*, 377–393.
Deaton, A., & Paxson, C. (1997). *Poverty among children and the elderly in developing countries* (Vol. 179). Princeton, NJ: Princeton University, Woodrow Wilson School of Public and International Affairs, Research Program in Development Studies.
Depner, C. E., & Ingersoll-Dayton, B. (1988). Supportive relationships in later life. *Psychology and Aging, 3*(4), 348.
Elder, Jr, G. H. (1994). Time, human agency, and social change: Perspectives on the life course. *Social Psychology Quarterly*, 4–15.
Elder, Jr, G. H., George, L. K., & Shanahan, M. J. (1996). *Psychosocial stress over the life course*.
Eliot, T. A. (1961). Social Security Administration Headquarters; Baltimore. The legal background of the Social Security Act, Speech delivered. Retrieved from http://www.ssa.gov/history/officials.html
Esplen, E. (2009). *Gender and care*. Overview report: Institute of Development Studies.
Folbre, N. (2014). The care economy in Africa: Subsistence production and unpaid care. *Journal of African Economies, 23*(suppl 1), i128–i156.
Folbre, N., & Nelson, J. A. (2000). For love or money—Or Both? *The Journal of Economic Perspectives, 14*(4), 123–140.
Friedemann-Sanchez, G., & Griffin, J. M. (2011). Defining the boundaries between unpaid labor and unpaid caregiving. Review of the health and social sciences. *Human Development and Capabilities, 12*(4), 511–534.
George, L. K. (1993). Sociological perspectives on life transitions. *Annual Review of Sociology, 19*, 353–373.
Gersen, E. M. (1976). On quality of life. *American Sociological Review, 41*, 793–806.
Goode, W. J. (1960). A theory of role strain. *American Sociological Review, 25*, 483–496.
Ghosh, S., Greenberg, J. S., & Seltzer, M. M. (2012). Adaptation to a spouse's disability by parents of adult children with mental illness or developmental disability. *Psychiatric Services, 63*(11), 1118–1124.
Gu, D., Dupre, M. E., & Liu, G. (2007). Characteristics of the institutionalized and community-residing oldest-old in China. *Social Science and Medicine, 64*(4), 871–883.

Gupta, R. (2009). Systems perspective: Understanding care giving of the elderly in India. *Health Care for Women International, 30*(12), 1040–1054.
Gupta, R., Rowe, N., & Pillai, V. K. (2009). Perceived caregiver burden in India implications for social services. *Affilia, 24*(1), 69–79.
Hareven, T. K. (1994). Aging and generational relations: A historical and life course perspective. *Annual Review of Sociology, 20*, 437–461.
Harper, S. (2006). *Addressing the implications of global ageing*. Dordrecht: Springer Netherlands.
Harper, S. (2010). The capacity of social security and health care institutions to adapt to an ageing world. *International Social Security Review, 63*(3–4), 177–196.
Hermalin, A. I., Ofstedal, M. B., Chang, M-C. (1992). Types of supports for the aged and their providers in Taiwan. In: K. Hareven Tamara (Ed.), *Aging and Generational Relations: Life-Course and Cross-Cultural Perspectives*. pp. 179–215. New York: Aldine DeGruyter.
Hirosima, K. (1997). Projection of living arrangements of the elderly in Japan: 1990–2010. *Genus, 53*, 79–111.
Ho, D. Y. F., & Lee, L. Y. (1974). Authoritarianism and attitude toward filial piety in Chinese teachers. *The Journal of Social Psychology, 92*(2), 305–306.
Hobfoll, S. E. (1988). *The ecology of stress*. Washington DC: Hemisphere Publication Services.
Hobfoll, S. E. (1989). Conservation of resources: A new attempt at conceptualizing stress. *American Psychologist, 44*, 513–524.
Hong, J., & Seltzer, M. M. (1995). The psychological consequences of multiple roles: The non-normative case. *Journal of Health and Social Behavior, 36*, 386–398.
Jamuna, D. (1995). Elder care: The developing Indian scenario. In *Challenges before the elderly: An Indian scenario* (pp. 257–272).
Jamuna, D. (1997). Stress dimensions among caregivers of the elderly. *The Indian Journal of Medical Research, 106*, 381–388.
Jamuna, D. (2003). Issues of elder care and elder abuse in the Indian context. *Journal of Aging & Social Policy, 15*(2–3), 125–142.
Jamuna, D., Lalitha, K., & Ramamurti, P. V. (2003). Future of elder care in India: Hope or despair some alternative core models. *Indian Journal of Gerontology, 17*, 147–156.
Kahn, R. L., & Antonucci, T. C. (1980). Convoys over the life course: Attachment, roles, and social support. *Life-Span Development and Behavior*.
Kincannon, C., He, W., & West, L. (2005). Demography of aging in China and the United States and the economic well-being of their older populations. *Journal of Cross-Cultural Gerontology, 20*(3), 243–255.
Kramer, B. J., & Lambert, J. D. (1999). Caregiving as a life course transition among older husbands: A prospective study. *The Gerontologist, 39*(6), 658–667.
Lamb, S. (2005). Cultural and moral values surrounding care and (in) dependence in late life: Reflections from India in an era of global modernity. *Care Management Journals, 6*(2), 80–89.
Lamb, S. (2013). In/dependence, intergenerational uncertainty, and the ambivalent state: Perceptions of old age security in India. *South Asia: Journal of South Asian Studies, 36*(1), 65–78.
Lilly, M., Laporte, A., & Coyte, P. (2007). Labor market work and home care's unpaid caregivers: A systematic review of labor force participation rates, predictors of labor market withdrawal, and hours of work. *The Milbank Quarterly, 85*(4), 641–691.
Lloyd-Sherlock, P. (2000). Old age and poverty in developing countries: New policy challenges. *World Development, 28*(12), 2157–2168.
Lloyd-Sherlock, P. (2002). Formal social protection for older people in developing countries: Three different approaches. *Journal of Social Policy, 31*(04), 695.
Lloyd-Sherlock, P. (2010). *Population ageing and international development: From generalisation to evidence*. Bristol, UK; Portland, OR: Policy.
Macmillan, R., & Eliason, S. R. (2003). *Characterizing the life course as role configurations and pathways* (pp. 529–554). US: Springer.
Marks, R. K. (1977). Multiple roles and role strain: Some notes on human energy, time and commitment. *American Sociological Review, 42*, 921–936.

Marks, N. F., Lambert, J. D., & Choi, H. (2002). Transitions to caregiving, gender, and psychological well-being: A prospective US National Study. *Journal of Marriage and Family, 64*(3), 657–667.
Martin, L. G. (1990). Changing intergenerational family relations in East Asia. *Annals of the American Academy of Political and Social Science, 510*, 102–114.
Menaghan, E. G. (1989). Role changes and psychological well-being: Variations in effects by gender and role repertoire. *Social Forces, 67*(3), 693–714.
Mendez-Luck, C., Kennedy, D., & Wallace, S. (2008). Concepts of burden in giving care to older relatives: A study of female caregivers in a Mexico City neighborhood. *Journal of Cross-Cultural Gerontology, 23*(3), 265–282.
Moen, P., Robison, J., & Dempster-McClain, D. (1995). Caregiving and women's wellbeing: A life course approach. *Journal of Health and Social Behavior, 36*, 59–273.
Moen, P., Robison, J., & Fields, V. (1994). Women's work and caregiving roles: A life-course approach. *Journal of Gerontology, Social Sciences, 49*, S176–S186.
Murray, C. J. L., & Lopez, A. D. (1997a). Global mortality, disability, and the contribution of risk factors: Global Burden of Disease Study. *The Lancet, 349*(9063), 1436–1442.
Murray, C. J. L., & Lopez, A. D. (1997b). Regional patterns of disability-free life expectancy and disability-adjusted life expectancy: Global Burden of Disease Study. *The Lancet, 349*(9062), 1347–1352.
Murray, C. J., Vos, T., Lozano, R., Naghavi, M., Flaxman, A. D., Michaud, C., et al. (2012). Disability-adjusted life years (DALYs) for 291 diseases and injuries in 21 regions, 1990–2010: A systematic analysis for the Global Burden of Disease Study 2010. *Lancet, 380*(9859), 2197–2223.
Neugarten, B. L., Moore, J. W., & Lowe, J. C. (1965). Age norms, age constraints, and adult socialization. *The American Journal of Sociology, 70*(6), 710–717.
Parker, G. (1993). *With this body: Caring and disability in marriage*. Buckingham, England: Open University Press.
Patel, V., & Prince, M. (2001). Ageing and mental health in a developing country: Who cares? Qualitative studies from Goa, India. [10.1017/S0033291799003098]. *Psychological Medicine, 31*, 29–38.
Pearlin, L. I. (2010). The life course and the stress process: Some conceptual comparisons. *The Journals of Gerontology Series B: Psychological Sciences and Social Sciences, 65*(2), 207–215.
Pearlin, L. I., Mullan, J. T., Semple, S. J., & Skaff, M. M. (1990). Caregiving and the stress process: An overview of concepts and their measures. *The Gerontologist, 30*, 583–594.
Repetti, R. L., Matthews, K. A., & Waldron, I. (1989). Effects of paid employment on women's mental and physical health. *American Psychologist, 44*, 1394–1401.
Rothbard, N. P. (2001). Enriching or depleting? The dynamics of engagement in work and family roles. *Administrative Science Quarterly, 46*, 655–684.
Rajan, S. I., & Kumar, S. (2003). Living arrangements among Indian elderly: New evidence from national family health survey. *Economic and Political Weekly*, 75–80.
Ruggles, S. (2007). The decline of intergenerational co-residence in the United States, 1850 to 2000. *American Sociological Review, 72*(6), 964–989.
Ruggles, S. (2010). Stem families and joint families in comparative historical perspective. *Population Development Review, 36*(3), 563–577.
Ruggles, S. (2011). Intergenerational coresidence and Family Transitions in the United States, 1850–1880. *Journal of Marriage and Family, 73*(1), 138–148.
Ruggles, S., & Heggeness, M. (2008). Intergenerational coresidence in developing countries. Amsterdam: Blackwell Publishing Ltd.
Salomon, J. A., Wang, H., Freeman, M. K., Vos, T., Flaxman, A. D., Lopez, A. D., et al. (2012). Healthy life expectancy for 187 countries, 1990–2010: A systematic analysis for the Global Burden Disease Study 2010. [Research Support, Non-U.S. Gov't]. *Lancet, 380*(9859), 2144–2162.

Seltzer, M. M., & Li, L. W. (2000). The dynamics of caregiving transitions during a three-year prospective study. *The Gerontologist, 40*(2), 165–178.

Settersten, R. A. (2003). *Invitation to the life course: Toward new understandings of later life*. Amityville, N.Y.: Baywood Pub. Co.

Shaji, K. S., Arun Kishore, N. R., Lal, K. P., & Prince, M. (2002). Revealing a hidden problem. An evaluation of a community dementia case-finding program from the Indian 10/66 Dementia Research Network. (10.1002/gps.553). *International Journal of Geriatric Psychiatry, 17*, 222–225.

Shaji, K. S., Smitha, K., Lal, K. P., & Prince, M. J. (2003). Caregivers of people with Alzheimer's disease: a qualitative study from the Indian 10/66 Dementia Research Network. *International Journal of Geriatric Psychiatry,18*(1), 1–6.

Shrestha, L. B. (2000). Population aging in developing countries. *Health Affairs, 19*(3), 204–212.

Sieber, S. D. (1974). Toward a theory of role accumulation. *American Sociological Review, 39*, 567–578.

Silverstein, M., & Giarrusso, R. (2010). Aging and family life: A decade review. *Journal of Marriage and Family, 72*(5), 1039–1058.

United Nations. (2005). Living arrangement of older persons around the world (Vol. ST/ESA/SER.A/240). New York, NY: United Nations.

United Nations. (2008). World population prospects: The 2008 revision population database, from http://esa.un.org/unpp/

United Nations Population Fund. (2011). Report on the status of elderly in selected states of India. Retrieved from http://www.isec.ac.in/AgeingReport_28Nov2012_LowRes-1.pdf

Vo, P. H., Penrose, K., & Heymann, J. (2007). Working to exit poverty while caring for children's health and development in Vietnam. *Community, Workand Family, 10*(2), 179–199.

Vos, T., Flaxman, A. D., Naghavi, M., Lozano, R., Michaud, C., Ezzati, M., et al. (2012). Years lived with disability (YLDs) for 1160 sequelae of 289 diseases and injuries 1990–2010: A systematic analysis for the Global Burden of Disease Study 2010. *Lancet, 380*(9859), 2163–2196.

Wachter, K. W. (1997). Kinship resources for the elderly. *Philosophical Transactions of the Royal Society of London. Series B, Biological Sciences, 352*(1363), 1811–1817.

Walker, A. J., & Pratt, C. C. (1991). Daughters' help to mothers: Intergenerational aid versus caregiving. *Journal of Marriage and the Family, 53*(1), 3–12.

Zhan, H., Liu, G., & Bai, H.-G. (2005). Recent development in Chinese elder homes: A reconciliation of traditional culture. *Ageing International, 30*(2), 167-187-187-167-187-187.

Zhan, H. J. (2002). Chinese caregiving burden and the future burden of elder care in life-course perspective. *The International Journal of Aging and Human Development, 54*(4), 267–290.

Zhan, H. J. (2005). Social-economic context of parent care—Explaining Chinese caregivers' psychological and emotional well-being. *Journal of Gerontological Social Work, 45*(4), 83–100.

Chapter 3
Social Psychology and Gerontology: Integrating Theory to Explain and Intervene in Age Discrimination Towards Older People in Europe

Sibila Marques, Christin-Melanie Vauclair, Hannah J. Swift, Christopher Bratt, Maria Luísa Lima and Dominic Abrams

> *We have enough research! We have enough theories! What we need are programs to help senior citizens in need!*
> (M. Kuhn 1983)

Abstract Today, age discrimination is one of the most fundamental forms of discrimination endured by Europeans. In an aging society this carries important consequences for the overall health and wellbeing of European citizens. This chapter discusses how integrating Social Psychology with Gerontology theorizing may contribute to the design of proper research and interventions dealing with this pressing social issue. To illustrate our discussion we present two case studies based on our findings from the "Experiences and expressions of ageism" module of the *European Social Survey*. Case 1 shows how the perception of age discrimination by older people mediates and helps to explain the effects of wealth inequality on older people's subjective health. Case 2 presents compelling evidence showing that, among older people, identifying with being an older person is associated with poor health outcomes, especially in countries where older people's status is lower. These findings are discussed in light of their implications for theory and practical intervention in this domain.

Keywords Ageism · Social psychology · Europe · Social identity · Health

---

S. Marques (✉) · C.-M. Vauclair · M.L. Lima
Instituto Universitário de Lisboa (ISCTE-IUL), CIS-IUL,
Av das Forças Armadas, 1649-026 Lisbon, Portugal
e-mail: Sibila.Marques@iscte.pt

H.J. Swift · C. Bratt · D. Abrams
Centre for the Study of Group Processes, School of Psychology,
University of Kent, Canterbury CT2 7NP, UK

T. Samanta (ed.), *Cross-Cultural and Cross-Disciplinary Perspectives in Social Gerontology*, DOI 10.1007/978-981-10-1654-7_3

This heart-felt comment from the founder of the Grey Panthers movement in the USA, Maggie Kuhn, at the Gerontological Society of America's annual conference, draws attention to the frustrations often felt by those working with and representing the people whose lives we are trying to change and improve with our research. It is a good example of the perceived gap between theory and its application to changing or developing policies and practices to enhance the lives of older people. Those who do applied research know that it can be effortful to disseminate and share research with end users of research, i.e. those who make practical use of the research findings. It is often difficult to identify and connect with relevant practitioners, charitable organizations and policy-makers who actually use research-based evidence to develop interventions and policy. It seems that theory has become devalued in general and it appears difficult for academics and researchers to communicate with people on the ground. Why did this happen? Bengtson et al. (2000) enumerate four factors that may have contributed to this state of affairs. First of all, theoretical approaches are perceived as lacking an overall structure and coherence. Instead of one theory that "explains best", most often the literature is filled with several approaches that do not seem to connect at first sight. This lack of a "grand theory", that could simply explain what is happening, creates a feeling of confusions that draws away the interest of practitioners. Second, the urgent need to solve real social problems creates pressure in practitioners that leads them to skip literature review. Often things happen as they were the first time such a problem has appeared, and only sometimes will practitioners have enough time and motivation to consider best practices in the field already out there and published in scientific journals. In fact, this lack of use of scientific proven theories and interventions is often explained by a third factor, which is the spread of post-modernist *Zeitgeist,* claiming that it is rather difficult to make generalizations and that all subjects should be treated as unique cases, with their specificities and needs. Science and scientific knowledge are thus questioned as the best foundations of human knowledge. Finally, another factor that may explain the lack of interest for theory-based gerontological interventions is the resistance to cross-disciplinary and interdisciplinary investigations in gerontology. If researchers in the field limit their investigations to just one narrow field of scientific research this probably creates an over-simplified explanation of phenomena, limiting their interest to explain real-life situations. In that case, the quest for an interrelation between levels of explanation and different fields in research could be a good way to increase the perceived interest of theories for people in the field.

One may ask: how can academics, using different research methods and approaches, provide joined and coherent evidence? Can this really make a contribution to what happens in social gerontology?

In this chapter our goal is to show that theory is in fact important as a basis to solid practical and policy interventions. In particular, we will try to demonstrate how social psychological research may contribute in a significant way to gerontological debates and evidence to reduce ageism experienced by older people. Based on the findings of the European Social Survey our goal is to present the case where findings from large surveys, built on solid theoretical grounds, can have important implications for exposing age discrimination regarding older people in the European context. We will also discuss the practical and policy implications this research may have for real-life intervention in the gerontological field.

3.1 "*There Is Nothing as Practical as a Good Theory*": Introduction to Theory-Based Social Gerontology Interventions

The above quote from the famous social psychologist Kurt Lewin at the Society for the Psychological Study of Social Issues conference in 1943 translates well the spirit of this chapter and of this book. In this sense, this would be an interesting answer to Maggie Kuhn's comment trying to show that if it is true that we need better interventions, the best theoretical evidence should also support them. In Lewin's vision, theory and practice should play an equal status and complementary role in the explanation of social phenomena. And why is theory important? Its importance is well translated in the very definition of theory.

Theories can be defined as "the construction of explicit explanations in accounting for empirical findings" (Bengtson et al. 2000, p. 3). By building knowledge through a systematic and cumulative way, theories provide a set of lenses through which we can make sense of what we find in research. The key process is thus helping *explain* and *predict* events taking in consideration a broader context of inquiry and one can find several examples where this process is important. For example, many studies in social psychology explore the fundamental role of social norms and social influence. A classic experience in 1954 shows how is it possible to predict situations of intergroup conflict and conflict resolution. In this famous study, a group of boys is asked to join a summer camp in Robber's Cave State Park, USA. Without the boys' knowledge, their parents agreed to let them collaborate in a study of intergroup conflict created by Muzafer Sherif and colleagues. When the boys arrive to the camp, they are divided into rival teams struggling to win a competition: the Eagles and the Rattlers. The entire situation is set to promote the competition between the groups thus creating some tension. The conflict escalates after the Eagles had legitimately won the tournament and the Rattlers decided to raid their cabins and steal the prizes. By dividing the boys into two groups Sherif and colleagues created two different group identities that their members strove to defend and gain relative advantage. In this experience, Sherif tests, in the field, the factors that may contribute to promote competition and conflict between groups.

In order to solve this conflict, the experimenters created a situation where members of the two groups needed to collaborate side-by-side in order to reach common goals: repairing the camp truck after it broke down or to join monetary efforts to rent a movie. The creation of these situations helped build a common and unique identity for both groups—those of the boys in the summer camp as a whole

—thus, mitigating the conflict. This is a case where theoretical-based predictions, based on realistic group theory, shows very practical results. This work is fundamental to understanding many real-life situations of conflict and escalation of violence and to increase the possibilities of increased cooperation between different groups.

These types of processes have also been recently studied in gerontological settings of practical intervention. In some studies, Haslam et al. (2014) have shown how building a common identity with senior residents in a home setting is fundamental to increasing their wellbeing and feelings of belonging. In an experimental study, 36 seniors were randomly allocated to three conditions: an intervention where they made decisions about lounge refurbishment as a group, a comparison condition where the staff made those decisions, or a no treatment control. Participants in the intervention condition showed improvements in cognitive performance and home satisfaction. Interestingly, there was also a significant increase in the level of identification with the group. Emphasizing group cohesion and identity turned out to be a key factor to increase senior's adaptation in this case.

In this chapter we argue that, as a theoretical-based approach, social psychology may have fundamental contributions to give to the gerontological field. We make our point by giving specific examples of how a "grand theory" in social psychology —the Social Identity Approach—may help understand the way older people are affected by age discrimination in the European context.

3.2 Introduction to Social Psychology: The Importance of Multilevel Explanations

One can think of social psychology as "the scientific study of the effects of social and cognitive processes on the way individuals perceive, influence and relate to others" (Smith and Mackie 2007, p. 3). First of all, social psychologists' studies are based on systematic methods of gathering information on a structured way. Moreover, social psychology is especially interested in understanding the way cognitive and social processes influence behaviours. These cognitive processes are the ways in which our perceptions, memories, thoughts and motives influence our understanding of the word, and thus guide our actions. Crucially, social influences exert their effects through cognitive processes. The way we think, feel and act towards other people all reflect the influences of our social contexts and influences. Thus, cognitive and social processes are inextricably linked.

Thinking about the object of social psychological studies, Smith and Mackie (2007) identify two basic ways in which people affect each other. On one hand, social psychologists are interested in studying the *individual in the group*. In this case, the main focus is on the way individuals are affected by others who are physically present (e.g. individuals who are leading, providing trustworthy information, waiting in a line). The effects of this physical presence depend heavily on the way individuals interpret the situation. On the other hand, social psychology also looks the other way around, that is, trying to understand the *group in the individual*, that is, how the representations and thoughts we have regarding the groups we belong to influence our individual actions and perceptions. Social psychology has shown that our group memberships are an essential part of our self-concepts and identity (Turner et al. 1987), influencing us even when other group members are absent. In fact, we are all aware that even when other fans are physically absent, we still defend our national football teams if we feel strongly about this membership. Just being a member of the group connects us to the group, creating a sense of join fate and making us willing to defend and act in defence of that group.

One of the fundamental assertions of social psychology is the notion that individuals play an active role in the construction of social situations (Taylor 1998). In fact, our responses to social environments are heavily dependent on the way we interpret them. As an older person, one can think of our place in society in different ways depending on the manner we perceive this category. As we shall see in more detail below (Case 2), if we identify ourselves with the group of older people, in societies where perceptions of the social status attributed to this age group is higher, we will have significantly better health than if we use the same type of process in societies where perceptions of old age are more negative. Hence, the effects of perceiving oneself as old depend on the meaning attributed to this age category in society. The study of this phenomenon holds important implications for interventions in different types of societies with varying perceptions of aging and old age.

Another fundamental characteristic of social psychological studies, that makes the link to gerontology especially interesting, is the fact that this discipline has in its nature an intrinsic and traditional focus on theory application to the understanding of social issues (Taylor 1998). Its very foundations and growth was marked by the need to create new solutions for practical problems such as the two world wars, the Great Depression, and the rise of Hitler. Since its roots, social psychologists have sought to understand the influences of social forces on human behaviour to increase predictive and protective power over social phenomena. Hence, the use of this theoretical framework to aging issues is a natural extension of this mode of thinking in the discipline.

Finally, yet another factor that makes the social psychological approach interesting is its potential for multilevel explanations. As we noted above, social psychology is concerned with the study of the feelings, thoughts and behaviours of individuals in social situations. There are various ways on how these social situations can be conceptualized. At the micro- and meso-level, they can constitute interactions and relations among individuals in a group (i.e. group processes) or individuals' self-concept that is inextricably linked to the in- and out-group with which they identify and distinguish themselves respectively (Ellemers et al. 2002; Simon et al. 1995; Turner et al. 1987). At the macro-level, these social situations go beyond the individuals and the groups and include larger institutions and social structures as well as culture and society. A common focus in these kinds of studies lies in identifying how these macro-level factors influence an individual's feelings, thinking and behaviour, although the study of reciprocal associations can also be of interest. For instance, concepts such as age stereotypes and age prejudice do not emerge in a social vacuum. Groups and whole societies tend to share similar stereotypes and prejudices (Schaller et al. 2002) which constitute a collective reality that has an impact on aging individuals' in terms of their self-concept and age identification. The roots of this perspective are grounded in what is called a methodological collectivism. Durkheim (1897/1951) was one of its advocators and argued that there are societal "realities external to the individual" (p. 37) which exert some power over the individual and neglecting them would mean gaining an incomplete understanding of human functioning.

Besides these collective social phenomena, there are also macrostructural ones such as economic and political systems (e.g. free-market capitalism vs. communism) and population structures (e.g. changing demographics) that can have an effect on older people. For instance, Frymer (2005) argues that in order to understand prejudice towards social groups in society, it is important to go beyond the assumption that prejudice is just a result of individual psychological attitudes (irrational biases) and to examine to what extent institutions encourage prejudice and discriminatory acts by motivating people to behave in a prejudiced manner. A prominent example would be if health care expenditures are cut substantially due to an economic recession and health practitioners are confronted with making treatment choices and use age boundaries as a guidance.

In sum, both objective macro-environments as well as cultural and collective phenomena relate to human mind and behaviour. The resulting epistemological perspective has also been coined "socio-ecological psychology" (Oishi and Graham 2010), which expresses the idea that there is an inextricable and reciprocal linkage between social psychological processes and the social and physical environment in which people live. As such, it is a highly interdisciplinary perspective that needs to take into account theories from other areas such as sociology, economics, political science and geography rendering it extremely relevant to gerontology.

One of the biggest challenges for social psychologists' attempts to take into account the macro-level was the lack of proper research methods. Yet, the development of multilevel modelling opened new doors to social psychologists. Multilevel modelling is a statistical regression approach that enables researchers to test multilevel theories. Although it has already been used for several decades in educational research, sociology, econometrics and biometrics, it has only recently been employed more frequently by social psychologists. Use of this type of methodologies has been really important to test important theoretical hypotheses in the field. In this chapter we will see in more detail how the use of these methods has been useful to understand ageism and age discrimination regarding older people in the European context. However, before presenting these evidences it is important to discuss in bit more detail the theoretical background underlying this research. In this regard, we give special emphasis to the role played by one particular theory in the social psychological field—the Social Identity Approach—as a good example of a theory that is fitted to be the base for real-life interventions in gerontology and other applied fields.

3.2.1 The Social Identity Approach

The social identity approach embodies a theoretical and research tradition that has been developed for more than a quarter of a century. Framed by the development of post-war European perspective to social psychology, it emerged from the need of Henry Tajfel and colleagues to offer an alternative to contemporary individualistic visions in social psychology (Allport 1924). The aim was to create a theoretical approach to social psychology that would be able to deal with the relationship between the individual and society: that is to study the *social* dimension of human behaviours (Hogg and Abrams 1988; Tajfel 1984).

From a meta-theoretical point of view (Hogg and Abrams 1988), Social identity assumes a psychological Marxist's vision of the world. The main idea is that, society is composed by social categories that stand in power and status relations to one another. These relationships are not static, instead they are in constant conflict between them causing changes in society's structure. Social categories become human groups because the individuals that compose them come to understand that they share a common belief system. In keeping with Marx and the symbolic interactionists (e.g. Mead 1934), the social identity approach considers that identity and self-conception mediates between social categories and individual behaviours. However, it goes further. It assumes a social psychological approach, and explores the psychological processes involved in translating social categories into human groups. In a very fundamental way, the social identity approach explores the "groups within the individual", as opposed to the "individuals in the group". Being part of a group confers individuals with a *social identity*, or a shared/collective representation of who one is and how one should behave (Hogg and Abrams 1988). Social identity is a central concept in this approach and according to Tajfel (1972) refers to the "the individual's knowledge that he belongs to certain social groups, together with some emotional and value significance to him of the group membership" (as cited in Turner 1982, p. 18). Hence, the social identity approach assumes a clear motivational perspective of human behaviours, considering individuals' motives attached with group belonging (Operario and Fiske 1999).

According to Tajfel and Turner (1979, 1986), social behaviours can vary along a continuum from interpersonal to intergroup. At the "interpersonal" extreme, the behaviours of individuals are determined by their personal relationships and their idiosyncratic personal qualities. On the other hand, at the "intergroup" extreme, the way individuals behave is fully determined by their belonging to different social groups or categories, by their social identities. Shift between the continuum (and the adoption of a specific social identity) varies in function of psychological and social factors.

Using the "minimal group paradigm" Tajfel and colleagues (Tajfel et al. 1971) showed that the mere social categorization of people into distinct groups was enough to produce intergroup behaviours in which participants favoured in-group over out-group members. According to Tajfel (1972), mere social categorization was enough to create a social identity for the subjects, in the sense that they accepted it as a relevant self-definition in the situation. Based on the results of the "minimal group studies", Tajfel and Turner (1986) derived three theoretical principles: (1) Individuals strive to achieve or maintain positive social identity; (2) Positive identity is based to a large extent on favourable comparison that can be made between the in-group and some relevant out-group; and (3) When social identity is unsatisfactory, individuals will strive to either leave their existing group and join some more positively distinct group and/or make their existing group more positive. Regarding this third aspect, Tajfel and Turner (1979, 1986) elaborated in more detail the possibilities available for members of devalued social groups to deal with the predicaments of a negative social identity. What happens in the case an individual belongs to a group judged in a negative manner?

Being a member of a subordinate group, may render individuals with a negative social identity and hence lower self-esteem. This is an unsatisfactory state and mobilizes individuals to change the situation. According to Tajfel and Turner (1979, 1986), members of devalued groups can choose to shift among more intergroup or interpersonal behaviours as a way to deal with the negative situation. However, this is dependent upon the way individuals perceive the nature and the structure of the relations between groups in their society, that is, depends on their *subjective belief structures*. Tajfel and Turner (1979, 1986) identified two major belief systems. The belief system of "social mobility" is based on the idea that the boundaries between groups are permeable, in the sense that individuals can leave their group ("exit"), in search of one that provides a more satisfactory identity ("pass"). The belief in social mobility leads subordinate group members to adopt individualistic strategies (individual mobility), that help to change one's personal position, but that leaves the group's position unchanged. On the other hand, the belief system of "social change" rests on the assumption that intergroup boundaries are impermeable, and that it is relatively impossible passing from a low to a high-status group. In this sense, negative consequences of group membership cannot be escaped simply by redefining oneself out of a group and into a dominant group. The only way to cope with this negative situation is to adopt group strategies, aimed to accomplish a relatively positive re-evaluation of the in-group (i.e. social creativity or social competition strategies).

Social Identity Theory is a solid theoretical trend within social psychology, tested in many different setting by diverse methods (e.g. correlational, experimental). In this chapter, our particular interest is to understand how this theoretical framework may be used to understand in specific age discrimination regarding older people. Next, we present what are age categorization and the factors that may have influenced ageism against this age group.

3.3 Age Categorization and the Origins of Ageism

Age as long been recognized as an important basis of social organization and social integration (Hagestad and Uhlenberg 2005). Both sociology and anthropology have shown the importance of age as a criterion for participation in society's division of labour. For instance, in their model of social stratification, Riley and Foner (1968) showed how age has a prominent role in the process of "matching people and roles". In the same vein, classical anthropological accounts of age grading (for a review please see Cain 1964) showed how age groups differed in their rights and responsibilities, and how rites of passage had such an important role marking the transition into older age groups. Hence, there is no doubt that age, like other social categorisations such as social class, race or gender, serves as an important social marker which is used to infer roles, status, power and social responsibilities. This structuring has several advantages because it promotes predictability of the life course and provides individuals a sense of belonging through peer ship (Garstka et al. 2004).

It is clear that age categorization is a fundamental tool to structure our societies. Age limits are established in a manner to guarantee predictability of the life course. Importantly, however, is that we need to take in consideration that this division into age groups is not free of content; in fact, each age category has its associated contents expressing guidelines for individual behaviours. The idea that "tells me your age, and I tell you what to do" does not carry only benefits. In fact, sometimes it may lead to "us versus them" distinctions, promoting prejudice and discrimination (Hagestad and Uhlenberg 2005). Butler (1969) was the first to notice the negative effects of this "classification" according to one's age group; in this case particularly against the older age group. Back in the 60s, the District of Columbia proposed to build a public housing project designed for poor seniors in the Maryland, USA. This project was highly controversial and local residents fought hard against its implementation based on the ideas that this would bring tax losses, cost, and zoning and property values. However, according to Butler, the financial concern was only part of the resident's feelings of irritation. Instead, Butler argued that there seemed to be, in fact, an underlying "ageism" against these older people which he defined as "the subjective experience implied in the popular notion that of the generation gap…a deep seated uneasiness on the part if the young and the middle-aged—a personal revulsion to and distaste for growing old, disease, disability, and fear of powerlessness, 'uselessness', and death" (Butler 1969).

Currently the most commonly accepted definition of ageism involves two basic ideas: (i) it is an "umbrella" concept (Braithwaite 2002) which refers to the several components of attitudes. Hence, several authors (Kite and Wagner 2002) adopt the traditional tripartite model of attitudes (Eagly and Chaiken 1993) and assume that ageism comprises three basic components which are an affective component, represented by prejudicial feelings (i.e. if I like the group), a cognitive component, represented by beliefs and stereotypes about age groups (i.e. what I think about the group), and a behavioural component, represented by behaviour and discrimination against the group (i.e. how I act towards the group).

Several international studies on ageism have shown that old age is perceived systematically in a more negative way than younger ages (Kite et al. 2005; Nelson 2002) and that these negative representations are often associated with prejudice and discrimination against older people. These studies showed that older people are perceived to be inferior to middle-aged people in aspects like power and social status, wealth, respect and influence (Cameron 1970; Foner 1984; Garstka et al. 2004; Pampel 1998; Youmans 1971). Hence, older people seem to be in fact, the lowest status group at least in our western contemporary societies (Garstka et al. 2004). Why is old age devalued in comparison with other age categories?

Several factors are frequently referred to explain these negative perceptions of older people (Cuddy and Fiske 2002). The first one is the fact that older people are not actively working. In a society that values productivity, this factor considerably diminishes their perceived value as it seems to impoverish their contribution to society and diminish their objective wealth and power (Branco and Williamson 1982; Nelson 2005). The second one refers to the fact that they are now considered a burden to society due to the amount of spending that governments make with aging programs (and this is linked with the fact that we witness an unprecedented growth in the number of older people who in need of assistance). Today there is a diffused idea that the funds are not distributed in an equitable way to all age groups and that older people are using much more than what they need at the expenses of younger generation (Binstock 2005). In this regard, a recent study has shown a significant interaction between the level of modernization of a country (measured as a compound index of measurable indicators of life expectancy, income, education and the level of country's urbanization) and the percentage of older people actively working on the perceptions of older people social status (Vauclair et al. 2015a). These results show that, especially in less modern countries, older people are perceived as a group of higher social status if they also still have an active working role in society. It seems that working status may contribute to improve the material situation of living of older people in a poor country, thus improving their perceived social status.

Finally, a third factor that seems to explain ageism against older people is the fact that they are perceived as the most threatening reminder to individuals of their inevitable mortality. According to *terror management theory* (Greenberg et al. 2002) younger generations fear their own fate of diminishing beauty, health, sensation and ultimately, death. In this sense, they tend to neglect older people and the aging process. All these factors seem to contribute to maintain ageism against older people.

Faced with this situation, how can older people cope with this type of stigmatization? Does age discrimination affect the health of older people?

In the next section we explore this issue by giving in more detail the example of two cases we analysed based on the data collected with the "Experiences and expressions of ageism" of the 2008/2009 Round 4 of the European Social Survey. We show how the social identity approach, as a theoretical framework, may be useful to understand real-life and complex phenomena as age discrimination against older people across European countries.

3.3.1 *Consequences of Negative Attitudes to Age and Age Discrimination for People Aged 70 and Over Across the European Region: Two Case Studies*

This work used data from the European Social Survey (ESS). The ESS is an academically driven cross-national survey that has been conducted every two years across Europe since 2001. The survey measures the attitudes, beliefs and behaviour patterns of diverse populations in more than 30 nations. The ESS employs the most rigorous methodologies to achieve cross-national comparability of the data. The survey consists of a fixed module and two or three rotating modules. Round 4 with data collection in 2008 and 2009 included the rotating module on *Experiences and Expressions of Ageism*. The module was designed by members of the EURAGE group led by Prof. Dominic Abrams (University of Kent), Prof Luisa Lima (CIS-IUL at ISCTE-IUL) and Prof. Genevieve Coudin (Université Paris V). It contains 55 items which were developed and pilot tested extensively within a framework that has been subjected to detailed scrutiny, peer review and evaluation by experts in the ESS Central Coordinating Team. The ageism module provides representative samples from 29 countries and over 55,000 individuals belonging to the European region (with the addition of Israel). The survey methodology is based on computer-based personal interviews, with national samples of between 1215 and 2576 people aged 15 years. The questions in the module reflect seven key domains which were guided by theoretical models from social psychology, i.e. social identity approach (which focus on the way people categorize one another, and hence who they are likely to stereotype), stereotype content theory (why particular groups are stereotyped in particular ways), intergroup threat theory (how different types of threat give rise to prejudice), and intergroup contact theory (the idea that friendship across group boundaries can reduce intergroup prejudice). It is a clear example of how theories in social psychology may be useful to understand wide-scale phenomena regarding ageism and aging across different countries. In this chapter we will give special emphasis to explanations based mainly on the social identity approach.

The research presented in this section followed multilevel modelling statistical techniques. Multilevel modelling is used when data have a hierarchical or so-called nested or clustered structure, which means that observations at one level of analysis are nested within observations at another level. In the ESS, individuals are nested within countries and we analysed this with two-level modelling; level 1 being individuals and level 2 being countries. The outcome variable is then measured at both levels and predictor variables will often differ between the individual- and the country-level. For instance, subjective health can be the outcome variable at both levels; individual predictors (e.g. age) are used at level 1, country-level predictors are used at level 2 (e.g. GDP). Beyond examining main effects of predictors at level-1 and level-2, multilevel modelling can also investigate interaction effects between predictors at different levels (see Case 2).

An important consideration is that multilevel research requires multilevel theories which entail an interdisciplinary approach. When researchers use context variables at the country-level as predictors for individual-level outcomes, they are conducting interdisciplinary research at the interface of sociology and psychology. Although there are numerous challenges inherent in this approach, as with any interdisciplinary research, testing multilevel theories with multilevel modelling can be very powerful, as we shall see in more detail later, beside its academic value it can also be highly informative for policy-makers (see, e.g. Abrams et al. 2011b). Next we present two case studies where we show how this method yields significant and important findings that may be the basis for solid gerontological policy and social interventions.

3.3.1.1 Case 1: Perceived Age Discrimination as a Mediator of the Association Between Income Inequality and Older People's Self-rated Health in the European Region

Self-perceived health among people aged 70 and older varies between countries, apparently dependent on income inequality within the country. In this study, we demonstrate that this association between income inequality and lower subjective health among older people can be explained by perceived age discrimination. The negative association between the income inequality of a country and poor health is known as the relative income hypothesis and has been widely researched by epidemiologists and sociologists since the 1990s. Until recently, much of the research has investigated this relationship between income inequality and health at the country-level (Ram 2006; Wilkinson and Pickett 2006, 2007). However, the multilevel approach we pursued allowed us to further research by exploring the process that explains how the income inequality of a country influences the self-perceived health of individual's aged 70 and over.

Taking into consideration the social psychological literature we noted that prejudice and discrimination against low status groups is more prevalent in unequal societies (Marmot and Wilkinson 2001; Wilkinson and Pickett 2007). As previously noted, older people are usually seen as a low status group relative to other age groups across Western and European cultures (Abrams et al. 2011a; Garstka et al. 2004). Thus, in more unequal societies older people belong to a social group that should be especially vulnerable to prejudice and experiences of discrimination. Evidence also shows that experiences of discrimination is a psycho-social stressor that can have quite profound negative impact on an individual's health (see Pascoe and Smart Richman 2009, for a meta-analytic overview) and this extends to perceived age discrimination (Luo et al. 2011; van den Heuvel and van Santvoort 2011; Vogt Yuan 2007). According to the social identity approach (Tajfel and

Turner 1986), belonging to a devalued group renders individuals with a negative self-evaluation that should be the basis for lower self-esteem. Also, the experience of discrimination incorporates both a social rejection and a largely uncontrollable event which are the two psycho-social stressors that have been found to be associated with the largest increase in stress hormones and the longest time of recovery (Dickerson and Kemeny 2004). Stress hormones, such as cortisol, are related with psychological, physiological and physical health functioning and can increase the risk of negative health outcomes (McEwen 1998). The common perception that older people have low social status, together with a societal context characterized by income inequality, are likely to increase older people's vulnerability to age prejudice. As prejudice is a stressor that chronically activates the physiological system with adverse health effects, it is likely to be an important psycho-social factor that explains how income inequality affects the health of older people.

We tested these pathways using ESS data by specifying a multilevel mediation model to explain self-perceived health, with income inequality specified at the country-level and perceived discrimination specified at the individual level (for further details see Vauclair et al. 2015b; data available in http://www.europeansocialsurvey.org/methodology/questionnaire/ESS4_rotating_modules.html). Even after controlling for relevant demographic variables (e.g. gender, age, education, subjective socio-economic status), perceived age discrimination significantly mediated the link between income inequality and self-perceived health in Europe. This finding strongly suggests that it is not only up to the individual to stay healthy in old age, but that the societal and social context matters too. A country's income inequality creates a form of 'social inequality' in which older people are more likely to be discriminated against. This finding is all the more concerning considering that income inequalities are predicted to increase in the future (OECD 2008), suggesting that prejudice and discrimination—important psycho-social stressors—may increase too. Population aging already puts a heavy strain on public and private budgets (International Monetary Fund 2012). However, these findings provide important insights to key challenges more developed countries face in how to prolong the healthy, active years in the aging population. Policy initiatives targeted at promoting health in later life need to take into account a multilevel perspective in order to be effective.

3.3.1.2 Case 2: "Being Old and Ill" Across Different Countries: Social Status, Age Identification and Older People's Subjective Health

In a second study we investigated the extent to which the relationship between age identification and self-perceived health varies depending on the societal evaluations of older people's social status. According to social identity theory, as already referred before, when an important social identity is associated with the membership of a group that is of low status, stigmatized or socially devalued by others, there are significant negative implications for self-esteem and life satisfaction

(McCoy and Major 2003), levels of depression and anxiety and sense of coping, self-efficacy and support (Kellessi et al. 2009). Accordingly, there is some evidence that identifying as an older person is related to worse self-perceived (*subjective*) health (Engle and Graney 1985). For example, Stephan et al. (2012) found that older individuals who perceived themselves as 'old' rated their health as poorer than older individuals who perceived themselves as younger. However, in a number of contexts, research has shown that group identification can serve as a buffer to counteract the negative effects of discrimination on health and wellbeing (Branscombe et al. 1999; Garstka et al. 2004; Ramos et al. 2012). Jones et al. (2011) demonstrated how individuals with acquired brain injury could protect themselves from some the adverse effects associated with their injury by identifying themselves as a group of survivors. Similarly, research revealed that older people in care homes show an increase in wellbeing after acquiring a common group identity (Haslam et al. 2010). This positive effect of social identification on health and wellbeing could be due to increased provision of social support, social ties and social capital (Levine et al. 2005; Iyer et al. 2009). However, we proposed that these mixed findings could be dependent on the value and status attached to people's in-groups (Tajfel and Turner 1986; Verkuyten 2009).

Drawing on social identity theory, we tested the hypothesis that the relationship between old age identification and health and wellbeing in old age should be linked to the perceived social status of the old age group in society (Hogg and Abrams 1988; Oakes et al. 1994; Tajfel and Turner 1986). To do this we explored the moderating role of perceived social status of people aged 70 and over (at the country-level) on the relationship between old age identification and self-perceived health of people aged 70 and over. In other words, using the ESS data we explored whether there was a significant interaction between the perceived social status of people aged 70 and over and individual's age identification in a multilevel model. The results showed that even after controlling for a host of individual difference variables and country differences in wealth as measured by GDP, we found a significant interaction between social status and age identification as predictors of self-perceived health. The positive relationship between age identification and subjective ill health was only present in countries where older people had a particularly low status (such as, Bulgaria, Croatia, Hungry, Poland, Slovakia and Ukraine). In countries where people over 70 had a higher status (such as Switzerland, the Nordic countries, United Kingdom, Belgium and Germany), older people's age identification was not related to subjective ill health (for more information see, Marques et al. 2015a; data available in http://www.europeansocialsurvey.org/methodology/questionnaire/ESS4_rotating_modules.html).

These findings represent a rare but highly meaningful and robust test of social identity theory's core predictions about the way societal status and identification relate to different social groups (cf. Abrams 2013). Specifically, the findings support the macro-social prediction that identifying with a negatively valued social group or category is likely to be an unpleasant state that has negative implications for the self-concept. Here we show that the implications extend to the health of individual members of the group. Interestingly, these results hold even when we control for the effects of the country's wealth and income inequality (measured by GDP and GINI), indicating that the interaction between age identification and social status is not an artefact of economic factors such as wealth or inequality within the country. Status differences between groups reflect the dominant ideology and widely cultural shared beliefs associated with group members (Abrams 2013; Hogg and Abrams 1988). Hence, distinct from the effects that more objective forms of material deprivation may have on an individual's health, *just* being a member of a devalued category with which one identifies has negative implications for subjective health.

3.3.1.3 Main Conclusions of the Two Case Studies

- Both case studies explore the negative impacts of ageism (via perceived discrimination and or negative attributions of status afforded to older people) on the self-perceived health of those aged 70 and older.
- Both case studies on looked at those who self-selected as 'old'.
- Case 2 shows the potential mitigating effects that strong age identity can have on health- but only when societies value older people.
- These studies reveal that tackling income inequality, perceived age discrimination and valuing older people by viewing them as higher status would all impact positively on older people's health.

3.4 How Can Social Psychology Inform Intervention and Policy-Making to Reduce Ageism and Age Discrimination

In this chapter we tried to make the claim that theory is important and a fundamental tool of analyses to guide our visions and interventions in the world. Interventions and policy-making in the gerontological field should be evidenced-based, preferably adapting the knowledge available through the solid testing of theories already established in the fields—such as social identity theory—to gain predictive and explanatory power to understand real issues that matter to older people.

In particular, in this chapter we have shown how ageism may hinder the health of older people, particularly in more unequal countries. Inequality within countries seems to be associated with an increase in the perceptions of discrimination in older people and this in turn adversely affects their health. Hence, in Case 1 we showed that belonging to a group that is negatively evaluated is associated with harmful effects for its individual members (in this case, older people).

However, in Case 2 we also show that this negative effect of ageism on health is very much dependent also on the way older people identify with being old. In fact, in countries where the social status of older people is perceived to be lower, more identification with this age group seems to lead to lower health levels. This relationship is not present in countries where the social status of older people is perceived to be higher. These results are completely in tune with social identity theory's claims, thus showing the significant predictive value of this theory in the explanation of ageism effects on older people across societies. Moreover, given their wide scope and representativeness, we argue that they also hold important and significant implications for interventions and policy-making in this field.

The strategy against ageism is primarily important because ageism violates human rights. In this regard it is important to recognize that, for instance, the article 21 from the Charter of Fundamental Rights of the European Union, already recognizes the discrimination of people because of their age as a violation of fundamental rights of the human being. As we shall see below, several countries have also already adopted such a legal perspective.

The increase in life expectancy has been a real human achievement. However, this is not just a matter of adding "more years to life"; we need a real change to improve also "more life into those years". Living longer, but better is what we should expect for present and future societies. If negative perceptions of old age have such a hindering effect on older people's health levels, they should be changed, and this change should occur at a broad ideological level.

On the other hand, in an aging society with an increasing demographic pressure of the older age groups, the fight against ageism is also a necessity. The proportion of persons aged 60 and over is expected to double between 2007 and 2050, and their actual number will more than triple, reaching 2 billion by 2050. In most countries, the number of those over 80 is likely to quadruple to nearly 400 million by then (United Nations 2015). In this scenario, the maintenance of older people active and healthy is an imperative for maintaining the economic growth and stability of the social security and pension systems.

The type of research described in this chapter is fundamental because it shows, in a scientific manner, that ageism matters and that it has significant effects on a broad sample of people across many countries. In this case, it is important that any prevention strategy should be created at a broad level.

For instance, the European Commission has already identified the need to address this issue. In a recent study, conducted in the realm of the FP7 project SIforAGE (www.siforage.eu) our team had the opportunity to explore the implementation of age discrimination laws (AADL's) across European countries (Marques et al. 2014). This analysis was developed based on two strategies: (1) documental analysis and (2) narrative interviews with key stakeholders. First of all, an analysis of the main documents and legislation regarding age discrimination in a sample of European countries (Austria, France, Italy, Poland and Portugal) was performed. This analysis was heavily based on the respective country reports elaborated by the European Network of Legal Experts in the non-discrimination field. This network of legal experts constitutes an important support to the European Commission by providing independent information and advice on relevant developments in the Member States in the non-discrimination field.

The main findings obtained in the documental analysis revealed that AADL's are already implemented in the legislation of all the 5 countries analysed. Most of these laws are related to the work field, prohibiting the discrimination based on age regarding the public and private employment. More specifically, these laws intend to promote equality regarding the access to job opportunities, career progression and salary increase to all employees independently of their age. Beyond work, most of the countries under analyses have also extended these anti-age discrimination legislations to cover other fields like social protection, social advantages, education, goods and services and housing.

Regarding the compliance of the anti-age discrimination laws there is, however, a gap between legislation and the practical implementation of these laws. This lower level of compliance can possible be related to the complexity of the legal framework, the low awareness and knowledge of legislation addressing discrimination issues and the absence of a specialized body on this field. More specifically, the legal experts from the five countries highlight the importance of developing a coordinated work between different institutions of important areas of action in society like ONG's, social scientists, public administrations and trade unions.

In a second phase, interviews were conducted with public administration employees responsible for the implementation of selected programmes in three areas in the aging field (health, labour and transport) at three levels of analyses (local, regional and national). The goal was to evaluate their knowledge of ADDL's, perceived relevance of such laws and actual compliance of their practices with such laws. A total of 50 interviews were obtained from 5 countries (Austria, France, Italy, Poland and Portugal) through a coordinated work developed between the SIforAGE partners involved in this task.

The results obtained revealed that the majority of the program planners interviewed share an awareness that age discrimination is a widespread phenomenon, affecting older people in several areas of their life such as employment, transportation, health, social media and within the family (mostly in the format of abuse or negligence against older people). Besides, the interviewees highlighted the relevance of the AADL's in order to promote the fight against ageism regarding older people, representing a step forward towards social change.

In accordance with the results obtained, our team suggested important guidelines as the major output of the work developed: avoid complex legislation that makes hard for actors in the practice domains to understand and apply; there should be an increased effort made by the governments to disseminate AADL's across society by promoting a coordinated work between different social actors (e.g. NGOS's and public offices); it is important to create a body of experts similar to European Network of Legal Experts in the non-discrimination field but that accompanies and evaluate the actual implementation of AADL's in the field; and finally, it is fundamental that ageism is addressed in a broader way, in order to promote a wider social change of mentalities.

But besides improving the legislation and its compliance, it is also important to undertake other type of initiatives. For instance, intergenerational activities that promote a closer cooperation between different age groups seem like an important route to take. Contact between different groups seems to be a promising avenue to change ageist perceptions, for instance, in children (Cunha et al. 2014). Also, within the context of the SIforAGE project, we have been working on the development of an intergenerational program to fight ageism in children (see the manual of the program imAGES for more details on the specificities of this program) (www.siforage.eu; Marques et al. 2015b).

These are only two examples of concrete actions that can be undertaken in the route to prevent ageism in our societies. Other types of initiatives of similar nature are, for instance, the training of professionals that interact with older people in health settings or the promotion of a true work age diversity ethics on organizational contexts. All these initiatives are still in their very early ages and should be further developed.

Finally, the studies described in this chapter are based on European data. Apart from the EUA, to our knowledge there are very few studies exploring ageism and age perceptions in other cultures. Given the expression of demographic aging in the world, we believe it would be important to extend the scientific study of this issue to a wider scale through, for instance, the application of a world survey such as, for instance, the International Social Survey Programme.

We hope this chapter showed the need to use theoretically fundamental research to legitimize the importance of this subject. Our main claim is clear: we need in fact more policies and interventions in the gerontological field. Given the urgency of this matter, we should start this enterprise of theoretically driven empirical research as soon as possible.

References

Abrams, D. (2013). Social identity and groups. In J. M. Levine (Ed.), *Group processes* (pp. 267–296). New York: Psychology Press.

Abrams, D., Russell, P. S., Vauclair, C.-M., & Swift, H. (2011). *Ageism in Europe and the UK. Findings from the European Social Survey*. A Report from EURAGE (European Research Group on Attitudes to Age) commissioned by Age UK.

Abrams, D., Vauclair, C.-M., & Swift, H. (2011b). *Predictors of attitudes to age across Europe*. London: Department of Work and Pensions.

Allport, F. (1924). *Social psychology*. Boston: Houghton-Mifflin.

Bengtson, V. L., Rice, C. J., & Johnson, M. L. (2000). *Are theories of aging important? Models and explanations in gerontology at the turn of the century*: SEDAP Research paper.

Binstock, R. H. (2005). Old-age policies, politics, and ageism. *Generations XXIX*, 73–78.

Braithwaite, V. (2002). Reducing ageism. In T. D. Nelson (Ed.), *Ageism: Stereotyping and prejudice against older persons* (pp. 311–337). Cambridge, MA: The MIT Press.

Branco, K. J., & Williamson, J. B. (1982). Stereotyping and he lyfe cycle: Views of aging and the aged. In A. G. Miller (Ed.), *In the eye of the beholder: Contemporary issues in stereotyping* (pp. 364–410). New York: Praeger.

Branscombe, N. R., Schmitt, M. T., & Harvey, R. D. (1999). Perceiving pervasive discrimination among African-Americans: Implications for group identification and well-being. *Journal of Personality and Social Psychology, 77*, 135–149. doi:10.1037//0022-3514.77.1.135

Butler, G. (1969). Age-ism: Another form of bigotry. *Gerontologist, 9*, 243–246. doi:10.1093/geront/9.4_Part_1.243
Cain, L. D. (1964). Life course and social structure. In R. E. L. Faris (Ed.), *Handbook of modern sociology* (pp. 272–309). Chicago: Rand-McNally.
Cameron, P. (1970). The generation gap: Which generation is believed powerful versus generational members' self-appraisals. *Developmental Psychology, 3*, 403–404. doi:10.1037/h0030050
Cuddy, A. J. C., & Fiske, S. T. (2002). Doddering but dear: Process, content, and function in stereotyping of older persons In T. D. Nelson (Ed.), *Ageism: Stereotyping and prejudice against older persons* (pp. 3–26). Cambridge, MA: MIT Press.
Cunha, F., Marques, S., & Borges Rodrigues, R. (2014). GerAções Lx: Pilot project to decrease ageism and promote a positive self-concept in youngsters and seniors. *Journal of Intergenerational Relationships, 12*(2), 184–191.
Dickerson, S. S., & Kemeny, M. E. (2004). Acute stressors and cortisol responses: A theoretical integration and synthesis of laboratory research. *Psychological Bulletin, 130*(3), 355–391. doi:10.1037/0033-2909.130.3.355
Durkheim, E. (1951). *Suicide: A study in sociology* (J. A. Spaulding & G. Simpson, Trans.). New York: Free Press (Original work published 1897).
Eagly, A. H., & Chaiken, S. (1993). *The psychology of attitudes*. Fort Worth, TX: Harcourt Brace.
Ellemers, N., Spears, R., & Doosje, B. (2002). Self and social identity. *Annual Review of Psychology, 53*, 161–186. doi:10.1146/annurev.psych.53.100901.135228
Engle, V. F., & Graney, M. J. (1985). Self-assessed and functional health of older women. *International Journal of Aging and Human Development, 22*, 301–348. doi:10.2190/C8EC-6B0R-TBTD-ULVL
Foner, N. (1984). *Ages in conflict: A cross cultural perspective on inequality between old and young*. New York: Columbia University Press.
Frymer, P. (2005). Racism revised: Courts, labor law, and the institutional construction of racial animus. *American Political Science Review, 99*, 373–387. doi:10.1017/S0003055405051725
Garstka, T. A., Schmitt, M. T., Branscombe, N. R., & Hummert, M. L. (2004). How young and older adults differ in their responses to perceived age discrimination. *Psychology and Aging, 19* (2), 326–335. doi:10.1037/0882-7974.19.2.326
Greenberg, J., Schimel, J., & Martens, A. (2002). Ageism: Denying the face of the future. In T. D. Nelson (Ed.), *Ageism: Stereotyping and prejudice against older persons* (pp. 27–48). Cambridge, MA: MIT Press.
Haslam, C., Cruwys, T., & Haslam, S. A. (2014). "The we's have it": Evidence for the distinctive benefits of group engagement in enhancing cognitive health in aging. *Social Science and Medicine, 120*, 57–66. doi:10.1016/j.socscimed.2014.08.037
Haslam, C., Haslam, S. A., Jetten, J., Bevins, A., Ravenscroft, S., & Tonks, J. (2010). The social treatment: Benefits of group reminiscence and group activity for the cognitive performance and well-being of older adults in residential care. *Psychology and Aging, 25*, 157–167. doi:10.1037/a0018256
Hagestad, G. O., & Uhlenberg, P. (2005). The social separation of old and young: A root of ageism. *Journal of Social Issues, 61*(2), 343–360. doi:10.1111/j.1540-4560.2005.00409.x
Hogg, M. A., & Abrams, D. (1988). *Social identifications: A social psychology of intergroup relations and group processes*. London: Routledge.
International Monetary Fund. (2012). *Global financial stability report: The quest for lasting stability*. Washington: Author.
Iyer, A., Jetten, J., Tsivrikos, D., Postmes, T., & Haslam, S. A. (2009). The more (and the more compatible) the merrier: Multiple group memberships and identity compatibility as predictors of adjustment after life transitions. *British Journal of Social Psychology, 48*, 707–733. doi:10.1348/014466608X397628
Jones, J. M., Haslam, S. A., Jetten, J., Williams, W. H., Morris, R., & Saroyan, S. (2011). That which doesn't kill us can make us stronger (and more satisfied with life): The contribution of

personal and social changes to well-being after acquired brain injury. *Psychology & Health, 26*, 353–366. doi:10.1080/08870440903440699
Kellessi, B., Reicher, S., & Cassidy, C. (2009). Surviving the Kosovo conflict: A study of social identity, appraisal of extreme events, and mental well-being. *Applied Psychology: An International Review, 58*, 59–83. doi:10.1111/j.1464-0597.2008.00382.x
Kite, M., Stockdale, G. D., Whitley, B. E., & Johnson, B. (2005). Attitudes toward younger and older adults: An updated meta-analytic review. *Journal of Social Issues, 61*, 241–266. doi:10.1111/j.1540-4560.2005.00404.x
Kite, M., & Wagner, L. S. (2002). Attitudes toward older adults. In T. D. Nelson (Ed.), *Ageism: Stereotyping and prejudice against older persons* (pp. 129–162). Cambridge: MIT Press.
Levine, R. M., Prosser, A., Evans, D., & Reicher, S. D. (2005). Identity and emergency intervention: How social group membership and inclusiveness of group boundaries shapes helping behavior. *Personality and Social Psychology Bulletin, 31*, 443–453. doi:10.1177/0146167204271651
Luo, Y., Xu, J., Granberg, E., & Wentworth, W. M. (2011). A longitudinal study of social status, perceived discrimination, and physical and emotional health among older adults. *Research on Aging, 34*, 275–301. doi:10.1177/0164027511426151
Marmot, M. G., & Wilkinson, R. G. (2001). Psychosocial and material pathways in the relation between income and health: A response to Lynch. *British Medical Journal, 322*, 1233–1236. doi:10.1136/bmj.322.7296.1233
Marques, S., Mendonça, J., Vauclair, C. M., Bernardes, S., Batel, S., & Lima, M. L. (2014). The SIforAGE project: Social innovation for healthy and active aging. *Revista Transcultural, VI*, 69–86.
Marques, S., Swift, H. J., Vauclair, C.-M., Lima, M. L., Bratt, C., & Abrams, D. (2015a). 'Being old and ill' across different countries: Social status, age identification and older people's subjective health. *Psychology & Health, 30*, 699–714. doi:10.1080/08870446.2014.938742
Marques, S., Vauclair, C. M., Rodrigues, R., Mendonça, J., Gerardo, F., Cunha, F., Sena, C., & Leitão, E. (2015b). *imAGES—Intervention program to promote positive images of aging in children/adolescents*. Lisbon: SCML & LEYA editors.
McCoy, S. K., & Major, B. (2003). Group identification moderates emotional responses to perceived prejudice. *Personality and Social Psychology Bulletin, 29*, 1005–1017. doi:10.1177/0146167203253466
McEwen, B. S. (1998). Protective and damaging effects of stress mediators. *New England Journal of Medicine, 338*, 171–179.
Mead, G. H. (1934). *Mind, self and society*. Chicago, IL: University of Chicago Press.
Nelson, T. D. (Ed.). (2002). *Stereotyping and prejudice against older persons*. Cambridge, MA: Bradford Book.
Nelson, T. D. (2005). Ageism: Prejudice against our feared future self. *Journal of Social Issues, 61* (2), 207–221. doi:10.1111/j.1540-4560.2005.00402.x
Oakes, P. C., Haslam, S. A., & Turner, J. C. (1994). *Stereotyping and social reality*. Oxford: Blackwell.
Oishi, S., & Graham, J. (2010). Social ecology: Lost and found in psychological science. *Perspectives on Psychological Science, 5*, 356–377. doi:10.1177/1745691610374588
Operario, D., & Fiske, S. T. (1999). Integrating social identity and social cognition: A framework for bridging diverse perspectives. In D. Abrams & M. A. Hogg (Eds.), *social identity and social cognition* (pp. 26–54). Oxford: Balckwell Publishers Ltd.
Organisation for Economic Cooperation and Development. (2008). *Growing unequal? Income distribution and poverty in OECD countries*. Paris, France: OECD Publishing.
Pampel, F. C. (1998). *Aging, social inequality, and public policy*. Thousand Oaks, CA: Pine Forge Press.
Pascoe, E. A., & Smart Richman, L. (2009). Perceived discrimination and health: a meta-analytic review.*Psychological bulletin, 135*(4), 531–554. doi:10.1037/a0016059

Ram, R. (2006). Further examination of the cross-country association between income inequality and population health. *Social Science and Medicine, 62*, 779–791. doi:10.1016/j.socscimed.2005.06.034

Ramos, M. R., Cassidy, C., Reicher, S., & Haslam, S. A. (2012). A longitudinal investigation of the rejection–identification hypothesis. *British Journal of Social Psychology, 51*, 642–660. doi:10.1111/j.2044-8309.2011.02029.x

Riley, M. W., & Foner, A. (1968). *Aging and society* (Vol. 1). New York: Russel Sage Foundation.

Schaller, M., Conway, L. G, I. I. I., & Tanchuk, T. L. (2002). Selective pressures on the once and future contents of ethnic stereotypes: Effects of the communicability of traits. *Journal of Personality and Social Psychology, 82*, 861–877. doi:10.1037/0022-3514.82.6.861

Simon, B., Pantaleo, G., & Mummendey, A. (1995). Unique individual or interchangeable group member? The accentuation of intragroup differences versus similarities as an indicator of the individual self versus the collective self. *Journal of Personality and Social Psychology, 69*, 106–119. doi:10.1037/0022-3514.69.1.106

Smith, E. R., & Mackie, D. M. (2007). *Social psychology* (3rd ed.). Philadelphia: Psychology Press.

Stephan, Y., Demulier, V., & Terracciano, A. (2012). Personality, self-rated health, and subjective age in a life-span sample: The moderating role of chronological age. *Psychology and Aging, 27*, 875–880. doi:10.1037/a002830

Tajfel, H. (1972). La catégorization sociale. In S. Moscovici (Ed.), *Introduction à la psychologie sociale* (Vol. 1). Paris: Larousse.

Tajfel, H. (Ed.). (1984). *The social dimension: European developments in social psychology* (Vols. 1 and 2). Cambridge and Paris: Cambridge University Press and Editions de la Maison des sciences de l'Homme.

Tajfel, H., Flament, C., Billig, M. G., & Bundy, R. F. (1971). Social categorization and intergroup behaviour. *European Journal of Social Psychology, 1*, 149–177. doi:10.1002/ejsp.2420010202

Tajfel, H., & Turner, J. C. (1979). An integrative theory of intergroup conflict. In W. G. Austin & S. Worchel (Eds.), *The social psychology of intergroup relations* (pp. 33–47). Monterey, CA: Brooks/Cole.

Tajfel, H., & Turner, J. C. (1986). The social identity theory of intergroup behavior. In S. Worchel & W. Austin (Eds.), *Psychology of intergroup relations* (pp. 33–48). Chicago: Nelson-Hall.

Taylor, S. E. (1998). The social being in social psychology. In D. Gilbert, S. Fiske, & G. Lindzey (Eds.), *The handbook of social psychology* (4th ed., pp. 58–95). New York: McGraw-Hill.

Turner, J. C. (1982). Towards a cognitive redefinition of the social group. In H. Tajfel (Ed.), *Social identity and intergroup relations* (pp. 15–40). Cambridge: Cambridge University Press.

Turner, J. C., Hogg, M. A., Oakes, P. J., Reicher, S. D., & Wetherell, M. S. (1987). *Rediscovering the social group: A self-categorization theory*. Basil: Blackwell.

United Nations (2015). *World population prospects: The 2015 revision, key findings and advance tables*. New York: United Nations, Department of Economic and Social Affairs, Population Division.

van den Heuvel, W. J. A., & van Santvoort, M. M. (2011). Experienced discrimination amongst European old citizens. *European Journal of Ageing, 8*(4), 291–299.

Vauclair, C. M., Marques, S., Lima, M. L., Bratt, C., Swift, H. J., & Abrams, D. (2015a). Subjective social status of older people across countries: The role of modernization and employment. *The Journals of Gerontology Series B: Psychological Sciences and Social Sciences, 70*, 650–660.

Vauclair, C.-M., Marques, S., Lima, M. L., Abrams, D., Swift, H., & Bratt, C. (2015b). Perceived age discrimination as a mediator of the association between income inequality and older people's self-rated health in the European region. *Journal of Gerontology Series B: Psychological Sciences and Social Sciences, 70*, 901–912. doi:10.1093/geronb/gbu066

Verkuyten, M. (2009). Assimilation ideology and situational well-being among ethnic minority members. *Journal of Experimental Social Psychology, 46*, 269–275. doi:10.1016/j.jesp.2009.11.007

Vogt Yuan, A. S. (2007). Perceived age discrimination and mental health. *Social Forces*, *86*(1), 291–311. Univ North Carolina Press. doi:10.1353/sof.2007.0113

Wilkinson, R. G., & Pickett, K. E. (2006). Income inequality and population health: A review and explanation of the evidence. *Social Science and Medicine, 62*, 1768–1784. doi:10.1016/j.socscimed.2005.08.036

Wilkinson, R. G., & Pickett, K. E. (2007). The problems of relative deprivation: Why some societies do better than others. *Social Science and Medicine, 65*, 1965–1978. doi:10.1016/j.socscimed.2007.05.041

Youmans, E. G. (1971). Generational perceptions of life cycle stages in an urban/rural area. *Gerontologist, 1*, 284–288.

Part II
Culture, Contexts and Aging

Chapter 4
Aging and Subjectivity: Ethnography, Experience and Cultural Context

Jason Danely

Abstract Anthropologists use the concept of subjectivity to describe the interplay between feeling, experience and social context. How can ethnography help researchers link theories of subjectivity to practices of working with older adults? This chapter brings together critical gerontology of global aging, narrative gerontology and anthropological theories of subjectivity to examine the experience of aging in contemporary Japan. In 2015, over one in four Japanese people were over the age of 65, and as pensioners enrolled in the national mandatory long-term care insurance programme, older Japanese adults, like those elsewhere in the world, feel pushed and pulled by a variety of interests as they attempt to manage interpersonal relationships, health and hopes. One narrative that has emerged from this context of longevity and care was a narrative of old age as being "burdensome". Using examples of this narrative from fieldwork with older adults between 2005 and 2014, I argue that these concerns reveal tensions between competing subjectivities. While many older people still aspire to maintain selves embedded in interdependent and reciprocal relationships, care services address them as if they were autonomous individuals. This chapter describes the frustration this brings for thinking about future possible selves in old age, and considers alternative cultural models of subjectivity.

Keywords Subjectivity · Japan · Care · Burden · Ethnography

4.1 Anthropology, Aging and Subjectivity

Since the late twentieth century, anthropologists influenced by critical cultural theories developed mainly by post-colonial, feminist and queer studies movements have shifted their focus from descriptions of mainstream social structures to more socially and politically marginalized voices, including those of children, disabled

J. Danely (✉)
Oxford Brookes University, Oxford, UK
e-mail: jdanely@brookes.ac.uk

T. Samanta (ed.), *Cross-Cultural and Cross-Disciplinary Perspectives in Social Gerontology*, DOI 10.1007/978-981-10-1654-7_4

persons, the ill and the old. While this shift has led to important contributions to our understanding of these previously under-recognized groups, and addressed ethically problematic omissions in the field, it has also become the subject of recent critiques for reinscribing the subject of anthropology as what Joel Robbins has referred to as the "suffering slot" (Robbins 2013). That is, marginal voices have been characterized as suffering subjects of dominant culture, at once part of the larger human experience of political subjection but also made into a subject for anthropology. Alternately, anthropology would do well, Robbins suggests, to also study "the good", hopeful, and wellbeing that can be found throughout the world (e.g. Corsin-Jiménez 2008; Suzuki 2013).

Not surprisingly, anthropological work on aging is following a similar trend, from a recognition of the once excluded voices and suffering of older people to a focus on care, empathy, hope and creative aging around the world (cf. Danely 2014; Graham and Stephenson 2010; Lynch and Danely 2013). This work asserts the need to challenge both the popular alarmist discourses on the economic and political consequences of global aging as well as universalist framing assumptions such as "successful aging" (Cosco et al. 2014; Gullette 2004; Lamb 2014; Martin et al. 2015; Rubinstein and de Medeiros 2014) that dominate the gerontological literature in health and social sciences. Anthropology aims to not only offer points for critical reflection, but also to establish a space for examining individual and cultural diversity in aging as a basis for a more inclusive, holistic perspective of everyday lived experience. This chapter argues the concept of *subjectivity* brings these aims together by locating aging in processes of self-formation in response to cultural and political contexts. Increased longevity and rapid, dramatic social change in many parts of the world now challenges local traditional understandings of age-based life-course trajectories (Meiu 2014), while technologies, globalization, and neoliberalism have given rise to new relationships of power between and within generations (Baars et al. 2006; Cole and Durham 2007; Otto 2013; Pols 2012; Vincent et al. 2006). These changes are not only external to and acting upon older persons, but become integrated into personal narratives and emotionally charged subjectivities. Robbins-Ruszkowski (2013a, b), for example, shows how 'Third Age' groups in Poland link moral ideals of self-knowledge and self-responsibility over aging in ways that articulate with older person's values and memories of collective and political identity while at the same time realigning those ideals towards new hopes for the future based on a more global, neoliberal vision. In a very different setting, Leibing (2014) shows how narratives of place and illness become intertwined with aging subjectivities and experiences of bodily suffering in a Brazilian favela. Studies like these illustrate the strength of ethnographic approaches for linking local, personal narratives to global developments in areas like health policy, medical research and citizenship.

These anthropological studies of aging share a common concern with subjectivity: dynamic inner processes that make up our selves (emotions, thoughts, embodied senses, e.g.) as they are shaped and constrained by cultural norms and power relations (Biehl et al. 2007; Luhrmann 2006, Ortner 2005; Parish 2008). Subjectivity has been described as "the emotional world of the political subject"

(Luhrmann 2006, 358), and a perspective that requires putting the "political at the heart of the psychological and the psychological at the heart of the political" (Good 2012). In this way, subjectivity differs from theories of the self as conceived purely as the internal and invisible "seat of the soul", and theories of the subject that hinge on the fundamental social determination of subjects (Lester 2005, 37–47). It differs as well from accounts of "identity" which assume a stable, coherent and self-aware individual. Subjectivity implies a process that is perpetually unfinished, inherently relational and forms our way of responding world.

Gerontologists and geriatric psychologists have written hundreds, if not thousands of research articles on associations between feelings of "subjective stress" and "subjective wellbeing" (which includes elements of both social and emotional wellbeing) and the physical, institutional and everyday life-styles of older adults. However, the results that emerge look more like Latour's (1987) dissected whale—something massive and mysterious to be cut up and measured scientifically—rather than a picture of culturally embedded, thinking, feeling and relational subjects. While feelings of subjective wellbeing or unease are important from both a clinical and humanistic standpoint, they do not in themselves reveal the ways cultural constraints (sanctions, values, stigma, e.g.) transform initial appraisals and self-reflective judgments that make up subjectivity. Critical and cultural gerontologists (particularly from the humanities) have started to bridge the divide between the psychological and the cultural processes affecting aging subjectivity (Twigg and Martin 2014). This work argues that investigating embodiment, narrative, agency and identity helps shed light on ethical and political puzzles of experience (Neilson 2012), gender and sexuality (Twigg 2013) and the ethics of care (Nortvedt 2003), and provide very promising opportunities for interdisciplinary cross-fertilization between the work of anthropologists and more mainstream gerontology.

In this chapter, I will illustrate how thinking about subjectivity can help us understand the experience of older adults in contemporary urban Japan, where I conduct fieldwork. Japanese adults, like those in many other places in the world, are living longer, healthier lives than at any time in history, and yet longevity has brought along new challenges as they find themselves pushed and pulled by a variety of interests (family, community, social welfare agencies, insurance groups, e.g.) while striving to manage interpersonal relationships, day-to-day stresses and maintaining meaning, hope and wellbeing (Danely 2016; Long 2012; Matsumoto 2011; Traphagan 2000). In my research, I was interested in how this new politics of old age in Japan affected not only the social and institutional frameworks, but also how it affected older adults' feelings about loss, dependence, self-worth and meaning in life. I was also interested in whether we could place these feelings in a broader context of Japanese values, aesthetics and concepts of mind/body interconnectedness (cf. Lock 1993; Schattschneider 2003). Exploring subjectivities of older adults in this way, I contend, could help us understand what kinds of lives these older Japanese adults feel possible for themselves, and what kinds of futures become disrupted, blocked, or foreclosed due to cultural and political circumstances.

4.2 Cultural Models of Aging and the Self in Japan

The nature of the Japanese self remains a matter of considerable discussion and debate (Rosenberger 1994). While acknowledging that any singular definition of the Japanese self will be both insufficient and unable to account for historical and cultural change, there is wide agreement that not only is the *cultural model* (D'Andrade 1995) of the self in Japan more interdependent and relationally construed (Doi 1985; Markus and Kitayama 1991; Shimizu 2001) but also that it has sharper divisions between sometimes dramatically contrastive public and private selves when compared to western cultural models of the whole, coherent, individual actor (Bachnik and Quinn 1994; Hendry 1993; Kondo 1990; Levy 1999). Children and adults are socialized to accept and act in a manner that locates the sense of self within a larger relational world that requires empathic awareness and emotional selectivity depending on the social context (Grossmann et al. 2015). This cultural model is reproduced and habitually reinscribed with each generation, embedded in language, habits and popular cultural activities. The nationally holiday "Sports Day" (*taiiku no hi*), for example, is observed by both children and older people, centring on community gatherings where participants compete in relay-races and other cooperative (rather than team or individual) sports events. To the extent that individual winners are recognized at community Sports Day events, they are always recognized first as a representative of their local block association, and not merely as a talented individual. These kinds of events emphasize an ethos of solidarity, discipline and egalitarianism that cuts across socioeconomic and generational divisions, organizing feelings and sensations of participating and incorporating them into subjectivity.

What might an event like Sports Day tell us about aging subjectivity? The last time I participated in one of these Sports Day events, in October of 2013, I noticed an event that I had not seen before, called "Guardian Corps" (*mimamoritai*). The game consisted of six children (three on each side of the elliptical track) running a short distance to pick up a number card laid on the track, then matching their number to those on signs held by older men and women standing further down the track. The children (two for each adult) held hands with the older man or woman between them and "ran" together to the finish line. Guardian Corps was at once a performance of positively valued, extra-familial, community-based intergenerational care and a thinly masked suggestion of elder dependence (everything depended on the actions of the young "guardians"). Those watching the event cheered the children as they escorted their partners across the finish line, even though most of them abruptly dropped the hand of the older person as soon as they crossed the finish line so that they could collect their prize of candy and return to their parents.

The kind of passive, dependent subjectivity performed by older adults in this event at first resembles what anthropologist Ruth Benedict (1946/1974) described as a culturally accepted role for older Japanese adults. Such descriptions, perhaps appropriate a century ago, have contributed to a common assumption that

Confucian values of filial piety, strong interdependence within the family (indicated by high rates of coresidence and care), and cultural traditions such as the veneration of the ancestors, have made older Japanese adults impervious to the kinds of emotional insecurity faced elsewhere in the world. Even in the early 1990s, Hashimoto (1996) characterized older Japanese adults as assuming a greater degree of dependence and decline in old age as well as a greater assumption that care is best provided through one's family relationships (1996, 75–76). Though moving beyond ideologies of the family and focusing on close analysis of individual narratives, Hashimoto nonetheless argues that norms of reciprocity within the Japanese family (and the shame of abandoning an older parent in need) supported a culturally shared pattern of "deservedness" that appeared much less pronounced in her self-reliant American sample (Hashimoto 1996, 83).

I would argue, however, that over the last generation, the narrative of old age and the subjectivity it composes have created greater opportunity to resist a narrative of decline and dependence. Several informants, for example, told me that while older family members used to deserve care and indulgence, changes in the family and in the implications of longevity mean greater concern about "becoming a burden" (*hito no sewa ni naru*) on others. This complaint was often used to explain the rejection of care or coresidence with adult children in old age. Such complaints could be interpreted as masked solicitations of care (*amae*) (Tomita 1994); polite fictions covering real desires with humble self-depreciation; expressions of pride and a desire for independence or a means of enforcing and regulating reciprocity (cf. Rosenberger 2009). These explanations follow a functionalist logic—the complaint allows older people to signal an awareness of anticipated dependence on others while retaining a sense of agency and dignity for the moment—but in what way might we say this strategy indicates older adults' subjectivity?

In order to answer this question, it is important to examine the emotional states associated with feelings of being a burden (not only shame and guilt, but also in some cases acceptance and relief), as well as the political context and broader discourses of the national burden of Japan's aging population. Older Japanese men and women were very aware that unlike their parents' generation, their cohort was likely to face a more prolonged period of dependence and old age, dying not of a single sudden incident, but multiple slowly progressing chronic causes. In this context, complaints of being burdensome express an effort to navigate and make moral sense of two contradictory self-orientations: one that values reciprocal interdependence and another that privileges individual self-reliance and social separation.

In the next sections, I will provide a brief description of the study methods followed by illustrations from ethnographic fieldwork on aging and grief in Kyoto, Japan. Finally, I will reflect on possible implications of this study and others like it for expanding our understanding of aging and wellbeing.

4.3 Ethnography, Narrative and Experience

Anthropologists examine subjectivity by immersing themselves in the worlds of their subjects, not only shadowing, but also participating alongside them in daily activities of the communities they are part of. This allows the researcher to gain a sense of how people go about their everyday lives as they narrate their experiences, motivations, aspirations, struggles and successes. Narratives have been recognized within anthropology as keys to understanding the ways people articulate experiences into forms that can be considered culturally meaningful (Bruner 1991; Jackson 2002). The capacity to form and reform these narratives across transitions in the life course contributes to wellbeing in old age (Baars 2012; McAdams 2005). Wilińska and Anbäcken (2013) note the urgent need for this kind of qualitative analysis in Japan, especially when it comes to understanding wellbeing and emotion. For the ethnographer, these narratives are more than idiosyncratic expressions of bounded selves, but should be examined within their cultural, historical and discursive contexts to derive broader inductive arguments about the influence of material and social conditions that make certain narratives possible while foreclosing on others. By concentrating on subjectivity, we might see how experiences such as aging and care, while universal, are also socially constrained by the particular meanings and values learned over a person's life course, generating the conditions in which individuals compose their own stories (Kenyon and Clark 2001; Phoenix et al. 2010).

This research employed a descriptive, qualitative approach. It is based on ethnographic participant observation and open-ended interviews conducted with older adults in Kyoto, Japan, an urban setting where over one-fourth of the population is over the age of 65 (25.9 %). I followed several key informants as they went about everyday activities in public social settings such as community centres, peer-group clubs, shopping trips, volunteer work and attending religious services. I also spent time in older people's homes, observing interactions with other family members, eating, doing housework and observing ceremonies for the household ancestors. This was essential for understanding the ways old age is embodied in practices in different environments.

I also conducted over one hundred hours of interviews with older adults over the course of 2 years, focusing on 12 key individuals from diverse backgrounds (Danely 2014, 15–18), Interviews included individual life histories as well as personal views on aging, family and bereavement. Observational and interview data from these twelve key informants were analysed for emergent themes, including culturally specific keywords and generational cohort experiences. Narrative coherence and structures were compared between informants of different backgrounds in order to explore the connection between subjectivity, agency, and wellbeing in later life. Initial interviews were conducted between 2005 and 2006 with follow-up interviews conducted with seven of the original ten informants in 2007, 2013 and 2014.

Although some older Japanese people can be quite gregarious, especially with friends and peers, most adhere to social norms of modesty and restraint when it comes to speaking with outsiders. Kyotoites are known to be particularly reticent. One colleague with years of experience across Japan had to give up on Kyoto after three months of attempts to find older people who would speak with her. Long-term engagement was therefore critical for contacting individuals and verifying the content of their narratives over time.

Noticing my frustration early in the fieldwork, one of my closest informants told me, "You have to tell [interviewees] that you are there to learn from them. If you say you are doing some kind of 'research', they will think you are like those people at the hospital or a civil servant from the ward office and you won't get anywhere!" As this woman indicates, older people in Japan are quite accustomed to playing the role of a research subject for various institutions. I realized this later when, after I offered a small gift card to one of my interviewees, she surprised me by adding it to a large envelope she kept full of identical cards received from filling out various hospital surveys. Others I spoke with told me that I couldn't expect anyone to speak to me unless I "*najimu*", a word that literally means becoming "soaked through" like a piece of dyed cloth or dissolved into one substance, like spices in a soup.

If participant observation was the soup, peppered with occasional improvisations and adjusted through constant tastings, then ethnography was the recipe. It was a place where I would write about the way things came together, a collaborative narrative that I would continually refine until I was satisfied that my Japanese readers could sense the familiar feelings and situations.

Gradually, I learned how to *najimu*, spending more time listening, following each individual's concerns as they were willing to share with me. I was lucky enough to meet some key individuals who then provided the first wave of introductions. From there, my network expanded across different areas of the city. When I found it becoming difficult to spend at least one day each month with each informant, I stopped expanding out and concentrated on in-depth narrative work with a cross-section of the overall group. The stories I gathered in turn provided key insights to conflicts, contradictions or confusions that sometimes puzzle large scale questionnaire-based researchers. It also provided case studies that forced me to rethink my assumptions about not only the interaction between nature and culture, or agency and hegemony, but also representation, mediation, practice, embodiment and memory.

4.4 Social and Institutional Context

In 2014, the ratio of the Japanese population over the age of 65 surpassed one in four, or 35 million people, each of whom were eligible to receive benefits under the national Long-Term Care Insurance system (LTCI) and access healthcare under the national health insurance programme. The Japanese health care insurance system is separate from LTCI financially and administratively, although both are important components supporting health and wellbeing, and persons employed in one sector often move to the other and retain important professional relationships as well as personal friendships with those in both systems. In addition, municipalities, wards (an administrative regional sub-section of a municipality) and local neighbourhood groups often held lectures, workshops, volunteer activities and social events that brought together doctors, nurses, care managers, residential care assistants and the whole spectrum of others involved in care work.

LTCI is paid for through a combination of taxes and insurance payments, which are mandatory for everyone over the age of forty and adjusted regularly (Campbell et al. 2010; Tamiya et al. 2011). In 2014, the average rate had, for the first time, exceeded 5000 yen per insured person per month. The LTCI plan also introduced a graded system of care needs assessed using a checklist based on Activities of Daily Living (ADL). Depending on how one scored, the insurance program covered 80–90 % of eligible costs for services such as in-home helpers, day-service respite or residential care. If the insured feels like they require more services either because of a change in condition or a perceived need, they would discuss this with a member of the community social welfare association, an assigned care manager or staff at a regional comprehensive centre (*chiiki hōkatsu sentā*) for older adults who could either refer private services or aid in requesting changes from the ward office overseeing LTCI.

Older adults are not only increasing in number, but they are also living longer, meaning an unprecedented number of people in Japan are over 85 and will likely have 6–9 years of dependence ahead of them. Only 12 % of those needing care (according to the LTCI checklist) list formal services as their main care provider. Over 75 % receive care from family, with 80 % of those coresiding with the family member who is offering care. 41 % of older adults coreside with an adult child (compared with about 15 % in the US, Tamiya et al. 2011, 71). These numbers may indicate instrumental variables at work, rather than the older person's stated preference (Takagi and Silverstein 2006), since overall survey results show a strong preference among older adults to live independently. Neither are older people averse to residential care institutions. LTCI eligible non-profit residential care institutions (nursing homes) have shown improvements in quality to adhere to insurance guidelines, and the general image is much less dreary than it was in the past. As of 2014, over 520,000 people were registered on waiting lists for special nursing care homes (*tokuyō rōjin hōmu*).

Despite Japan's comparatively accessible and affordable long-term care system, there are indications that it will not be able to maintain the current benefits and cost structure as Japanese citizens continue to age. In April of 2014, for example, the latest major revision raised the co-pay on care benefits from ten to twenty percent for those meeting a certain income threshold (one that most living in the city found far too low). At the same time, eligibility requirements for receiving many benefits were changed, especially for those with relatively low need. The consequence for many families is that care responsibilities have become difficult to avoid. For older adults, the feeling of being burdensome is a response to a situation where they are faced with a long life, insecure care and possible friction in the family. They are even seen as a burden on the national economy; the generation who worked hard for some of Japan's most economically prosperous years are now the generation responsible for the economic drag of social care costs.

4.5 Burden Complaints and Emotion

Mr. Hasegawa and his wife live alone in a narrow two-story home in central Kyoto. I had known the Hasegawas for almost 10 years, having first met them at a Senior Community Centre, where Mr. Hasegawa was a well-liked member of a few clubs, but rarely joined his wife, who was fonder of social dance. Now in their mid-eighties, the Hasegawas have stopped attending the centre, citing various physical limitations. As she slowly peeled a persimmon Mrs. Hasegawa sighed, "When you get over 80", "every year feels like you've aged five!"

A heavy smoker whose demeanor was generally anxious and jittery, Mr. Hasegawa slumped back in his chair beside her. He had recently been in the hospital for a serious case of herpes which he had let go for several weeks before his wife finally convinced him to go. He frowned and held his gaze downward as she told the story, gently chiding him as if he were a small child. "He was just impossible when we did end up going", she continued, "He didn't want to stay there and insisted I take him home. Just so impossible!"

After a moment, Mr. Hasegawa, whose speech had become garbled over the years from various gum and tooth problems, managed to grumble out "I don't like hospitals. They just want to take your money and don't let you do anything". His temper was flaring up and his speech became even faster and more difficult to understand. When I asked him what level of LTCI care need (*yōkaigodo*) he qualified for, trying to understand if he was perhaps not taking advantage of some of the resources available, he replied "I've never taken that test!" spitting out the words like rotten fruit. "It's all politicians who say they are taking care of people, but they're just taking your money!"

Mr. Hasegawa was not the only one who felt this way about being cared for, but he was much less reserved about his anger. For others, rejecting care is a quieter matter of social withdrawal. One community leader I spoke with talked about how he has had trouble getting members of his neighbourhood to comply with the long-term care insurance guidelines to assess care eligibility because so many felt that they didn't want to be a burden:

> You know, Japanese people, it is one of our kind of defects but we don't like to have other people take care of us, you know? They just think I don't want to be a burden. I think you should say the truth, that you can't go to the bathroom by yourself or something like that, but then people just say 'I don't want to have others taking care of me'. They used to have the community leaders report who was in need of care and everything, but now it is up to the individual themselves to report… Japanese people don't like to 'let other people in'. That's the difference between Japanese people and foreigners.

Mr. Hasegawa not only rejected the health insurance system, but also care from his sons, both of whom were married and lived elsewhere in Japan. Unlike the government and health system, which he felt was predating on vulnerable older individuals, Mr. and Mrs. Hasegawa often spoke glowingly of their children, who come to visit every month. They were happy and proud to have had successful children with families of their own, but burdening them with care would threaten to disrupt their children's lives and make them feel even more shame. When they spoke about this, Mrs. Hasegawa gave another resigned sigh. Mr. Hasegawa brushed the question aside, dismissing the possibility. It was hardly a decision that made them happy, but one, at least, that made them feel supportive of their children, even at the potential cost of their own health.

While the Hasegawas still had each other (they referred to themselves as "*rōrōkaigo*", or the old caring for the old), feelings of being a burden came up among older people living alone as well. Mrs. Tanaka, for example, had been living alone since her husband died less than one year earlier. While she was still able to do most things herself, she also had emergency information written down and taped to her refrigerator and an emergency button pendant she would wear in case she had an accident and needed assistance. I asked if she was interested in moving, now that she lived alone. To my surprise, she told me,

> My daughter has been asking me to live with her since one of her sons has moved out and she has a little extra space. I think she is worried about me. But I don't want to burden her. I know it is just so Japanese [to say that]. Besides, I don't think our tastes really match. It wouldn't be very comfortable.

After a few moments, however, she looked back down to her lap, where her hands were nervously picking at the ends of her sweater sleeves. "Really, what should I do? I think [my daughter] is busy enough as it is. I don't think I am ready".

Placing the decision to seek care in abeyance was a popular strategy for many, and yet when I followed up with individuals to see if they had taken any concrete steps to organize or arrange a care plan, most had not. Even making a plan, it seemed, was too uncomfortable to begin.

For some, and especially for men, the solution to avoiding becoming a burden lay in forging a life of independence. A 70-year-old widower, Mr. Sato invited me to his flat on the second floor of his son's family's house to see how well this can be done. "I have my own door, even my own key, so I don't have to see my daughter-in-law! Sometimes no one even knows that I am here", he told me proudly, easing himself onto a couch in front of a coffee table covered with various educational magazines. Mr. Sato did not appear worried about being a burden, nor did he expect to live his last years in long-term care. In his words, he wanted to "die alone", without a funeral, since that would be a costly and unnecessary burden. In his mind, living alone was best because he "could die in here and no one would know!"

> When I asked Mr. Sato if he was lonely, he quickly and flatly rejected the idea at first, saying

> You probably wouldn't understand this sort of thing, but I have been living on my own for a while since my wife died, so I am just used to it. The important thing is not being a burden on other people. Nowadays things like 'eldest son' [taking care of you] don't matter [J: don't matter?] Not at all. I mean, for some people it does matter, if you need a successor, but what do I have? I don't have any business or anything, nothing at all to pass on!

When I interviewed him later, little appeared to have changed in his life. He still lived in the same flat, and still attended clubs and activities so he would not grow frail or senile and have to rely on "strangers" for his care. Yet when I asked him about close relationships with friends and family, he replied "My wife is gone, my wife's mother [who lived with us] is gone. I am lonely".

While some like Mr. Sato, saw little point in continuing affective bonds or links of interdependence, others who felt that they couldn't rely on family, turned to organizations in the community. Rather than feeling a burden, participation in these groups offers an opportunity to practice interdependence. Mr. Nakamura, who still ran a small business in his early seventies, participated in the local community organizations, but was concerned about lack of participation of younger families who he felt had less appreciation for community ties:

> We have a community organization to take care of other older people, but when I ask someone who is in their thirties if they want to join, they say, 'what's so interesting about that?'! It isn't about being interesting! It's about doing something for other people. After all, someday you might need some help from other people, right? Young people think about themselves, their own family above everything else. They say they have to take care of their kids or their wife or something so they can't go do community things. That's how young people are. I'm not judging and saying that's wrong or anything…

Mr. Nakamura shrugged his shoulders as he said this, then rubbed his forehead and smoothed back a few wisps of long white hair across his head. He always stopped short of saying the attitudes of younger people were wrong, preferring to adopt an attitude of resignation. This was the case as well when he spoke of the inability to convince his son to move back to the old family home.

Feelings of being a burden are expressed sometimes with sad sighs, sometimes with anger, sometimes with humour. One woman I met went so far as to consult with a diviner in order to find out what to do about her care, telling me "I don't want to live very long and be a burden on everyone. 86 is just fine by me". Last time I saw here in 2014, she had turned 88 and appeared to be living happily with her two daughters.

In any of these cases, it is not difficult to see how reluctance to seek or accept help, and to feel one's own life to be burdensome can be potentially disruptive to formal and informal care providers. In extreme cases, staunch refusals of care may lead to suicide (Popp and Wilhelm 2009; Traphagan 2006). More commonly, refusing care to avoid becoming a burden on others isolates the older person in what has become a widespread phenomenon of precarious solitude, no longer limited to those without children or resources. Many other older people become cognitively impaired in the course of delaying their care plans, a situation that can make arranging formal care, end of life decisions and intergenerational transfers difficult, costly and contentious.

4.6 Narratives of Possible Future Selves

One way to understand these narratives and the subjectivity they represent is to think of them as efforts to manage future possible selves (Markus and Herzog 1991; Parish 2008). That is, when older adults say that they do not want to be cared for, they are imagining the emotional consequences to their future self of disrupting the lives of those around them, of the delicately balanced bonds of interdependence.

If one cannot rely on interdependence in the household, there are typically few other options than to depend on state-provisioned long-term care. This offers another kind of shame—the shame of being abandoned. Long-term care insurance and geriatric care logics operate in an individuating narrative form, beginning, perhaps with the care certification form (the voluntary checklist to assess care need, often manipulated to receive more or less care services), and culminating in many cases with a long-term hospitalization. Entering the domain of long-term care, and especially medical care, was considered "outside of the community" rather than an extension of some shared sphere of the 'public'. In both cases, however, the consequences of an extended period of old age left one dependent, either on one's family or on the social support system, and the most one could then do in this situation would be to try and uphold a moral future self. One could imagine a future self who remains integrated within relationships of mutuality without the need to become the object of care (or abandonment). But maintaining this possible future self has psychological costs, as the emotional strain, worry and anger of older adults like Mr. Hasegawa reveals. Managing the self in the face of uncertainty, change and shifting social positions is an ongoing work that faces additional complications in old age.

Interestingly, many older adults in Japan found hope in an alternate narrative of the future self based in cultural beliefs and practices of caring for the spirits and ancestors (Danely 2014). As I have argued elsewhere, the spirits of the departed are experienced as ever-present, watchful and protective presences that shared an intimate connection to the pains of loss as well as to a hopeful promise of transcendence and continuity beyond age (Danely 2014, 2016). In caring for the spirits through offerings and memorials, older adults could forge a narrative that was moral, vulnerable and dependent. Indeed, older adults, irrespective of formal religious belief, told me that they viewed their life and choices to be largely credited to their ancestors, and so one is naturally dependent upon them, the same way a child is dependent on its parents.

By integrating culturally meaningful narratives such as ancestor memorial into one's own narrative, older adults produced a shift in subjectivity. Suffering and wellbeing were situated into a larger framework of the hitherto and heretofore, as the individual gives way to a greater sense of linked lives, present and invisible. One woman called this simply the 'blue sky', which for her was at once peaceful, clear and without fear for oneself. As one came to inhabit this subjectivity, each encounter and each loss felt meaningful in an intuitive way, and for those who had dreams or visions of parents, siblings and spouses who had passed away, these encounters were usually profoundly comforting.

4.7 Conclusion: Subjectivity as a Bridge

Examining age and subjectivity requires that we take into account the ways older adults compose cultural narratives of wellbeing involving modes of moral experience. These enrich our understandings of the possibilities that might be cultivated, even in the midst of feelings of loss or abandonment. But spiritual subjectivity should not be removed from other cultural values or political relationships. What I have tried to show is that powerful narratives such as being a burden in old age are difficult to separate from sources of meaning and wellbeing. The strain of being a burden on one's family or the state finds a particular elaboration in the dependence on the spirits and ancestors. As life changes, emotions fluctuate, offering chances for reappraisals of subjectivity. Research into the ways this occurs in different settings among different groups is a good starting place for greater dialogue between gerontologists and anthropologists.

References

Baars, J. (2012). *Aging and the art of living*. Baltimore: Johns Hopkins University Press.

Baars, J., Dannefer, D., Phillipson, C., & Walker, A. (2006). *Aging, globalization and inequality: The new critical gerontology*. Amityville, NY: Baywood Publishing Co., Inc.

Bachnik, J. M., & Quinn, C. J. (1994). *Situated meaning: Inside and outside in Japanese self, society, and language*. Princeton: Princeton University Press.

Benedict, R. (1946/1974). *The chrysanthemum and the sword: Patterns of Japanese culture*. New York: New American Library.

Biehl, J. G., Good, B., & Kleinman, A. (2007). *Subjectivity: Ethnographic investigations*. Berkeley: University of California Press.

Bruner, J. (1991). The narrative construction of reality. *Critical Inquiry, 18*(1), 1–21 (1991). doi:10.1086/448619

Campbell, J. C., Ikegami, N., & Gibson, M. J. (2010). Lessons from public long-term care insurance in Germany and Japan. *Health Affairs, 29*(1), 87–95 (2010). doi:10.1377/hlthaff.2009.0548

Cole, J., & Durham, D. L. (2007). *Generations and globalization: Youth, age, and family in the new world economy*. Bloomington: Indiana University Press.

Corsin-Jiménez, A. (Ed.). (2008). *Culture and well-being: Anthropological approaches to freedom and political ethics*. Ann Arbor, MI: Pluto Press.

Cosco, T. D., Prina, A. M., Perales, J., Stephan, B., & Brayne, C. (2014). Whose 'successful ageing'?: Lay-and researcher-driven conceptualisations of ageing well". *The European Journal of Psychiatry, 28*(2), 124–130.

D'Andrade, R. G. (1994). *The development of cognitive anthropology*. New York: Cambridge University Press.

Danely, J. (2014). *Aging and loss: Mourning and maturity in contemporary Japan*. New Brunswick, NJ: Rutgers University Press.

Danely, J. (2016). Hope in an ageing Japan: Transience and transcendence. *Contemporary Japan, 28*(1), 13–31. doi:10.1515/cj-2016-0002

Doi, T. (1985). *The anatomy of self: The individual versus society*. Tokyo: Kodansha USA.

Good, B. J. (2012). Theorizing the "subject" of medical and psychiatric anthropology. *Journal of the Royal Anthropological Institute, 18*(3), 515–535.

Graham, J., & Stephenson, P. (2010). *Aging and Loss*. Contesting the Dominant Paradigm. Toronto: University of Toronto Press.
Grossmann, I., Huynh, A. C. & Ellsworth, P. C. (2015). Emotional complexity: Clarifying definitions and cultural correlates. http://psycnet.apa.org/psycinfo/2015-57060-001/
Gullette, M. M. (2004). *Aged by culture*. Chicago: University of Chicago Press.
Hashimoto, A. (1996). *The gift of generations: Japanese and American perspectives on aging and the social contract*. Cambridge: Cambridge University Press.
Hendry, J. (1993). *Wrapping culture: Politeness, presentation, and power in Japan and other societies*. Oxford: Clarendon Press.
Jackson, M. (2002). *The politics of storytelling: Violence, transgression, and intersubjectivity*. Portland, OR: Museum Tusculanum Press.
Kenyon, G. M., & Clark, P. G. (2001). *Narrative gerontology: Theory, research, and practice*. New York: Springer Publishing Company.
Kondo, D. K. (1990). *Crafting selves: Power, gender, and discourses of identity in a Japanese workplace*. Chicago: University Of Chicago Press.
Lamb, S. (2014). Permanent personhood or meaningful decline? Toward a critical anthropology of successful aging. *Journal of Aging Studies, 29*, 41–52. doi:10.1016/j.jaging.2013.12.006
Latour, B. (1987). *Science in action: How to follow scientists and engineers through society*. Milton Keynes, England: Open University Press.
Leibing, A. (2014). Heterotopia and illness: Older women and hypertension in a Brazilian favela. *Anthropology & Aging Quarterly, 34*(4), 225–237 (2014). doi:10.5195/aa.2014.2
Lester, R. (2005). *Jesus in our wombs: Embodying modernity in a Mexican convent*. Berkeley: Univeristy of California Press.
Levy, B. R. (1999). The inner self of the Japanese elderly: A defense against negative stereotypes of aging. *International Journal of Aging and Human Development, 48*(2), 131–144.
Lock, M. M. (1993). *Encounters with aging mythologies of menopause in Japan and North America*. Berkeley: University of California Press.
Long, S. O. (2012). Bodies, technologies, and aging in Japan: Thinking about old people and their silver products. *Journal of Cross-Cultural Gerontology, 27*(2), 119–137.
Luhrmann, T. R. (2006). Subjectivity. *Anthropology Theory, 6*(3), 345–361.
Lynch, C., & Danely, J. (Eds.). (2013). *Transitions and transformations: Cultural perspectives on aging and the life course*. Oxford: Berghahn Books.
Markus, H. R., & Herzog, A. R. (1991). The role of the self concept in aging. *In Annual review of gerontology and geriatrics, 11*, 110–143.
Markus, H. R., & Kitayama, S. (1991). Culture and the Self: Implications for Cognition, Emotion, and Motivation. *Psychological Review, 98*(2), 224–253. doi:10.1037/0033-295X.98.2.224
Martin, P., Kelly, N., Kahana, B., Kahana, E., Willcox, B. J., Willcox, D. C., et al. (2015). Defining successful aging: A tangible or elusive concept? *The Gerontologist, 55*(1), 14–25. doi:10.1093/geront/gnu044
Matsumoto, Y. (2011). *Faces of aging: The lived experiences of the elderly in Japan*. Stanford, CA: Stanford University Press.
McAdams, D. P. (2005). Studying lives in time: A narrative approach. *In Towards an interdisciplinary perspective on the life course: Advances in life course research, 10*, 237–258.
Meiu, G. P. (2014). "Beach-boy elders" and "young big-men": Subverting the temporalities of ageing in Kenya's ethno-erotic economies. *Ethnos*, 1–25. doi:10.1080/00141844.2014.938674
Neilson, B. (2012). Ageing, experience, biopolitics: Life's unfolding. *Body & Society, 18*(3–4), 44–71.
Nortvedt, P. (2003). Subjectivity and vulnerability: Reflections on the foundation of ethical sensibility. *Nursing Philosophy, 4*(3), 222–230. doi:10.1046/j.1466-769X.2003.00120.x
Ortner, S. B. (2005). *Subjectivity and cultural critique. Anthropological Theory, 5*(1), 31–52.
Otto, L. (2013). Negotiating a healthy body in old age: Preventative home visits and biopolitics. *International Journal of Ageing and Later Life* (advance access) http://www.ep.liu.se/ej/ijal/2013/v8/i1/13-111/ijal13-111.pdf

Parish, S. M. (2008). *Subjectivity and suffering in American culture: Possible selves*. New York: Palgrave Macmillan.
Phoenix, C., Smith, B., & Sparkes, A. C. (2010). Narrative analysis in aging studies: A typology for consideration. *Journal of Aging Studies, 24*(1), 1–11.
Pols, J. (2012). *Care at a distance on the closeness of technology*. Amsterdam: Amsterdam University Press.
Popp, J., & Wilhelm, J. H. (2009). Altern und suizidalität im heutigen Japan. *Godzik, 7*, t3–t93.
Robbins, J. (2013). Beyond the suffering subject: Toward an anthropology of the good. *Journal of the Royal Anthropological Institute, 19*(3), 447–462. doi:10.1111/1467-9655.12044
Robbins-Ruszkowski, J. C. (2013a). Shifting moral ideals of aging in Poland: Suffering, self-actualization, and the nation. In C. Lynch & J. Danely (Eds.), *Transitions and transformations: Cultural perspectives on aging and the life course*. Oxford: Berghahn Books.
Robbins-Ruszkowski, J. C. (2013b). Challenging marginalization at the universities of the third age in Poland. *Anthropology & Aging Quarterly, 34*(2), 157–169. doi:10.5195/aa.2013.18
Rosenberger, N. R. (1994). *Japanese sense of self*. Cambridge: Cambridge University Press.
Rosenberger, H. (2009). Complaint discourse, aging and caregiving among the Ju/'hoansi of Botswana. In J. Sololovsky (Ed.), *The cultural context of aging: Worldwide perspectives* (3rd ed., pp. 30–52). London: Bergin&Garvey.
Rubinstein, R. L., & de Medeiros, K. (2014). Successful aging', gerontological theory and neoliberalism: A qualitative critique. *The Gerontologist, August,*. doi:10.1093/geront/gnu080.
Schattschneider, E. (2003). *Immortal wishes: Labor and transcendence on a Japanese sacred mountain*. Durham, N.C: Duke University Press Books.
Shimizu, H. (2001). Japanese adolescent boys' senses of empathy (omoiyari) and Carol Gilligan's perspectives on the morality of care: A phenomenological approach. *Culture & Psychology, 7*(4), 453–475. doi:10.1177/1354067X0174003
Suzuki, N. (2013). Preface. In N. Suzuki (Ed.), *The anthropology of aging and well-being: Searching for the space and time to cultivate life together*, *Senri Ethnological Studies* (Vol. 80, pp. i–ix). Osaka: National Museum of Ethnology.
Takagi, E., & Silverstein, M. (2006). Intergenerational coresidence of the Japanese elderly are cultural norms proactive or reactive? *Research on Aging, 28*(4), 473–492.
Tamiya, N., Noguchi, H., Nishi, A., Reich, M. R., Ikegami, N., Hashimoto, H., et al. (2011). Population ageing and wellbeing: Lessons from Japan's long-term care insurance policy. *The Lancet, 378*(9797), 1183–1192.
Tomita, S. K. (1994). The consideration of cultural factors in the research of elder mistreatment with an in-depth look at the Japanese. *Journal of Cross-Cultural Gerontology, 9*(1), 39–52.
Traphagan, J. W. (2000). *Taming oblivion: Aging bodies and the fear of senility in Japan*. Albany: State University of New York Press.
Traphagan, J. W. (2006). Intergenerational ambivalence and suicide in rural Japan. *Journal of Intergenerational Relationships, 8*(1), 21–37.
Twigg, J. (2013). *Fashion and age: Dress, the body and later life*. London: A&C Black.
Twigg, J., & Martin, W. (2014). The Challenge of Cultural Gerontology. *The Gerontologist,*. doi:10.1093/geront/gnu061
Vincent, J. A., Phillipson, C., & Downs, M. (2006). *The futures of old age*. Thousand Oaks, Calif: SAGE Publications Ltd.
Wilińska, M., & Anbäcken, E.-M. (2013). In search of the everyday life of older people in Japan: Reflections based on scholarly literature. *Journal of Cross-Cultural Gerontology, 28*(4), 435–451. doi:10.1007/s10823-013-9204-7

Chapter 5
Neoliberalism and Resilience Among Older Yoruba People in a Semiurban Community, South West Nigeria

Ojo Melvin Agunbiade and Akanni Ibukun Akinyemi

Abstract Growing older in some communities in Africa is becoming more challenging and worrisome. The situation is compounded by an increase in neoliberal based reforms and the absence of state-funded pension schemes that could cushion the everyday challenges that comes with old age. By adopting an interactionist perspective, this chapter examines the consequences of neoliberal social policies by focusing on experiences of ageism and adoption of resilience among Older Yoruba People in a Semiurban Community, South West Nigeria. Using a purposive sampling approach, 37 elderly Yoruba men and women from two post-conflict communities participated in semi-structured face-to-face interviews. The findings revealed resilience as essential to surviving elderly abuse, marginalization and neglect in old age. Among this social category of older Yoruba people, resilience is functional to achieving healthy aging experiences. This is coming up within a cultural context devoid of formal social support in meeting aging needs and challenges. In confronting some of the daily challenges, the participants narrated their adoption of resilience in coping with different life situations. Ageism did not emerge as a major issue at a personal level but perceived as an increasing problem for other elderly in their communities. However, some of the narratives confirmed different dimensions of ageism as participants' recounted abuse and dwindling quality in intergenerational relations. Hence, prompt identification and promotion of resilience across the diverse social categories of the elderly might be instrumental in negotiating aging experiences and coping with the inherent challenges of liberalization and chatting clear directions in later life within a social context.

O.M. Agunbiade (✉)
Department of Sociology and Anthropology, Obafemi Awolowo University, lle-Ife, Nigeria
e-mail: ojomelvin@yahoo.com

O.M. Agunbiade
Department of Sociology, University of the Witwatersrand, Johannesburg, South Africa

A.I. Akinyemi
Department of Demography and Social Statistics, Obafemi Awolowo University, lle-Ife, Nigeria

T. Samanta (ed.), *Cross-Cultural and Cross-Disciplinary Perspectives in Social Gerontology*, DOI 10.1007/978-981-10-1654-7_5

Keywords Ageism · Neoliberalism · Elderly people · Resilience · Nigeria

5.1 Introduction

Across different cultural settings, there are variations in the ways old and older people are treated, and this may range from reverence and respect to abandonment and deprivation (Calasanti 2005). Within time and space, each dimension of these possible experiences has its consequences on older people themselves and the society. Abandonment and deprivation have negative consequences on the aging experiences of old and older people. There are challenges in arriving at a consensus on a particular terminology that can capture the deprivation and marginalization that comes with the aging process. A somewhat encompassing terminology that captures age-based stereotyping and discrimination is the concept of ageism (Kite et al. 2004). As a social construct, ageism describes "a set of social relations that discriminate against older people" (Minichiello et al. 2000; Kite et al. 2004).

Stereotyping and age-based discrimination has psychosocial (Garstka et al. 2004; Levy 2009; Coudin and Alexopoulos 2010), and health implications on the aged (Grant 1996; Wagner and Wagner 2003). These consequences are different as there are variations in the experiences and possible implications of age-based discrimination across cultures (Cruikshank 2013; Rippon et al. 2014). Accounting for the variations includes prevailing political and economic ideology, cultural beliefs, values and practices (Barber et al. 2014; Adeniyi-Ogunyankin 2012; Aléx 2010; Calasanti 2005). These factors among others also affect the quality of the networks of support, an important social variable that can mitigate the negative aspects of the aging process, improving experiences and quality of life (Thomas 2010; Hill et al. 2014).

In some traditional African communities, old age is something that is desirable, and it serves as evidence of approaching contact with the spirit world (Cohen 1994; Togunu-Bickersteth 1988). As such, there is a relatively favourable social disposition to aging signs such as greying, thinning hair and wrinkling of the skin. As a reinforcement, old age in some Yoruba communities also comes with chieftaincy titles, increased engagement in rituals and a compassionate inclusion in other traditional duties (Togunu-Bickersteth 1988; Ajala 2006). However, over the years, and similarly to changes in many developing countries, the social values attached to old age are somewhat decreasing mainly with the potential effects of tailored neoliberal reforms. These changes, among other factors, have increased the vulnerability and marginalization of old and older people thereby worsening their aging experiences (Adeniyi-Ogunyankin 2012). At the individual level, resilience is an important individual variable that mitigates the negative experiences and reactions to age-based discrimination, marginalization and perceived vulnerability in old age.

Socio-economic and political challenges affect resilience at personal and social levels thereby predisposing old and older people to abuse, discrimination and marginalization. Also, elderly abuse thrives where either a high degree of poverty, inadequate social response by way of appropriate policies or the lack of a robust informal network of support. With the growing socio-economic and political challenges in sub-Saharan Africa, high levels of poverty and neglect of the elderly make the aging process challenging and unrewarding (Aboderin 2011). Available empirical evidence indicates that the absence of a formal and informal network of support in old age increases the predisposition and vulnerability to a series of health challenges (Mégret 2011; Shetty 2012). Several studies have examined ageism and the effects on quality of life and wellbeing among elderly people, but few studies are exploring the meanings, experiences and strategies of coping with ageism (Minichiello et al. 2000; Calasanti 2005; Lagacé et al. 2012). In the absence of a quality support network, old and older people as active co-constructors of their realities adopt different coping measures by becoming resilient and in minimizing the negative aspects of aging (Lagacé et al. 2012).

With the increasing emphasis on reforming economies through liberalization and the absence of state-funded pension schemes for the aged, there is the possibility that abuse, vulnerability and survival challenges may become common in the everyday experiences of old and older people. As an ideological framework, neoliberalism promotes individuality and competition as vital to attaining competitive advantage. Such emphasis will ensure a shift in focus from mutual support to individual survival and possible depreciation in the fragile informal support for the aged and other marginal social categories. In the Yoruba worldview, old age constitutes a period in which old people depend on their family members for survival (Togunu-Bickersteth 1988). This places a heavy burden on family members and children, in particular, to care for the wellbeing and survival of their elderly relatives (Togonu-Bickersteth 1989; Dokpesi 2015). The gradual adoption of neoliberal policies in a mixed economy like Nigeria has led to a reduction in policies that promote subsidizing of some essential services for the general populace. Against this backdrop, marginalized and vulnerable groups, in particular, will become disadvantaged because of the limited or absent social safety nets that could promote healthy aging experiences.

By adopting an interactionist perspective, this chapter examines the consequences of neoliberal social policies with a focus on ageism and resilience in the everyday living of older Yoruba people in a semiurban community in South West Nigeria. This chapter starts with an overview of cultural scripts around older adult caregiving and support in Yoruba culture. Drawing from available studies, we focused on the basics of neoliberalism and a possible situation of aging experiences, vulnerability and resilience within the framework of neoliberalism. The framework provides an opportunity to understand how old and older people employ their agency in negotiating survival challenges through the aging process. Using a qualitative approach that relies on face-to-face interviews, it was possible to generate empirical and situational evidence on the aging experiences and survival strategies among old and older Yoruba people. The findings show variations in agency and resilience within a relatively diverse social category of elderly. In conclusion, we argue that prompt identification and promotion of resilience strategies might be instrumental in improving aging experiences and coping with the inherent challenges of liberalization in low-resource settings.

5.2 Cultural Scripts Around Older Adult Caregiving and Support in Yoruba Culture

Cultural scripts around older adults caregiving and support are the guidelines that influence individual perceptions, response and relations with old and older members of the society (Fajemilehin 2001; Sokolovsky 2009). By nature, these scripts are dynamic, but transmittable and recycled from generations to generations. Broadly, cultural scripts around older adult caregiving and support exist on three levels, namely: the cultural descriptors, interpersonal and individual conviction (Sokolovsky 2009). In the Yoruba cultural contexts, cultural descriptors include non-verbal expressions and verbal statements encrypted in proverbs and traditional wise sayings that characterized expectations and outcomes from relations within and among different social categories (Olajide 2012). Through this process, the when, where, how and benefits of providing support and caregiving around old age become established (Olajide 2012). In many African traditions, proverbs are cultural wisdom employed in conversations and everyday expressions that promote conformity to social expectations and practices (Richmond 1987).

> Ọmọdé k. í w. Ṣọṣọ níbùjók.ó agba
> Etiquette demands that youth avert its eyes from age or, at least, look deferentially at it (Owomoyela 2005).

> Eetán lelégbè ẹyin; ọmọ bíbí inú ẹni lelégbè ẹni.
> Young palm fruits are the support of ripe ones; one's children are one's support (Owomoyela 2005: 304).

Like in many cultural contexts, the mutual rights and duties of the young to the aged are transmitted in diverse ways and promoted through various networks of relations to ensure continuity of support for the elderly. While the actual manifestations and availability of the needed support differ in practice, cultural descriptors provide an adaptable framework in a various network of relations and outcomes (Olajide 2012; Fajemilehin 2001). Caregiving and supports for older people are reciprocal and is based on the expectation that such elderly people during their productive years have given the needed support to their children or significant others. Among the Yoruba people, such scripts are gendered and context dependent (Oyewumi 2009; Makinde 2004). The interpersonal scripts also include the cultural rights and duties of children to their aged parents. At the interpersonal level, rehearsals of and possible adoptions of the cultural descriptors around providing support for the elderly or aged are invoked in the ways individuals or social categories talk or relate with their aged in public and interpersonal relations (Olajide 2012). For individuals, such cultural scripts include cognitive interpretations and adoption of these social expectations in their social relations with their parents and other old and older people. For individuals, the performance of and adherence to these duties and responsibilities could produce tensions, which could, however, be overlooked for their benefits.

For several reasons, deviations from such cultural obligations and rights are possible. Some individual factors and experiences could promote deviations on the parts of parents to their children and vice versa. Partly, this may depend on structural factors such as religious beliefs, socio-economic and political factors that produce systemic failures and tensions in society and at the household level. As such, the philosophy of caregiving and support from adult children to their aged parents or the elderly is changing gradually among the Yoruba people (Togunu-Bickersteth 1988). This change may also be attributed to many factors including modernization and adoption of new orientations to familial relations and responsibilities (Togonu-Bickersteth and Akinyemi 2014). This growing shift in the performance of duties and responsibilities affects both the old and the young, parents and children.

In summary, cultural descriptors provide a guide on the micro and macro network of relations in the household, community, workplace and other public places. With the growing influence of neoliberal based reforms in Nigeria, it will be useful to understand the implications of this structural change in aging experiences and performance of duties and rights as defined in cultural scripts of caregiving and support provision. Old age comes with vulnerability and dependency, which call for support at different levels. There are health, material and nonmaterial conditions that increase the need for support in old age. Despite the availability of empirical evidence showing the need for qualitative support in old age (Liu et al. 2014; Reichstadt et al. 2010), socio-economic and political challenges keep altering the availability and accessibility of support and subjective wellbeing in old age (George 2010). The situation is worrisome in societies with inadequate or absent social pensions for the aged.

5.3 Theoretical Framework

Neoliberalism as a terminology attracts different interpretations and portrays itself in diverse forms. In recent times, neoliberalism represents socio-economic and political ideologies that emphasize the roles of market forces and entrepreneurship as hallmarks of setting and realizing personal and societal goals and fulfilments (Peck et al. 2009). Within this framework, market forces power and state agents are given the right to dictate the process and distribution of scarce resources within given time and space (Harvey 2005). The key aim is to ensure that profit and capital accumulation gets into the hands of individuals and groups that will further improve their wellbeing and overall that of others. As a philosophy, competition and marketization ensure the "survival of the fittest" (Harvey 2005). As neoliberal reforms progress, individuals and social categories imbibe values and practice that emphasize freedom and personal survival. The expectation is that through adherence to these competitive values governed by market forces, high productivity and protection against abuse are readily encouraged and sustained. However, with the volatility of the market system and its interconnectivity with other systems at the global level (Peck et al. 2009), survival within sociocultural settings that are devoid of quality and responsive network of support becomes daunting.

From the viewpoint of neoliberalism, reforms will ensure efficiency and enhance the quality of life for everyone. However, neoliberal reforms in Nigeria have also brought hardships and the dwindling of networks of support for the aged (Adeniyi-Ogunyankin 2012). Rather than protecting the elderly, as a social category, elderly people are more dependent, disadvantaged and weak, and exposed to abuse and neglect. Compulsory retirement age and the intermittent provision of state pensions and welfare services for the retirees make old age a period even more vulnerable and marginalized. The situation is more challenging for old and older people who never had any formal employment and had had to work late into old age just to earn a living. Part of the challenges of surviving economic liberalization includes the ability to carve out a niche through competitive advantage. A local economy that is predominant agriculture, the thriving of commercial farming will imply the death of subsistence farming as subsidy regimes are altered to pave the way for market forces. When the foundation for liberalizing the Nigerian economy was laid, there were inadequate or inexistent relevant measures put in place to cope with the latent functions of neoliberalism. Then it is okay to see a worsening of the situation of the elderly especially those dwelling in cities or urban areas where individualism and efficient measures prevail over common forms of relations that characterized rural lives.

Despite the devastating effects of neoliberal reforms, some studies have revealed that older people with their physical weaknesses deploy different strategies including resilience in surviving poverty, ageism and neglect (Aléx 2010; Manning 2012). While in traditional Yoruba culture, old age is cherished and attracts respect from the younger generation, this perception is changing especially with the growing emphasis on materialism as the hallmark of success and the idea of competition and individual accumulation of wealth. From a micro-level gerontological theory, this chapter explores the aging experiences of this social category of elderly Yoruba people. It situates the individual experiences of this social category of adults within a neoliberal stance. Against this backdrop, it is possible to understand how old and older people utilize their agency in coping with the inherent contradictions within the neoliberalism framework.

5.4 Methodology

5.4.1 Research Design

In this study, the use of exploratory research design provided an opportunity to investigate ageism as embodied realities that cut across other spheres of the aging process. Exploratory designs are not novel in aging studies despite the scarcity of such studies in the literature. As such, Williams and Mohammed (2009) in a recent review call on researchers to adopt more of this type of design in understanding and addressing discrimination and inequalities in accessing important social resources especially among the growing elderly population across the globe. The use of such designs has tremendous advantages, among which the opportunity to understand the situation of events and resolutions from the perspectives of the old and older population. It also gives researchers a better opportunity to look into the world and challenges of aging. For the study participants, it develops a sense of belonging to the research process. It also provides needed information that could serve as indicators of new directions in probing into other areas of research interests. With a focus on aging experiences, an exploratory research design that consists of 37 face-to-face interviews was conducted among elderly Yoruba men and women recruited from two post-conflict communities (Ile-Ife and Modakeke-Ife).

5.4.2 Study Setting

The Yoruba people are one of the largest ethnic groups in Nigeria. Others include the Igbo and the Hausa people. There are differences among the Yoruba people despite the commonality of their history, and different sub-ethnic groups within the Yoruba group. These sub-ethnic groups consist of dominant groups like the Akoko, Awori, Egba, Ekiti, Ife, Ijebu, Ijesa, Okun, Oyo and some other marginal groups that are closely linked with the dominant groups (Falola 1999: 1). Despite the variations, there are close similarities in beliefs and consistency in practices around the care and support for the elderly among the various groups. This provides an opportunity to focus on a sub-ethnic group to gain insights into the commonality of the existing beliefs and practices and support for the elderly.

Thus, for this study, we recruited elderly Yoruba people aged 60 years and above from the two Yoruba communities. Both communities also have histories of clashes that could have worsened the availability of support for the elderly and a probable need for more resilience to confront the everyday challenges of living in post-conflict communities (Adisa and Agunbiade 2010). There are cases of conflicts in many communities in Nigeria (Onigu-Otite and Albert 1999), but the two communities were selected for this study for their proximity to the authors. A focus on the situation of the elderly in these two post-conflict Yoruba communities is not representative to capture ageism and neoliberal challenges in Nigeria. Nevertheless, a focus on the aging experiences of elderly people from these two communities provides an opportunity to gain insights into the daily challenges of living and survival strategies. Within the study settings, it becomes possible to understand how this social category of elderly Yoruba people negotiate and approach possible implications of existing neoliberal policies on their aging experiences. Similar to some African countries (Ferguson 2006), the Nigerian society has embraced neoliberal ideology in its economic and political reforms agenda (Adeniyi-Ogunyankin 2012). A major assumption and aim of neoliberalism are that emphasis on individual responsibilities and efforts encourages freedom and empowerment of different social categories and a possible improvement in their overall wellbeing (Ferguson 2006; Adeniyi-Ogunyankin 2012).

The two Yoruba communities purposively selected for this study are the Ile-Ife and Modakeke-Ife. Both communities are neighbours and share a long history of inter-communal conflicts. Also, both communities share boundaries, cultural beliefs and practices in several ways. The Ile-Ife, the proclaimed source of origin for the Yoruba ethnic group, is described as early settlers and the original inhabitants. From historical sources, the Ife-Modakeke people migrated from a geographical area to the present Ile-Ife centuries ago (Toriola 2001; Agbe 2001). This single factor among others has remained vital in dictating the dimensions and quantity of relations among member of both communities (Agbe 2001). Agriculture and trading at subsistence level remain the primary economic activities in both communities. Access to farmland comes through patrilineal inheritance and in recent times through sale or gifts from landowners. However, the past differences between indigenous people and migrants remain a vital source of conflicts and survival among members of both communities (Adisa and Agunbiade 2010; Soyinka-Airewele 2003). Issues around access to farmlands and ownership keep reoccurring, as it is vital to the identity and economic survival of both communities (Soyinka-Airewele 2003).

Thus, the post-conflict resolutions efforts have revolved around access to farmlands, identity protection and reduction of vulnerability of different social categories including the elderly. These efforts have produced some meaningful results especially the return of relative peace to the communities in recent times. However, it is noteworthy to emphasize that large-scale post-conflict reconstruction efforts are lacking within the two communities (Adisa and Agunbiade 2010). Many of the empowerment programmes and protection of vulnerable groups are from self-help groups, community development associations, and non-governmental organizations. Examples include small loan support for traders, empowerment for artisans and seedling support for farmers. The focus of these groups and initiatives are on empowering economically viable businesses run by men and women who are actively providing support for their households (Fabiyi et al. 2012). The success and possible impacts of these initiatives also depend on other factors within the local economy and other factors within the larger Nigerian economy. The logic behind such decision is sound, but it creates room for marginalization and possible neglect of vulnerable social categories like the elderly. Moreover, it also increases the dependability and isolation of the elderly. The loss of one or two household members to the crises over the years might likely increase their loneliness and possibility of experiencing other health challenges (Adisa and Agunbiade 2010). How then are the elderly surviving the everyday challenges in their old age in both communities?

5.4.3 Recruitment Procedure

Using a purposive sampling approach, 37 elderly Yoruba men and women were recruited from the two communities. Prior to the voluntary recruitment, all the participants had opportunities to listen to the objectives and aims of the study. Questions and answers were also provided to unclear areas. Through these efforts, both male and female volunteered personal information and experiences. However, personal identifiers are omitted from the findings in this chapter for confidentiality purposes.

More elderly men (22) than women (15) participated in this study. Some of the socio-demographics are stated in Table 5.1. In this study, elderly are classified as those who are 60 years and above, they are residents and indigenous to either community earlier or not later than 1997. With a long history of recurring communal conflicts between both communities, it is presumed that the conflicts, especially between 1997 and 2001, would have affected elderly people in diverse ways. Before both periods, there have been other incidences. By focusing on elderly people aged 60 years and above, we are not assuming homogeneity in what constitutes old age. For instance, the meanings of old age or older people transcend chronological order or number. The focus on age and experiences agreed with the key research question and the aim to understand resilience and aging experiences of this social category of the elderly within their social context. Hence, we presumed that being an elderly person either after or before the recent clashes would provide some unique survival challenges, hence the focus on this experience as part of the inclusion criteria.

Table 5.1 Socio-demographic profile of the interviewed elderly men and women

Characteristics	Male (n = 22)	Female (n = 15)
Age categories (years)		
60–69	7	6
70+	15	9
Marital status		
Married	9	8
Widowed	13	7
Type of marriage		
Monogamy	6	3
Polygyny	3	5
Religion		
Christianity	14	9
Islam	8	6
Educational status		
None	3	6
Primary	15	8
Post-primary	4	1

Having mentioned the inclusion criteria and aim, the initial identification of potential participants commenced from a purposive and systematic position. First, we identified the core residential areas in the communities. Such areas are occupied by a predominant proportion of the inhabitants. This was facilitated by the assistance of two youth leaders in both communities. The youth leaders assisted us with four community leaders who introduced the lead author to some of the potential participants. This initial contact provided an opportunity to speak to five men and three women. The lead author interviewed the men while a female field assistant interviewed the women. This was in order to make the interviewees more at ease. The initial interviewees also facilitated the opportunity for recruiting other participants. All the interviews took place in locations preferred by the interviewees. Thirty of the interviews took place in the homes of the participants while the remaining seven were conducted outside their homes as desired by the interviewee. The average interview lasted between 32 and 57 min. With cues from the literature, all the interviews were audiotaped and transcribed verbatim (Barbara Lent et al. 2013; Agunbiade 2014) in the Yoruba language. All the transcripts were then translated from Yoruba into English by two expert linguists in both languages. Also, part of the transcripts was shared with three men and five women among the interviewees. During the validity process, all the participants gave their consent to the taping of all the interview sessions. These steps were taken for the purpose of reconfirmation of some information after the data analysis may have commenced.

5.4.4 Data Collection and Analysis

A semi-structured interview guide was used in focusing the data collection. The early part of the discussion revolved around what old age is and the everyday challenges of living with an elderly person. These questions provided opportunities to investigate how they survive these challenges, sources of and the available quality of support. The subsequent questions also looked at their perceptions of the nature and quality of interpersonal and intergenerational relationships. Other topics were the ways elderly people are treated in public places like markets, bus, stops streets and neighbourhoods and religious circles, and what are the perceived forms of ageism in interpersonal relationships. Throughout the interviews, participants were encouraged to share personal and other experiences without any fear. Because of the process of recruitment, it was easier to gain rapport with the interviewees. The structuring of the questions around thematic issues provided an opportunity to explore the experiences of the elderly around issues relating to everyday challenges of living, perceived ageism and challenges of coping with these issues within the study context.

The transcripts were repeatedly read for familiarity with convert and divergent views on core issues. After that, inductive and deductive approach to qualitative coding was adopted in searching for salient themes. This was done to ensure retention of participants' experiences and meanings around their everyday challenges as older people. Three salient themes and two subthemes that emerged are discussed under the findings in relation to existing relevant evidence in the literature. We also provide excerpts from the individual interviews to provide contexts and insights into the themes and subthemes. Such excerpts contained personal experiences that were shared voluntarily among the participants. However, no personal identifiers of the participants of the informants were used to maintain confidentiality.

5.5 Findings

5.5.1 Socio-economic Profile of Interviewees

Twenty-two males and fifteen female older people aged 60–93 years participated in this study. The average age of the participants was 71 years. Three out of the 22 male participants were in polygynous marriage, and six were into monogamous marriage. Among the female interviewees, seven were widowed, and the remaining eight were married. All the female interviewees are engaged in some form of economic activity.

The majority of them are involved in small commerce, especially of foodstuffs. Among the males, four of them are artisans with the oldest aged 67 years. Three of the male interviewees are retirees, and nine own small businesses. Among the rest, four are involved in part-time farming, and five are religious leaders. The majority of the participants are Christians (23) and this was followed by Muslims (14). On average, more males (15) than females (8) had a minimum of the primary form of education. Eight male participants had up to ten adult children while ten out of the fifteen female interviewees had six adult children each. Only one out of the remaining five female interviewees has three adult children. None of the interviewees are childless; however, the emphasis given by interviewees was on fruitfulness in marriage and children's wellbeing.

5.5.2 Marginalization and Abuse in Old Age

Marginalization as a social phenomenon is believed to characterize aging experiences and affects wellbeing and quality of life (Wiles and Jayasinha 2013). Different terms such as social exclusion and social neglect abound in the literature focusing on investigating the dimensions of marginalization of the aged in society (Gilleard and Higgs 2010). Marginalization of the elderly depends on various factors: structural factors like socio-economic, political, religious and technological advancements obtainable within a social setting (Gilleard and Higgs 2010). Other factors include and individual factors such as biological and psychological differentials, gender and personal experiences. All these factors could facilitate the occurrence and the ramification of marginalization that could occur within and across different social categories (Phillipson 2007). Whether within the same social class of individuals or across social categories, the likely occurrence of marginalization also depends on gender and social class and ethnic affiliations especially in multiethnic or racial groups. This makes it possible to variation in degree and experiences of social categories even within a given social setting.

Old age comes with physical and psychological challenges that may cause the possibility of being marginalized and the experiences that follow. For some the participants, marginalization varies and depends on other factors. For instance, within the study setting some old and older people claimed that the involvement in community activities had reduced regarding ramifications of involvement, but believed that their social value has increased over the years. For this category of old and older people, the subjective health of the individual and the social capital defined in terms of value creation and ability to provide useful suggestions and solutions to community and group related challenges are critical indicators.

In the same vein, financial status and material security of the elderly also define the likely occurrence of marginalization and abuse in old age. Regarding material security, gender variation was obvious in the participants' description of what material security entails. From the perspectives of the male participants, these include possession of landed property (through either inheritance or personal acquisition), housing, the source of income, successful adult children and support from wives and family members. Access to landed properties, housing and availability of spouse were absent from the descriptions are given by the female participants.

> Marginalization is common among older people that failed to work hard in their youthful days (An elderly male, aged 78)

As a social phenomenon, marginalization could also occur irrespective of hard work or material security as expressed by some other participants. Among this category of participants, no matter what, old age will deny an individual the opportunity to take responsibilities, actively engage in social activities and community development. As humans, old age comes with physiological changes that could impair decision-making abilities and mobility difficulties. This would occur earlier among some individuals than others, and once this comes, marginalization becomes inevitable.

> *Ojo ogbo jee igba ti eniyan kiyo lee opolopo awon ikan mo.* (Old age is a period of loss of strength and difficulties in attending to activities) (An elderly woman, aged 79)

The appeal to biological explanation within this category of participants does not imply ignorance of how social context also determines the differences that exist in degree and experiences of marginalization in old age. However, their argument rests on the uniqueness and variations that come because of biological differences and susceptibility to living challenges and difficulties in old age.

Similar to biological factors, lifestyles and indulgence in socially undesirable activities could also reduce or encourage marginalization of an older person.

Examples of such behaviours include alcoholism, extramarital affairs, sexual abuse, gambling, and support for criminals either with weapons, spiritual powers or information and sale of family lands.

> There are different categories of older people. Some have finished already even though they are still living, they are dead already, and there is no how responsibilities can be committed to them. If you try it, they will mess up and make nonsense of the whole thing, and part of the reasons is that they are useless; they take other people's wives, drink and sell other people farms to make money (An elderly woman, aged 68)

From these extracts, participants attributed the occurrence of marginalization and associated experiences to socio-economic value and relevance as well as lifestyles and conformity to acceptable forms of behaviour. The attribution of marginalization to the creation of socio-economic relevance and lifestyle differentials takes the responsibilities of avoiding marginalization and promotion of social inclusion from the social domain to the individual level. This has its benefits as it encourages alertness of individuals to value creation and building of social relevance through conformity and public acceptance. However, it could also create challenges, and the possibility of stigmatization for older people who perhaps based on their lifestyles or biological differentials are unable to attain such social standards and choose to act differently.

Furthermore, the shift in responsibilities from communal ownership to individualism and personal achievements is gradually taking its toll among the participants in this study. This may be attributable to developments within the larger Nigerian society and the growing emphasis on materialism as social indicators of a healthy life. The rationalization of marginalization in old age among the participants in this study is also consistent with the literature. Some studies have shown older people hardly perceive their marginalization as deliberate or intentional as the physiological process of aging is often seen as a critical determinant (Chrisler et al. 2016; Minichiello et al. 2000). While biological factors such as physiological changes remain critical, other factors such as economic and political ideology could worsen the degree of marginalization alongside with the possible experiences and how the effects are handled within a given social setting.

A further probe into the quality of relations and experiences with others revealed the occurrence of elderly abuse in various forms. This phenomenon attracted different interpretations and reservations among the participants. As a social reality, irrespective of the definition and the problem around what constitutes elderly abuse, participants confirmed the incidence of elderly abuse and attributed it to deterioration of social values and respect for the elderly. This development was viewed as unacceptable and disheartening. The only exception when such experience was considered somewhat acceptable included when an elderly person behaves far below social expectations. Under such circumstances, an elderly person could suffer justifiable abuse due to his or her shortcomings. To many of the participants, the phenomenon of elderly abuse is also attributable to economic religious and political developments that have altered the quality of interpersonal relations, especially at the household and community levels. The roles of structural factors like economic challenges and religious practices like Pentecostalism have fuelled the occurrence of elderly abuse within the community. For some of the participants, economic challenges have altered the quality of relations expected of some elders as some have ventured into socially undesirable or even illegal activities like providing support for armed robbers and sale of the family landed properties. As exemplified in the words of two participants

> Survival challenges and neglect by one's children have turned some older people to act against their beliefs. Suffering and poverty can make old men and women to steal and sell family lands and farms. Selling of lands is a serious business now, and some elders have supported and killed other family members as a result (An elderly man, aged 62).

> Recently, there are cases of some older men supporting rituals just because of money. As spiritualists, they are supposed to use their powers to help and not to kill or use others for rituals (An elderly woman, aged 72)

The undesirability of elderly abuse emerged strongly and was disapproved by all the participants. The involvement of older persons in such activities is not novel: such occurrences have existed in the past. However, the frequency and dynamics of such practices among the elderly are worrisome especially with the increasing involvement of the elderly in rituals and socially unacceptable activities for monetary benefits. To some of the female interviewees, more men are involved in these practices than women, and they believed that many older men also engage in such practices due to their greed for money.

Similarly, elderly abuse might be justified when an older woman, in particular, is involved in witchcraft and uses her spiritual powers to harm and do evil. Older women and men that use spiritual powers for evil are viewed as not worthy of carrying the title *agbalagba* as they are unworthy role models for others. However, the possibility that some of the elders were only acting based on instructions and desires of their religious groups was not ruled out. Whatever the reasons for their actions, the social expectations are that older people will protect and guide younger ones and their children in particular so that they also support the elderly in old age when they become more frail and dependent. The interviewees lamented the unexplainable occurrences of witchcraft activities within the communities especially shortly after the end of the 2001 crisis in the study location. To some the interviewees, this was expected as the use of spiritual powers including sorcery and witchcraft was commonly accepted during the crisis period. To now abolish the use of such powers will be challenging.

The influx of Pentecostalism into the communities also received the attention of some of the interviewees. To this group of interviewees, the activities of some religious groups especially the Pentecostal Christians have fuelled the neglect of some older men and women perceived to possess witchcraft spirits. To These participants consider the possibility that these churches may not be encouraging elderly abuse, but the belief that witches and wizards are the ones contributing to mishaps and family problems including their children's may encourage neglect and abuse of older people in some situations.

> A 78-year-old woman in this community was accused of witchcraft by one of her daughters who had been married for 15 years and could not conceive. We have known her to be a Christian because of the problem. The woman came home recently and wanted to kill her aged mother. It took the intervention of neighbours, and that was the last time I saw her (An elderly woman, aged 68)

The acceptance of marginalization and elderly abuse as dictated by circumstances is not in favour of the elderly. Both situations can cause undesirable psychosocial problems in old age. Abuse, neglect and loneliness are part of the conditions that may cause depression and other psychological disorders. The prevailing economic challenges confronting many households have reduced the safety nets of older people. It is possible that the attention of adult children supposed to support their aged families would be divided between meeting their needs and those of their parents. In all ramifications, whether marginalization and elderly abuse is perceived as a problem or not, its consequences are beyond the elderly themselves. They affect every aspect of the society.

5.5.3 Interpersonal Relations and Perceived Ageism

Independence in old age is more challenging for elderly people living in social contexts where physical strength is much needed to meet the daily challenges of living (Ory et al. 2003; Hurd 2000). This encourages a high level of dependency on others for survival in many respects. Studies indicate that interpersonal relations and dependency exposes different social categories of individual to age-related discrimination. Ageism is not restricted to any single society: as a stereotype, ageism crosses national and cultural boundaries and, in fact, Vitman et al. (2014) described ageism is a pan-cultural phenomenon. As a nested social reality, the quality of interpersonal relation is central to whether an ageist experience may occur including the possible interpretations that may be associated with the occurrence.

Studies have reported cases of abuse, neglect, marginalization and destitution among older people in Nigeria and other communities in Africa (Fajemilehin et al. 2007). However, insight into how these experiences vary from one elderly person to others within the same community or similar social settings is lacking. Hence, much of the narratives around age-related discrimination and experiences emerged from old and older people's relations with others. From the different narratives, the interviewees decried the increasing incidence of maltreatment among some older people in their communities. Thus, we focused additional attention on the various forms of interpersonal relations, from the household to community levels. The narratives of the participants in this study revealed the existence of ageism despite their willingness to attribute such experience to their network of relations outside their households and neighbours. As such, variations abound regarding what, how and why old and older people have been treated differently even within their households and their current social settings.

5.5.3.1 Household Level

Some of the interviewees expressed familial bonds and good relationship with their immediate family members. At the individual level, we focused on decisions about food and finance for contextualized experiences. From the narratives, some old and older people enjoyed good relationship with their immediate family members, while others did not.

> Well, my relationship with my immediate family members is good. I have supported them in the past by training them to a level of independence. Hence, I expect good relations (An elderly male, aged 70).

> I am happy to have many of my relatives around, and none of them has treated me badly. I would say I have enjoyed a cordial relationship with them though we fight sometimes, and we settle it as well (An elderly woman, aged 78).

Decisions about what to eat and when foods are served provided an opportunity for a contextualized understanding of the aging experiences of the participants. Interviewees do not like eating the same food on a daily basis. Hence, they expressed a preference for local varieties, which, unfortunately, could be lacking. Among the interviewees, access to some cheap local meals and other varieties was easy as they could easily secure the items needed to prepare their favourites within their neighbourhood and nearby daily markets. However, this could be difficult when there is a health challenge or lack of money. For some of the women participants, especially those over 60 years of age, the presence of a grandchild, especially an adolescent, could prove useful despite the challenges of rearing an adolescent as a grandmother. The situation could become worse when such a grandmother depended on the adolescent for many things including fetching of water, access to energy and food on a daily basis. Two of the interviewees that are in their mid-70s recounted some experiences living with a dependent husband:

> To have some of these young around you can be interesting and frustrating sometimes. I earn a living through trade in little things and one my granddaughters' lives with me. As young as this girl is, she gives more trouble than I bargained for in life. Nevertheless, I prefer to have her around than staying alone (An elderly woman, aged 68).

> At my age, I cook and support my husband. He is more fragile and need more support, with or without people around; I'm determined until God calls him or me home (An elderly woman, aged 79).

Unfortunately, as much as some of them will prefer a level of dependence, it could become frustrating when there is a husband to cater for and when the husband in question is suffering from a health challenge. Based on cultural values and practices, women even in their old age provide different forms of support for people around even when they need similar support from others.

The situation is a bit different for men. The choice regarding what to eat and access to varieties depends on others and personal income. Concerns about personal money emerged dominantly in the narratives of the male interviewees. Similarly, eating local foods was preferred, and some of the male interviewees confessed a preference for eating local foods whether prepared at home or by commercial food vendors in their neighbourhoods. However, men that are poor lamented the difficulties around choice making as they eat whatever they can afford

> Growing up as an aged man without the support and personal income is a disaster. I have friends that are over 70 years and are still going to farm just because of what they would eat. It is sorrowful for a man to work and continue working until he cannot work again (An elderly male, aged 78)

The ability to choose what to eat and when was also similar to decision-making on health-related issues. To some of the participants, health and decision-making are areas of major concern. For the men, there had been occasions when they could make a decision on what to do but gave up the right due to financial incapacitation. For such periods, a participant described it as like hell for him

> It was so challenging for me the last time I was ill. I wanted to visit a nearby hospital but could not as there was no money. It was my aged wife (73 years) that took money from her business to buy me drugs from a nearby medicine vendor. I appreciated her so much since none of our children was around to give me money. (An elderly male, aged 82 years)

The daily challenges of living can make old age stressful and unpleasant. Inability to satisfy basic needs such as access to food and timely access to desired local varieties could also expose old and older people to abuse and discrimination. However, minimal work in old age keeps the body active and guarantees a source of income to meet basic needs like food. Farming in old age requires extra physical strength that could be detrimental to health in old age. Hence, much dependence on children and neighbours may become inevitable for some old and older people living in low economic resource settings also devoid of formal support from the government.

5.5.3.2 Public Sphere—Religious Involvement

From a religious perspective, some of the interviewees expressed the need for more teachings that will protect and encourage reverence for age and the aged in the society. The practice of dishonouring old and older people based on perceived age-related deficiencies is detrimental to the future interests of younger people themselves and society. From a cultural position, there are advantages and benefits for those who protect the aged and accord them the needed respect in social relations.

> Based on my understanding, young people should respect and bow before older people so that they can obtain mercy because the Bible says to honour your father and your mother so that you may live long and prosper (An elderly male, aged 70).

> We must honour older people with reverence and respect even the Bible says to honour your father and mother, and this goes with a promise of long life and not only your father but whosoever that is older than you (An elderly male, aged 71).

> Any young person that wants to live long must respect the older people even it is recorded in Islam that whoever dishonour old people will not live long (An elderly male, aged 62).

One of the quick points of reference in the narratives borders on leadership positions in modern religion. To the participants, new movements within Christianity and Islam have sprung up in recent times with ideas that promote discrimination against the aged. For instance, occupying religious positions in such movements depends on the charisma and perceived possession of unique gifts irrespective of the age and experience of the individuals. To some of the interviewees, effective leadership comes with age and experience that some of these organizations have failed to appreciate. Some of these organizations often die before their time as they lack experienced leaders. In the same vein, some of the interviewees also expressed concerns about the inadequate attention given to the elders in many of the Pentecostal churches in the area. Instead, many of the Pentecostal churches in the town encourage youths to occupy such positions and would only allow aged people to take passive positions. This was unlike the practice in the Orthodox churches. In this direction, some of the interviewees opined that the elderly would be better occupying some roles such as advisor, coordinator and overseers. These are leadership positions that require experience, courage, endurance, wisdom and knowledge that could address inevitable complex situations and issues. Often, young people lack these qualities and, therefore, need to take cues from elders in such positions.

> Certain jobs require experience and understanding that comes with age. In recent times, many of our younger ones have become so impatient and want to take everything under the pretence that older people are slow and incompetent to handle complex situations (An elderly male, aged 60).

> Older people should be given honourable responsibilities in the organization such as adviser, leadership position, coordinator and overseer because of their experiences, wisdom, endurance, courage and knowledge (An elderly woman, aged 68).

As a move away from exclusive rights of the elderly to some positions of responsibility to shared responsibilities with younger people, some of the interviewees argued that no age group be a reservoir of knowledge and, therefore, preferred shared responsibilities. One of the interviewees argued further, with the adage, "what is good for the goose is also good for the gander".

> No position can be reserved exclusively for older people in an organization because nobody is a reservoir of knowledge (An elderly male, aged 60).

At an individual level, three of the interviewees argued that age-based discrimination and marginalization in some churches in the community has discouraged them from becoming full worshippers. In the same vein, one interviewee explained that some of the current teachings in these churches emphasize individuality and craving for personal success at the detriment of providing support for the elderly. To this participant, this practice deviates from the traditional Yoruba belief system that promotes group success and communal responsibilities and achievements. Sharing promotes unity, solidarity and balances the craving to acquire more and more for oneself. Two other interviewees, however, took the blame away from the church and espoused the growing trend of individualism and struggle for personal recognition and competition as traits of modernism. In the opinion of these interviewees, quick success has overtaken the interests of many young people, and they would hardly listen to suggestions from older people because their suggestions often need more time and patience to be implemented.

5.5.4 Resilience and Surviving the Everyday Challenges of Living

The daily challenges of living and surviving in old age require some attitudinal disposition. Part of the problems includes dependence on others and the possible reoccurrence of some health challenges. In old age, resilience as an individual trait has remained useful in helping older people to cope with physiological, psychological and physical challenges associated with the aging process. Resilience is conceptualized as the potential will and determination to continue living despite some challenging or life threatening situations (Biggs 2004). The functionality of resilience for the individual differs as it provides opportunities to appreciate older people and possible ways to make their aging experiences interesting. In espousing the view further, the interviewees emphasized these differences about experiences and expectations

> Dying in old age is very easy if one gives up hope for life and is idle. Some of us are still selling a few things to earn a living (An elderly woman, aged 66)

> Old age differs from one individual to another, and this depends on how such individual spent his young stage of life. Some of our friends are just too old than their ages. Some of them worked too hard while young and cannot work now again and some it was too much pleasure (alcohol and sex) now they are something else, yet they have given up their hope of living (An elderly man, aged 71).

In this study, the participants narrated the adoption of different survival methods in old age. These include taking up paid employment such as security services as night watchmen, trading in small household items, and farming. In contrast, some of the participants also narrated the undesirable indulgence of some older people in activities that are socially degrading or even criminal, just to earn a living.

The involvement of the elderly in socially unlawful or illegal activities (like stealing, provision of support for armed robbery, and unlawful sale of landed properties) to earn a living is gradually becoming rampant in some Nigerian communities. Part of their explanations revolved around the growing difficulties in earning a living and neglect for adult children and other family members. In all ramifications, participants recounted the increasing difficulties in going through successful aging experiences as individuals; neglect and abuse are now normative. With grown-up adult children living in urban areas and parents in semiurban areas, they expect frequent and meaningful visits. However, when it becomes difficult for a child to provide the needed support, praying for the safety of their adult children and blessings becomes critical and taken as a call to duty. However, this was not the case as some the participants expressed personal resolutions to fend for themselves and ensure that they do not become a nuisance to their children and society.

5.6 Conclusions

This study presents qualitative evidence on aging, ageism and resilience within an interactionist gerontological framework. Particular attention was focused on the possible implications of the neoliberal-based social policies on the everyday experiences of older Yoruba people aged 60 years and above. This study adopts an interactionist approach in situating the aging experiences of old and older Yoruba people within a neoliberal theoretical framework.

Studies have shown the increasing occurrence of elder abuse and neglect in Nigeria and other countries in sub-Saharan Africa. Recent studies have shown how elder abuse and neglect is gradually becoming the norm in many ethnic groups in Nigeria, including the Yoruba. Against this backdrop, this chapter focused on the growing difficulties in living as an old or older person and the possibilities of adopting different measures that could be useful in achieving healthy aging experiences. The findings revealed that whether ageism was perceived and reacted to as defined in the literature, the reality of ageism remains critical to the possibility of experiencing healthy interpersonal relations. The particularity of this experience and the context count regarding what qualifies or is recognized as an ageists experience or occurrence. In the private and public spheres, some of the older people became resilient and devised means that probably minimized the effects of ageism in their everyday life experiences.

The study further revealed that resilience at the individual level and the availability of minimal support from various sources could be meaningful in diverse ways. This indicates the urgent need to reduce the growing challenges of aging social settings where poverty remains high and with the absence or inadequacy of a network of support. Second, it also shows a high level of sensitivity to existing inadequacies and challenges that adult children encounter in supplementing the existing low quality of support networks that a few older people are experiencing. Against this background, personal struggles and determination to work and earn a living were seen as a partial but useful way of keeping body and soul together. Unfortunately, some older people have taken the direction of unlawful means to earn a living. Whether the means are lawful or not, this also echo the need to consider old and older people in current efforts to protect vulnerable people and empower them, to maintain peace and stability in the society. An encouraging effort in providing state pension scheme for the aged is gradually taking shape in Ekiti and Osun States, two states in southwestern Nigeria. An expansion of these initiatives to cover all old and older people within the states and at the national level would guarantee a basic source of income. It has been argued that neoliberal reforms require a level of sensitivity to the needs of marginal and vulnerable groups in the society. These groups are also heterogeneous and particular in diverse ways.

These characteristics make it difficult to adopt a "straitjacket" approach to identifying and meeting the needs of marginalized and vulnerable groups.

The current policy gaps and limited knowledge of existing needs and experiences of being marginalized or vulnerable may increase inequalities and increase further the need for social safety nets that could promote healthy aging experiences. Within the Nigerian context, the impact of neoliberal reforms is taking a toll on religious values, beliefs and socio-economic relations. These developments may be changing the dependence patterns and increase the need for revitalizing agency and resilience in old age. Hence, prompt identification and promotion of resilience across the diverse social categories of elderly people might be instrumental in several ways. Such benefits could be in the ability to navigate aging experiences and cope with the inherent challenges of liberalization. It could also provide and increase personal sensitivity to identifying options and alternatives in charting clear directions in later life within a social context.

References

Aboderin, I. (2011). *Intergenerational Support and Old Age in Africa*. New Jersey, United States of America: Transaction Publishers.

Adeniyi-Ogunyankin, G. (2012). "When will I get my rest?" Neo-liberalism, women, class and ageing in Ibadan, Nigeria. *Agenda, 26*, 29–36.

Adisa, A. L., & Agunbiade, O. M. (2010). Aborigines' and Migrant Settlers' crisis and reconstruction efforts in two Yoruba Communities in Southwestern Nigeria. In D. S. Miller & J. D. Rivera (Eds.), *Community disaster recovery and resiliency: Exploring global opportunities and challenges* (pp. 287–302). London: CRC Press, Taylor and Francis Group.

Agbe, G. (2001). The Ife/Modakeke crisis an insider view. *Ife Psychologia, 9*, 14–20.

Agunbiade, O. M. (2014). Spirituality in Knowledge Production and the Practice of Traditional Herbal Medicine among the Yoruba People in Southwest Nigeria. In: E. Gemi-Iordanou, S. Gordon, R. Matthew, E. McInnes, & R. Pettitt (Eds.), *Medicine, Healing and Performance* (pp. 176–191). Oxford & Philadelphia: Oxbow Books.

Ajala, A. S. (2006). The changing perception of aging in Yoruba culture and its implications on the health of the elderly. *Anthropologist, 8*, 181–188.

Aléx, L. (2010). Resilience among very old men and women. *Journal of Research in Nursing, 15*, 419–431.

Barber, S. J., & Mather, M. (2014). Stereotype threat in older adults: When and why does it occur and who is most affected? In: P. Verhaeghen & C. Hertzog (Eds.), *The Oxford Handbook of Emotion, Social Cognition, and Problem Solving in Adulthood* (pp. 302–319). New York: Oxford University Press.

Biggs, S. (2004). Age, gender, narratives, and masquerades. *Journal of Aging Studies, 18*(1), 45–58.

Barbara Lent MA, M., & Herbert, C. (2013). Surgical culture in transition: gender matters and generation counts. *Canadian Journal of Surgery, 56*(3), 153.

Calasanti, T. (2005). Ageism, gravity, and gender: Experiences of aging bodies. *Generations, 29* (3), 8–12.

Chrisler, J. C., Barney, A., & Palatino, B. (2016). Ageism can be hazardous to women's health: Ageism, sexism, and stereotypes of older women in the healthcare system. *Journal of Social Issues, 72*, 86–104.

Cohen, L. (1994). Old age: Cultural and critical perspectives. *Annual Review of Anthropology, 23*, 137–158.
Coudin, G., & Alexopoulos, T. (2010). 'Help me! I'm old!' How negative aging stereotypes create dependency among older adults. *Aging & mental health, 14*, 516–523.
Cruikshank, M. (2013). *Learning to be old: Gender, culture, and aging*. Plymouth, United Kingdom: Rowman & Littlefield.
Dokpesi, A. O. (2015). The Future of Elderly Care in Nigeria: Borrowing a Leaf from a Foreign Land. *Ageing International, 40*(2), 81–97.
Fabiyi, O. O., Thontteh, O. E., & Borisade, P. (2012). Spatial and Social Dimensions of Post Conflict Urban Reconstruction Programme in South Western Nigeria. The case of Ile-Ife, Nigeria. *Journal of Settlements and Spatial Planning, 3*(2), 163.
Fajemilehin, B. (2001). Familial expectation and social support for the elderly in Yoruba communities of South Western Nigeria. *Journal of Social Sciences India, 51*, 31–37.
Fajemilehin, B. R., Ayandiran, E. O., & Salami, K. K. (2007). Elderly destitution in Ile-Ife community of Osun State, Nigeria. *International Journal of Nursing Practice, 13*, 161–165.
Falola, T. (1999). *Yoruba Gurus: Indigenous production of knowledge in Africa*: Africa World Press.
Ferguson, J. (2006). *Global shadows: Africa in the neoliberal world order*. Duke University Press.
Garstka, T. A., Schmitt, M. T., Branscombe, N. R., et al. (2004). How young and older adults differ in their responses to perceived age discrimination. *Psychology and Aging, 19*, 326.
George, L. K. (2010). Still Happy After All These Years: Research Frontiers on Subjective Well-being in Later Life. *The Journals of Gerontology Series B: Psychological Sciences and Social Sciences*. doi:10.1093/geronb/gbq006
Gilleard, C., & Higgs, P. (2010). Aging without agency: Theorizing the fourth age. *Aging & Mental Health, 14*, 121–128.
Grant, L. D. (1996). Effects of ageism on individual and health care providers' responses to healthy aging. *Health and Social Work, 21*, 9–15.
Harvey, D. (2005). *A brief history of neoliberalism*. Oxford: Oxford University Press.
Hill, P. L., Payne, B. R., Jackson, J. J., et al. (2014). Perceived social support predicts increased conscientiousness during older adulthood. *The Journals of Gerontology Series B: Psychological Sciences and Social Sciences, 69*, 543–547.
Hurd, L. C. (2000). "We're not old!": Older women's negotiation of aging and oldness. *Journal of Aging Studies, 13*, 419–439.
Kite, M. E., Wagner, L. S., & Nelson, T. (2004). Attitudes toward older adults. In T. D. Nelson (Ed.), *Ageism: Stereotyping and prejudice against older persons* (pp. 129–161). Massachusets, USA: The Massachusets Institute of Technology Press.
Lagacé, M., Tanguay, A., Lavallée, M.-L., et al. (2012). The silent impact of ageist communication in long term care facilities: Elders' perspectives on quality of life and coping strategies. *Journal of Aging Studies, 26*, 335–342.
Levy, B. (2009). Stereotype embodiment a psychosocial approach to aging. *Current Directions in Psychological Science, 18*, 332–336.
Liu, L., Gou, Z., & Zuo, J. (2014). Social support mediates loneliness and depression in elderly people. *Journal of Health Psychology*, 1359105314536941.
Makinde, T. (2004). Motherhood as a source of empowerment of women in Yoruba culture. *Nordic Journal of African Studies, 13*, 164–174.
Manning, L. K. (2012). Navigating hardships in old age: Exploring the relationship between spirituality and resilience in later life. *Qualitative Health Research, 23*, 568–575.
Mégret, F. (2011). The human rights of older persons: A growing challenge. *Human Rights Law Review, 11*, 37–66.
Minichiello, V., Browne, J., & Kendig, H. (2000). Perceptions and consequences of ageism: Views of older people. *Ageing and Society, 20*, 253–278.
Olajide, F. (2012). *My culture-Yoruba*. Leicestershire, UK: Troubador Publishing Ltd.
Onigu-Otite, K., & Albert, I. O. (1999). *Community conflicts in Nigeria: Management, resolution and transformation*. Spectrum Books Limited.

Ory, M., Kinney Hoffman, M., Hawkins, M., et al. (2003). Challenging aging stereotypes: Strategies for creating a more active society. *American Journal of Preventive Medicine, 25*, 164–171.

Owomoyela, O. (2005). *Yoruba proverbs*. University of Nebraska Press.

Oyewumi, O. (2009). Abiyamo: Theorizing African Motherhood. *JENdA: A Journal of Culture and African Women Studies,* (4). Retrieved from http://www.africaknowledgeproject.org/index.php/jenda/article/view/79

Peck, J., Theodore, N., & Brenner, N. (2009). Neoliberal urbanism: Models, moments, mutations. *SAIS Review of International Affairs, 29*(1), 49–66.

Phillipson, C. (2007). The 'elected' and the 'excluded': Sociological perspectives on the experience of place and community in old age. *Ageing and Society, 27*, 321–342.

Reichstadt, J., Sengupta, G., Depp, C. A., et al. (2010). Older adults' perspectives on successful aging: Qualitative interviews. *The American Journal of Geriatric Psychiatry, 18*, 567–575.

Richmond, E. B. (1987). Utilizing proverbs as a focal point to cultural awareness and communicative competence: Illustrations from Africa. *Foreign Language Annals, 20*, 213–216.

Rippon, I., Kneale, D., de Oliveira, C., et al. (2014). Perceived age discrimination in older adults. *Age and Ageing, 43*, 379–386.

Shetty, P. (2012). Grey matter: Ageing in developing countries. *The Lancet, 379*, 1285–1287.

Sokolovsky, J. (2009). *The cultural context of aging: Worldwide perspectives*. West Port, CT, USA: Praeger Publishers.

Soyinka-Airewele, P. (2003). When neutrality is Taboo. Navigating institutional identity in protracted conflict settings—The Nigerian Ife/Modakeke Case. *African and Asian Studies, 2*, 259–305.

Thomas, P. A. (2010). Is it better to give or to receive? Social support and the well-being of older adults. *The Journals of Gerontology Series B: Psychological Sciences and Social Sciences, 65*, 351–357.

Togonu-Bickersteth, F. (1989). Conflicts over caregiving: A discussion of filial obligations among adult Nigerian children. *J Cross Cult Gerontol, 4*, 35–48.

Togonu-Bickersteth, F., & Akinyemi, A. I. (2014). Ageing and national development in Nigeria: Costly assumptions and challenges for the future. *African Population Studies, 27*, 361–371.

Togunu-Bickersteth, F. (1988). Perception of old age among Yoruba aged. *Journal of Comparative Family Studies*, 113–122.

Toriola, J. O. (2001). The Ife/Modakeke crisis: An insider view. *Ife Psychologia, 9*, 21–29.

Vitman, A., Iecovich, E., & Alfasi, N. (2014). Ageism and social integration of older adults in their neighborhoods in Israel. *The Gerontologist, 54*, 177–189.

Wagner, L. S., & Wagner, T. H. (2003). The effect of age on the use of health and self-care information: Confronting the stereotype. *The Gerontologist, 43*, 318–324.

Wiles, J. L., & Jayasinha, R. (2013). Care for place: The contributions older people make to their communities. *Journal of aging studies, 27*, 93–101.

Williams, D. R., & Mohammed, S. A. (2009). Discrimination and racial disparities in health: Evidence and needed research. *Journal of Behavioral Medicine, 32*, 20–47.

Chapter 6
Social Capital, Interrupted: Sociological Reflections from Old Age Homes in Ahmedabad, India

Tannistha Samanta and Jagriti Gangopadhyay

Abstract A burgeoning literature has celebrated the complementary roles of bonding, bridging and linking potential of social capital. A disproportionately smaller set of studies have identified the less desirable consequences of social capital in terms of social control, exclusion and restrictions on freedoms have pointed out the limits of the (positive) functional value of social capital. In particular, we build on the latter critique of social capital and argue that moral-cultural scripts, social hierarchies and lived experiences of older Indians in elder care institutions trump the motivation to unlock the benefits of the much celebrated potential of social capital. To establish this argument, we have adopted a social constructionist perspective (Gergen and Gergen 2000) focusing on the cultural meanings of age and aging among 15 older adults living in two elder care institutions in the rapidly urbanizing city of Ahmedabad (Gujarat, India). Through in-depth interviews we illustrate that despite similar life circumstances, fate, shared histories, self-consciousness and agency, older people in these alternative care sites are often in the pursuit of a "meaningful decline" that frustrates the potential building/bridging capacities of social capital. We contend that moral-cultural scripts, belief systems and practices cripple the motivation to actively construct or benefit from social ties, interdependence and association. Overall, we extend previous studies on social networks of older adults in India and argue that an examination of the *limits* of social capital is crucial in understanding the lived experiences of older people and their access to networks and resources in non-familial settings. This is particularly useful as India experiences dramatic transformations in family structure marked by a gradual, yet steady, movement away from multigenerational settings to the individual, the market and the state, as emerging sites of aging and elder care.

T. Samanta (✉) · J. Gangopadhyay
Humanities and Social Sciences, Indian Institute of Technology Gandhinagar, Gandhinagar, India
e-mail: tannistha@iitgn.ac.in

J. Gangopadhyay
e-mail: jagriti.gangopadhyay@iitgn.ac.in

T. Samanta (ed.), *Cross-Cultural and Cross-Disciplinary Perspectives in Social Gerontology*, DOI 10.1007/978-981-10-1654-7_6

Keywords Aging · Social capital · Old age homes · Social constructivism · Cultural scripts · India

6.1 Introduction

A burgeoning literature has celebrated the complementary roles of bonding, bridging and linking potential of social capital (Bisung et al. 2014; Yip et al. 2007; Cannuscio. et al 2003; Colletta and Cullen 2000; Putnam 1993). A disproportionately smaller set of studies have identified the less desirable consequences of social capital in terms of social control, exclusion and restrictions on freedoms (Bloch et al. 2007; Cleaver 2005; Portes 1998; Waldinger 1995) and have pointed out the limits of the (positive) functional value of social capital. In particular, we build on the latter critique of social capital and argue that moral-cultural scripts, social hierarchies and lived experiences of older Indians in elder care institutions trump the motivation to unlock the benefits of the much celebrated potential of social capital. To establish this argument, we have adopted a social constructionist perspective (Gergen and Gergen 2000) focusing on the cultural meanings of age and aging among 15 older adults living in two elder care institutions in the rapidly urbanizing city of Ahmedabad (Gujarat, India). Through in-depth interviews we illustrate that despite similar life circumstances, fate, shared histories, self-consciousness and agency, older people in these alternative care sites are unable to tap the "bonding" potential of social capital to secure improved wellbeing. We argue that moral-cultural scripts, belief systems and practices cripple the motivation to actively construct or benefit from social ties, interdependence and association. We extend previous studies on social networks of older adults in India (Berkman et al. 2012; van Willingen and Chadha 2008) and contend that an examination of the *limits* of social capital is crucial in understanding the lived experiences of older people and their access to networks and resources in non-familial settings. This is particularly useful as India experiences dramatic transformations in family structure marked by a gradual, yet steady, movement away from multigenerational settings to the individual, the market and the state, as emerging sites of aging and elder care (Lamb 2013; Kalavar and Jamuna 2011).

The study location of Ahmedabad in the western Indian state of Gujarat, provided a useful site to understand the implications of an urbanizing society on the changing nature of filial obligations and intergenerational reciprocal transactions that are traditionally expected in India. Surprisingly, the upwardly mobile urban Gujarati middle class society known for their closed knit household units organized around paternalistic family-run businesses (Gupta et al. 2007), is also witnessing a rise in "old age homes" as alternative spaces of caring for the aged. In fact, the city of Ahmedabad has experienced a sharp increase in the number of older age homes in the last decade with the number rising from 11 in 2005 to 26 in 2010 (Times of India 2011). These trends are, however, new as living in multigenerational "joint" (or undivided) families still captures the popular imagination as the quintessentially

Indian way of life, morality and tradition (Lamb 2009). Given this background, one of the goals of this study was to understand how older adults in elder care institutions are (re)constructing the social meanings of aging and negotiating with the changing social realities. We find that older adults feel contrary pulls in the process of negotiating with these changing cultural forces, which in turn de-capitalizes the formation of social capital in non-familial settings.

Majority of the gerontological literature in India has focused on issues relating to living arrangements, social security and health outcomes of older adults (Samanta et al. 2015; Bloom et al. 2010; Pal 2004; Rajan and Kumar 2003). Recently, surveys and descriptive studies on old age homes have been gaining currency (Kalavar and Jamuna 2011; Liebig 2003; Rajan 2008). Curiously, only a handful of studies in India have examined questions surrounding social capital of older adults, a topic that has been elaborately studied in the West (Koutsogeorgou et al. 2014; Cramm et al. 2012; Cannuscio et al. 2003). Exceptions include, Berkman and Colleagues (2012) and Samanta (2014); however, the focus of these studies has been primarily on the facilitating action of social capital. Interpretive approaches (Glaser and Strauss 1967) adopted as part of this study guided our understanding to build an alternative framework of social capital. We suggest that the heuristic power of the concept of social capital will generate more "emancipatory knowledge" (Moody 1993) once the limits of the concept are critically examined. As such, the mainstream understanding of social capital with a singular focus on the positive aspects will most likely generate a weak policy tool to ensure alternative forms of caregiving and support for older people.

We begin by articulating the concept of social capital and the contested nature of its applications in different disciplines. Next we discuss our study and the methods adopted to examine the boundaries and limits of social capital. Later, drawing insights from the narratives, we engage in a discussion on the culturally unique meanings of age and aging. Finally, we summarize reflecting on gender, aging and the life course.

6.2 Social Capital: Theories and Debates

Portes (1998) contends that social capital has become one of the most popular exports from sociological theory to development policy, public health, education, community life, democracy, governance and organizational research (Vikram et al. 2012; Cornwell 2011; Lim and Putnam 2010; Carpiano 2007; Adler and Kwon 2002; Colletta and Cullen 2000; Gabbay and Zuckerman 1998). Yet, an application of this concept in the Indian gerontological literature is limited. Crudely put, social capital refers to the primordial features of social life, namely the goodwill related social ties, networks and trust that can be utilized for achieving support-moral, instrumental and material. "Like any other form of capital" (e.g. economic or human), social capital aids future productivity of individuals and groups in civil society and it has as its "conceptual cousin, community" (Putnam 1993). Taking these dimensions together, social capital is complex network of ties, associations or relations that bind people together as a community via certain norms and psychological capacities (e.g. trust) that are essential to achieve overall wellbeing and productivity (Farr 2004)

Though a more systematic contemporary analysis began with the influential works of Bourdieu (1985), Coleman (1988) and Putnam (1993), social capital's intellectual origins can be traced back to the nineteenth century with Durkheim's (1897) emphasis on group life as an antidote to *anomie* and suicide and Marx's (1894) analysis of class consciousness and solidarity in the industrial proletariat. In the course of this review, we restrict our discussion to the contemporary emergence and application of the concept. Bourdieu's (1985) treatment of the concept is instrumental. He defined social capital as "the aggregate of the actual or potential resources which are linked to possession of a durable network of more or less institutionalized relationships of mutual acquaintance or recognition", asserting social capital as a vital enabling resource and not a natural given and therefore must be constructed through investment strategies geared towards formalization of group relations (Portes 1998). This concept paved the way for Coleman (1988) who examined the role of social capital in the creation of human capital but at the same time rejecting the "extreme individualistic premises that often accompany it" (i.e. orthodox economic theories) (Coleman 1988: S95). Coleman's emphasis was on understanding the forms and the social structural conditions of social capital formation. According to him:

> Social capital is defined by its functions. It is not a single entity but a variety of entities, with two elements in common: they all consist of some aspect of social structures, and they facilitate certain actions of actors, within the structure…Like physical capital and human capital, social capital is not completely fungible but may be specific to certain activities. *A given form of social capital that is valuable in facilitating certain actions may be useless or even harmful for others.* [Coleman 1988: S98, emphasis added]

Though Coleman's focus is on the facilitating potential of social capital, he acknowledges the boundaries of this appropriable resource. We also find in Coleman's analysis the contradictory tensions of social capital through the group enforcement of norms. Coleman asserts, "a prescriptive norm within a collectivity that constitutes an especially important form of social capital is the norm that one should forgo self-interest and act in the interests of the collectivity". However, he also adds "this social capital, however, not only facilitates certain actions, it constrains others" (Coleman 1988: S104). Restrictions associated with strongly tied group norms were further elaborated by several other studies that highlighted the tension between community solidarity and individual freedom (Andrist 2008; Putnam 1993, 1995; Waldinger 1995; Rumbaut 1977). In particular, Putnam (1995, 2000) who persuasively demonstrated a decaying civic engagement and eroding social capital (e.g. trust, associational memberships, volunteering) in America heralding a less vigorous democracy, acknowledged the "dark side" of social capital: "social inequalities may be embedded in social capital" (Putnam 2000: 358). We will return to the discussion of social control and norm enforcement in our later reflections on the shared normative structure and mobility outcomes of older women in elder care institutions.

In an incisive criticism of the mainstream celebratory views of social capital, Cleaver (2005) argued that in development policy social capital is often used "to sidestep issues of inequality without proper consideration of the negative aspects of social life, or the structural constraints on the empowerment of the poor" (p. 894). Drawing from her ethnographic study in four Tanzanian villages, she concluded that the chronically poor are caught in interlocking disadvantages (e.g. fragile families, unstable marital arrangements and ill health) that constrain them to manoeuvre social relationships and collective action to their advantage. Cleaver's analysis clearly highlights the *limits* to the emancipatory potential of social capital. In a later economic study of reciprocity and mutual help in large groups, Bloch and Colleagues (2007) showed that people in well-coordinated, cooperative groups might find it beneficial to deviate in smaller groups rendering the original group fragile and unstable. The study emphasized the limits of cooperation (a form of social capital) in mutual help groups.

These examples illustrate that social capital on its own accord does not necessarily empower the actors. In a helpful review, Adler and Kwon (2002) articulate the need to consider the opportunity, motivation and ability dimensions of actors to determine the facilitating potential of social capital. And in the opportunity-motivation-ability schema, the authors argue, a movement from one stage to the other might not be automatic. Often missed is Portes' (1998) conceptualization of motivations; he contended that motivations are based on deeply internalized norms, engendered through socialization in childhood or experience later in life by the experience of a shared destiny with others. We build on this argument to illustrate how the bonding and trust potential of social capital formation is interrupted in old age homes, despite opportunity and agency among actors. In particular, we explain this "interruption" by highlighting the moral-cultural constructions and expectations surrounding aging and the life course.

6.3 Aging in a Non-family Setting: Old Age Homes in India

Though elder care institutions are not entirely a new phenomenon in India (for example, the Venkatagiri Chaultries of old Madras in 1872; see Nair 1995), the pay and stay western style residences (commonly known as "old age homes") have ushered a new era of modernity that is associated with urban residence, nuclear families, individualism and weakening family ties (Kalavar and Jamuna 2011; Lamb 2010; Liebig 2008). From a western perspective, these individuated living arrangements (in old age homes) may represent independence, personhood and "successful" aging (Rowe and Kahn 1997). However, Indian multigenerational families are organized around the ancient beliefs of *dharma* (moral-spiritual law that guides righteous living) and *seva* (respectful care and service), where caregiving of the elders is expected to sustain intimate ties of kinship and relationality

(Lamb 2013, Vera-Sanso 2004). It is therefore, not surprising that majority of older Indians co-reside in multigenerational households (Samanta et al. 2015; Sathyanarayana et al. 2012). In fact, there are demonstrated instrumental and health gains associated with multigenerational living for older persons in settings where filial traditions and intergenerational reciprocal ties are stronger (Samanta et al. 2015; Ku et al. 2013; Chen and Short 2008).

Studies have suggested that disintegration of the "joint" (or undivided) family system and erosion of filial obligations, growing affluence of the Indian middle class, higher rates of migration and job mobility, multiple income earners have contributed to the emergence of old age homes as alternative elder care spaces (Lamb 2013; Kalavar and Jamuna 2011; Liebig 2008). Since panel data on family structure and aging is limited in India, it is perhaps difficult to empirically ascertain the causes of such a shift in the caregiving arrangements. However, what seems to be clear is the feeling of loneliness, disengagement and abandonment among the residents in such homes. This is revealed both from national-level studies (e.g. Liebig 2008) as well in-depth ethnographic narratives on old age homes (Lamb 2009, 2013). Drawing from her intensive anthropological field work in Bengal, Lamb notes that the sense of abandonment among residents was analogous to being "thrown away" (discarded as trash); and this "throwing away" was not just of an individual but of a moral culture that upholds idealized notions of care and duty (Lamb 2013). Furthermore, the sense of social withdrawal and disengagement is reflected in the smaller network sizes that residents in old age homes have as compared to their joint family counterparts (Kalavar and Jamuna 2011; van Willingen and Chadha 2008). We will return to the discussion of social networks of the elderly in old age homes at a later point. Yet, studies have also reported how older Indians in such homes have accepted their fate and the irrevocable transience of a life in its decline (Lamb 2013; Prince, et al. 2007; Bhat and Dhruvarajan 2001; Dandekar 1996). Narratives and interviews from these studies have demonstrated that many have interpreted their stay in old age home as being comparable to the traditional Hindu forest-dwelling, spiritually contemplative phase of *vanaprastha*, as part of the four stages (*ashramas*) in the life course as suggested by the ancient Hindu *dharmasastras* (socio-religious principles of righteous living) (Olivelle 2009). Arguably, older Indians in these alternative care institutions, negotiate the contrary tensions of familial abandonment and individual denunciation of material wealth and binding ties. We argue that this moral-cultural paradox de-capitalizes the formation of interdependence, ties, trust and networks in old age homes.

6.4 The Study

The insights presented in this paper are derived from in-depth interviews conducted as part of a pilot study on two old age homes in the city of Ahmedabad. The HelpAge India office in Ahmedabad had supplied us with a list of old age homes operating in the city. Based on that list, initial stages of this study included informal interviews with the trustees, managers and a few residents of several old age homes in the city. Elder care institutions are rare in the rural areas of Gujarat, a phenomenon which has been observed in other studies as well (Lamb 2013; Liebig 2008; Rajan 2008). Hence for the most part of our study, we interviewed middle class elderly who were all Hindus. Elder care institutions in post-colonial India have been largely established and managed by Christian missionaries who catered to the poor and the destitute (Lamb 2009); contemporary India has seen a proliferation of old age homes that are run either by non-profit organizations, private entrepreneurs or by the Government. Studies on old age homes, though scarce, have often reported fewer number of Muslims as residents in these homes (Kalavar and Jamuna 2011; Liebig 2008; Rajan 2008). Since India has a stratified society and social distances in terms of residential segregation, language, identity, culinary practices and occupations still separate upper and lower castes as well as Hindus and Muslims (Beteille 1992; Bayly 2001; Drèze and Sen 2013), it is perhaps not surprising to find a lack of social group heterogeneity in our interviews. Additionally, political scientists have pointed out a collective overpowering disgust and intolerance among the Hindu middle class Gujaratis to non-vegetarianism, often associated with Muslims (Jasani 2014; Ibrahim 2007). In our initial round of interviews with managers at several old age homes, difference in dietary practices was often cited as a reason for not admitting Muslim elderly.

In the current study, interviews with the residents were followed up by focus group discussions. The selection procedure was based on the nature of the homes (one is a pay and stay facility and the other one is a government sponsored free facility) as well as on the ease of permissions obtained by the researchers. During our initial survey of old age homes in the city, gaining access to the residents for interviews and discussions were unsuccessful. Hence we decided to focus on the two homes where access was granted and the participants consented to our repeated visits. Ethical clearance for this study has been obtained from the Intuitional Ethics Committee of IIT-Gandhinagar. The interviews were conducted by the authors during the summer of 2014. The language of interaction with the study participants was primarily in Hindi and some Gujarati.

6.5 Methods

Our questionnaire design was guided by the social constructionist strategies (Silverman 2006; Dannefer and Perlmutter 1990; Glaser and Strauss 1967). As noted earlier, we were interested in understanding how older people in these homes are creating and maintaining social meanings of age and aging in their everyday lives. As Dannefer and Perlmutter (1990) explained how "these social processes of interaction" can be seen as dialectical-individual behaviour produces a "reality" which in turn structures individual lives (p. 90; also, Bengtson et al. 1997). For this purpose we had four distinct sections on the questionnaire: (1) socio-demographic information, (2) social resources, (3) health status and (4) general perception questions- that included a combination of Likert scale based questions and intensive interviews. By combining quantifiable questions with in-depth interviews, we believe that we were able to achieve "objectivity, not as an absolute or inherently meaningful condition to which humans react but as an accomplished aspect of human lived experience" (Dawson and Prus 1995: 113).

In particular, the "social resources" section of the questionnaire included questions that map the social capital of the participants (i.e. old age home residents) in terms of ties that they have developed inside the home, family support, networks, trust and organizational affiliations. For example, the section included questions such as whether or not the participants had friends *outside* the home or whether they had anyone *inside* the home on whom they can trust and confide. Additionally, it included questions on the frequency of family members or friends visiting the old age home residents in the last 3 months and the participants' own perception of how often someone (family or friend) should visit them. The "actual" and the "perceived" information were separated out to understand if there is a deficit in expectations and *seva*. Furthermore, we also asked our participants about the social activities that they engaged in and their perception of nurturing a link between old age homes and the society at large. Participants were asked to reflect on the notion of "successful aging" (Rowe and Kahn 1997) that is often characterized by sustained social engagement through volunteering, new skill learning and mentoring as well as an emerging "new" culture that endorses biomedicine and technology driven products (e.g. anti-aging creams, cosmetic surgery, hair dyes and special diets) to erase or delay age markers (Gergen and Gergen 2000). We find that the construction of "successful aging" in old age homes is marked by the cultural belief of "*maya*" that upholds social disengagement and stands in contrast to the (western) mainstream understanding of the concept.

In the "health" section of the questionnaire, participants were asked to report and rank their current physical and emotional health conditions. We also introduced a longitudinal element in mapping the participant's level of loneliness by asking if they felt that same 10 years back. This was done to examine if any life course event has significantly influenced their emotional status. Finally, participants were requested to voluntarily participate in focus group interactions and reflect on the social meanings of aging, filial obligations, kinship ties, individual agency and cultural norms guiding their behaviour in older ages. Due to the small number of study participants, both older men and women were part of these discussions. Thus, borrowing Silverman's (2006) idea, we believe that through our interviews our participants were not just constructing narratives, but also their social worlds and the cultural frames within which they were embedded.

6.6 Observations: Socio-demographic and Social Profiles of the Study Participants

We interviewed 10 women and 5 men for this pilot study. Mean ages of our participants were 73 (women) and 69 (men) years. The average length of stay in the old age homes was about 4.5 years. All women (excepting one) in our study were widowed and the men were either widowed or never married. Half of the respondents were high caste Brahmins and all of them belonged to the Hindu religion (as noted earlier). Gender differences in education, employment and control over economic resources were noted. For example, none of the older women in our study had completed high school as opposed to older men's education ranging from 12 to 15 completed years of schooling. Again, all men in our study reported having employed earlier in their lives, the corresponding number for women was 40 % (4 out of 10). Majority of the women (8 out of 10) didn't have any economic assets (e.g. land or home ownership, pension income, etc.). And those did, had limited control over the utilization of such incomes. These observations are not surprising given the complex interplay of patriarchy and gender-based hierarchies in India that restrict women's education and employment opportunities and their ability to control economic resources (Agarwal 1998; Desai 1994; Chen and Drèze 1992).

Nine out of fifteen study participants reported having at least one adult child. When asked about the reasons for moving into an elder care institution, the three top reasons mentioned by the study participants were widowhood, intergenerational conflict and absence of a male child. To our questions on social capital, none of the older women reported having friends outside the old age home, while 2 out of 5 older men reported having friends. Interestingly, though most of the study participants had access to a phone, older women seem to use them more (8 out of 10) frequently to connect to a female member in their families (daughter or a sister) than older men (1 out of 5) in our study. This finding is consistent with previous studies on network compositions of men and women (Chadha et al. 1990). In particular, women tend to have higher number of "blood relatives" than men. These observations might be explained by the fact that women are often perceived as "kin keepers" (Hagestad 1986) in most cultures suggesting role continuity across the life course.

Another interesting finding was the incongruence in the "actual" frequency of visits by friends or family members and the "expected" frequency (Fig. 6.1). Though the modal finding on the frequency of actual visits was that of "at least once in two months", older men and women expressed very subdued expectations from their friends and family members in terms of frequency of future visits. This finding is incompatible with the notion of *seva* [respectful service paradigmatically provided by the young to the elderly] where it is "normal and appropriate" for elders to be "dependent on kin for material, emotional and bodily support" (Lamb 2013). Lamb (2009, 2013) notes "appropriate and purposeful dependence" of older people on the young from her ethnographic work both in old age homes and families in Kolkata. Our study findings, instead, demonstrate a purposeful denunciation of ties and associations.

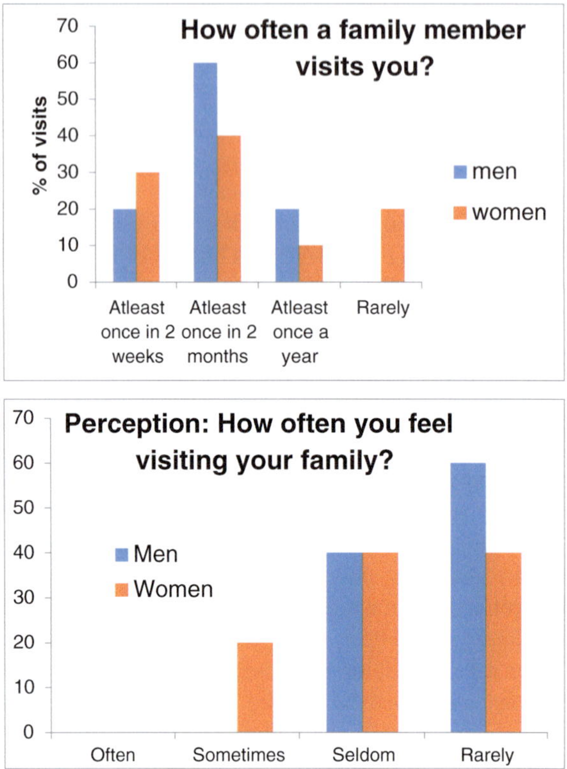

Fig. 6.1 Frequency of visits: actual and expected. *Source* Based on authors' interviews with the study participants, summer 2014

None of the study participants reported having anyone inside the old age home on whom they can trust and confide. Ten out of 15 of them reported feeling lonely "most of the time". On the question whether they would have had the same response if asked ten years back, 10 out of 15 participants reported "no". This response is perhaps not surprising, as most women in our study indicated widowhood as being the reason for moving into an old age home. Widowhood in India is not only marked by a definitive, tragic moment in a woman's life but a profound gendered transition into a life that is shrouded in incapacity, powerlessness and asexuality (Mastey 2009; Lamb 1997). Finally, none of them reported having any organizational affiliations/memberships (e.g. religious, charitable or political) highlighting the participants' very limited interaction with the world outside the homes.

6.7 Social Capital, Interrupted: Trust, Ties and Network Formation in Old Age Homes

Previous studies on old age homes in India have often reported lack of interaction among the residents and smaller network sizes when compared to those living in families (Kalavar and Jamuna 2011; Willingen and Chadha 2008; Arora and

Chadha 1995). However, these studies are limited in their exploration of the "why" question. In this paper, we have attempted to understand why despite being bounded by similar life circumstances, fate and shared living, social capital (e.g. trust, ties, interdependence and association) formation among older men and women in institutions is incomplete. Echoing Cleaver's (2005) critique of the mainstream understanding of social capital, we argue that trust might not automatically emerge from repeated interaction thereby disrupting the normative view of a "social family". In a situation where older men and women have accepted the fundamental impermanence of life, investing in collective action and new social relationships might not be automatic products of living together. All older men and women in our study seemed to be guided by the belief of *maya* (translated often as "illusion" but a more common usage of the term suggests ties, love and attachment). Over the life course, through emotional, instrumental and bodily transactions people may develop maya for their immediate family members, homes, places, experiences and memories. Hence, maya is expected to be heightened as one grows older. However, the ephemerality of life poses challenge to the nurturance and celebration of maya in later ages. Lamb's (2000, 2010, 2013) description of the paradoxical dimensions of *maya* is noteworthy. She asserts that while people are embedded within the ties of maya, "the emotional-physical ties of maya keep the soul or *atma*, literally "bound" to the body, making it difficult for the soul to leave. Having a lot of maya in old age can also cause one to linger on in a decrepit body past the natural time for dying" (Lamb 2010: 68). In our study, participants suggested a purposeful loosening of maya (ties, associations, bindings) before dying to be a realistic strategy to embrace death peacefully. The following quotes are illustrative of this overpowering cultural ideology:

> "I have stopped caring for the society. I moved here because I wanted to die in peace", (female respondent, 85 years); "I miss my family but I do not want to interfere any more. I have seen a lot in my life. I try not having any expectations from anyone now" (female respondent, 76 years); "I spend time reading on my own. After all, no one is really anyone's. I would like to die in a few years…..it is a frightful thought to live for another ten years" (male respondent, 65 years).

In the pursuit of a "meaningful decline" the potential capacities of social capital are frustrated. We contend that unless the cultural perceptions, values and belief systems governing the construction of old age among the elderly in institutions are acknowledged, any external initiative or effort to encourage the building of social capital might not be necessarily helpful. In other words, the mainstream positive view of building trust through repeated association is ruptured in situations where the benefits of reciprocal transactions of care and support are no longer expected. As Cleaver (2005) sharply puts it from her study on the limits and poverty of social capital: "Cooperation and collective action are arguably less motivated by strategic rationality, more by deeply held social norms and perceptions of the "right way of doing things" (Cleaver 2005: 900). In our study, for most of the participants, the "right way of doing things" was to reflect and confront the "fundamental transience of human life…hence, elder institutions becomes, in certain respects, an ancient Hindu way of ordering the life course" (Lamb 2013: 185)

6.8 Concluding Reflections: Gender and the Life Course

In closing, we reflect on a few observations that we drew from the narratives of older women and men. We highlight these examples as they seem to stray and confirm at the same time to the symbols of patriarchal social order. For example, most women in our study valued the relationship that they have with their adult daughters. We have noted earlier that older women in our study reported having phone conversations with their daughters at least once or twice a week, even when they had an adult son. In a country where the social institution of patriarchy contributes to persistent son preference and a systematic discrimination and devaluation of girls throughout the life course (as demonstrated powerfully in several sociological and demographic research on India, see, e.g. Das Gupta et al. 2003), this finding was unexpected. In fact, one of the well-documented reasons for son preference is old age security, especially for widows given their limited opportunities for inheritance and employment (Vlassoff 1990). Daughters, however, seem to provide emotional support and companionship to older parents, even when they are located at a physical distance. While we are unable to establish an empirical validity of this finding in the Indian context owing to limited research, but recent studies in the United States echo similar observations (Scientific American 2014). In contrast, a few older men lamented on the fate of India as (gender and age) hierarchies were challenged by their "modern" daughter-in-laws. One of them explained "She [daughter-in-law] would not give me tea on time. She is also careless about my grandchildren". India, like many other Asian societies, care-giving is traditionally commanded from women who continue to provide these services throughout their life course. Significantly, gerontological literature in India has documented caregiver burden being borne disproportionately by daughter-in-laws in Indian joint households (Gupta et al. 2012).These findings led us to reflect on the shifting Indian values surrounding family and *seva* and how they intersect with the normative gender role expectations.

Finally, we examine how patriarchy and group enforcement of norms restrict older women's mobility in old age homes. Older women from both the elder care institutions in our study reported fewer numbers of trips outside as compared to older men. We found that prior "permission" is required for women to leave the old age homes. Subsequently, the permission is reviewed by the management, and approvals are made based on the perceived importance and relevance of the trip. For example, a permission request for a trip to a local temple has a higher likelihood of being approved than visiting a non-family member. The management personnel in both these places argued that mobility restrictions on women is done ostensibly for "their (i.e. women's) own good", thereby constructing older women as helpless victims of age and deficiency. Controlling and disciplining a woman's body is not new in the cultural discourses of gender and sexuality in India (Puri 1999; Derne 1994). However, since older (widowed) women are regarded as naturally asexual in their post-reproductive lives (Lamb 1999), we were struck by the life course continuity of this gendered pattern. Older women themselves did not seem disappointed with this arrangement. Though some expressed longing to go out more often to buy small items of grocery and personal hygiene, overall they were contented with the permission policy of the old age homes. We interpreted this as the bounded solidarity of the collective that encourages enforcement of norms (Coleman 1988). We argue that a socially entrenched patriarchal structure normalizes social control and affects the lived experiences of women all along the life course. By drawing attention to the complex inter linkages of cultural ideology, normative structure and the limits of social capital, we expect that our work provided a useful next-step for future studies on alternative non-familial caregiving and a critical rethinking on the efforts of integrating the elderly with the larger society.

Acknowledgments This research is part of a larger research project entitled "Research Initiative on Social Gerontology", supported by the Indian Institute of Technology, Gandhinagar (IITGN). The lead author would like to thank Anusmita Devi, William Joe, Irudaya Rajan, U.S. Mishra and Kaushik Gopalan for their helpful suggestions on an earlier presentation.

References

Adler, P. S., & Kwon, S. W. (2002). Social capital: Prospects for a new concept. *The Academy of Management Review, 27*(1), 17–40.

Agarwal, B. (1998). Widows versus daughters or widows as daughters? Property, land and economic security in rural India. *Modern Asian Studies, 32*, 1–48.

Andrist, L. (2008). Social capital's dark side and patriarchy in India. Working Paper # 7, India Human Development Survey. Accessed: http://www.ihds.umd.edu/ihds_papers/scdarkside.pdf

Arora, M., & Chadha, N. K. (1995). Social support and life satisfaction of institutionalized elderly. *Indian Journal of Gerontology, 9*, 74–82.

Bayly, S. (2001). *Caste, society and politics in India from the eighteenth century to the modern age*. Cambridge University Press.

Bengtson, V. L., Burgess, E. O., & Parrott, T. M. (1997). Theory, explanation and a third generation of theoretical development in social gerontology. *Journals of Gerontology Series B: Social Sciences, 52*(B), S72–S88.

Berkman, L., Sekher, T. V., Capistrant, B., & Zheng, Y. (2012). Social networks, family and caregiving among older adults in India. In J. Smith, & M. Majumdar (Eds.), *Aging in India: Findings from new and emerging data initiatives*. Washington, DC: National Academies Press. Accessed at: http://www.ncbi.nlm.nih.gov/books/NBK109220/

Beteille, A (1992). Caste and Family: In representations of Indian society. *Anthropology Today, 8* (1), 13–18

Bhat, A. K., & Dhruvarajan, R. (2001). Ageing in India: Drifting intergenerational relations, challenges and options. *Ageing and Society, 21*(5), 621–640.

Bisung, E., Elliot, S., Schuster-Wallace, C. J., Karanja, D. M., & Bernard, A. (2014). Social capital, collective action and access to water in rural Kenya. *Social Science and Medicine, 119*, 147–154.

Bloch, F, Genicot, G and Ray, D (2007). Reciprocity in groups and limits to social capital. *American Economic Review Papers and Proceedings, 97*, 65–69.

Bloom, D. E., et al. (2010). Economic security arrangements in the context of population ageing in India. *International Social Security Review, 63*(3–4), 59–89. doi:10.1111/j.1468-246X.2010.01370.x

Bourdieu, P. (1985). The forms of capital. In J. G. Richardson (Ed.), *Handbook of theory and research for the sociology of education*. New York: Greenwood.

Cannuscio, C., Block, J., & Kawachi, I. (2003). Social capital and successful aging: The role of senior housing. *Annals of Internal Medicine* (Supplement), 139(5), No. 2.

Carpiano, R. M. (2007). Neighbourhood social capital and adult health: An empirical test of a Bourdieu-based model. *Health & Place, 13*(3), 639–656.

Chadha, N. K., Aggarwal, V., & Mangla, A. P. (1990). Social network and aging. *Spectra of Anthropological Progress, 12*, 15–32.

Chen, M., & Drèze, J. (1992). Widows and health in rural India. *Economic and Political Weekly, 27*, 81–92.

Chen, F., & Short, S. (2008). Household context and subjective well-being among the oldest old in China. *Journal of Family Issues, 29*, 1379–1403.

Cleaver, F. (2005). The inequality of social capital and the reproduction of chronic poverty. *World Development, 33*(6), 893–906

Coleman, J. S. (1988). Social capital in the creation of human capital. *The American Journal of Sociology, 94*. Title of Supplementary Material: Organization and Institutions: Sociological and Economic Approaches to the Analysis of Social Structure, S95–S120.

Cornwell, B. (2011). Independence through social networks: Bridging potential among older women and men. *The Journals of Gerontology, Series B: Psychological Sciences and Social Sciences, 66*(6), 782–794.

Cramm, J. M., Dijk, H. M., & Nieboer, A. P. (2012). The importance of neighbourhood social cohesion and social capital for the well-being of older adults in the community. *The Gerontologist, 52*(1), 142–150.

Colletta, N. J & Cullen M. L. (2000). *The nexus between violent conflict, social capital and social cohesion: Case studies from Cambodia and Rwanda*. Social Capital Initiative, Working Paper No. 23. The World Bank.

Dandekar, K. (1996). *The elderly in India*. New Delhi: Sage

Dannefer, D., & Perlmutter, M. (1990). Development as a multidimensional process: Individual and social constituents. *Human Development, 33*, 108–137.

Das Gupta, M., Zhenghua, J., Bohua, L., Zhenming, X., Chung, W., & Hwa-ok, B. (2003). Why is son preference so persistent in East and South Asia? A cross-country study of China, India and the Republic of Korea. *Journal of Development Studies, 40*(2), 153–187.

Dawson, L. L., & Prus, R. C. (1995). Postmodernism and linguistic reality versus symbolic interactionism and obdurate reality. *Studies in Symbolic Interaction, 17*, 105–124.

Derne, S. (1994). Hindu men talk about controlling women: Cultural ideas as a tool of the powerful. *Sociological Perspectives, 37*(2), 203–227.

Desai, S. (1994). *Gender inequalities and demographic behaviour: India*. New York: The Population Council.

Drèze, J., & Sen, A. (2013). *An uncertain glory: India and its contradictions*. Princeton University Press.

Durkheim, E. (1897) [1951] Suicide: a study in sociology. The Free Press.

Farr, J. (2004). Social capital: A conceptual history. *Political Theory, 32*(1), 6–33.

Gabbay, S., & Zuckerman, E. W. (1998). Social capital and opportunity in corporate R&D: The contingent effect of contact density on mobility expectations. *Social Science Research, 27*, 189–217.

Gergen, M. M., & Gergen, K. J. (2000). Qualitative inquiry: Tensions and transformations. In N. Denzin, & Y. Lincoln (Eds.), *Handbook of qualitative research* (2nd ed.). Thousand Oaks, CA: Sage

Glaser, B. G., & Strauss, A. L. (1967). *The discovery of grounded theory: Strategies of qualitative research*. New York: Aldine.

Gupta, V., Levenburg, N. M., Moore, L. L., Motwani, J., & Schwarz, T. V. (2007). Organization model of the southern Asia cluster family business. *The South East Asian Journal of Management, 1*(2), 125–142.

Gupta, R., Pillai, V. K., & Levy, E. (2012). Relationship quality and elder caregiver burden in India. *Journal of Social Intervention: Theory and Practice, 21*(2), 39–62.

Hagestad, G.O. (1986). Dimensions of time and family. *American Behavioral Scientist, 29*, 679–694.

Ibrahim, F. (2007). Capitalism, multiculturalism and tolerance: A perspective on "Vibrant Gujarat". *Economic and Political Weekly, 42*(34), 3446–3449.

Jasani, R. (2014). Book review, July 6. *Journal of Church and State.* Review of Ghassem-Fachandi, Parvis (2012). *Pogrom in Gujarat: Hindu Nationalism and Anti-Muslim Violence in India.* Princeton University Press.

Kalavar, J. M., & Jamuna, D. (2011). Aging of Indian women in India: The experience of older women in formal care homes. *Journal of women & aging, 23*(3), 203–215.

Koutsogeorgou, E., et al. (2014). *Social capital and self-rated health among older adults: A comparative analysis of Finland.* Poland and Spain: Ageing and Society, January. doi:10.1017/S0144686X13001025

Ku, L. J. E., Stearns, S. C., Houtven, C. H. V., Lee, S. D., Dilworth-Anderson, P., & Konrad, T. (2013). Impact of caring for grandchildren on the health of grandparents in Taiwan. *Journals of Gerontology, Series B: Psychological and Social Sciences, 68*(6), 1009–1021. doi:10.1093/geronb/gbt090

Lamb, S. (1997). The making and unmaking of persons: Notes on aging and gender in North India. *Ethos, 25*(3), 279–302.

Lamb, S. (1999). Aging, gender and widowhood: Perspectives from rural West Bengal. *Contributions to Indian Sociology, 33*(3), 541–570.

Lamb, S. (2000). *White Saris and Sweet Mangoes: Aging, gender and body in North India.* University of California Press.

Lamb, S. (2009). *Aging and the Indian Diaspora: Cosmopolitan families in India and abroad.* Bloomington, IN: Indiana University Press.

Lamb, S. (2010). Love and aging in Bengali families. In D. P. Mines & S. E. Lamb (Eds.), *Everyday life in south Asia.* Bloomington: Indiana University Press.

Lamb, S. (2013). Personhood, appropriate dependence and the rise of eldercare institutions in India. In C. Lynch, & J. Danely (Eds.), *Transitions and Transformations: Cultural perspectives on aging and the life course.* Berghahn.

Liebig, P. S. (2003) [2008]. Old-age homes and services: Old and new approaches to aged care. *Journal of aging & social policy, 15*(2–3), 159–178.

Lim, C., & Putnam, R. (2010). Religion, social networks and life satisfaction. *American Sociological Review, 75*(6), 914–933.

Marx, K. (1894). *Capital* (Vol. 3). New York: International.

Mastey, N. (2009). Examining empowerment among Indian widows: A qualitative study of the narratives of hindu widows in north Indian Ashrams. *Journal of International Women's Studies, 11*(2), 191–198.

Moody, H. R. (1993). Overview: What is critical gerontology and why is it important? In T. Cole, et al. (Eds.), *Voices and visions: Towards a critical gerontology.* New York: Springer.

Nair, T. K. (Ed.). (1995). *Care for the elderly: Directory of organizations caring for the elderly in India.* Madras: Madras Institute of Ageing.

Olivelle, P. (2009). *The law code of Manu.* Oxford World's Classics: Oxford University Press.

Pal, S. (2004). *Do children act as old-age security in rural India: Evidence from an analysis of elderly living arrangements.* Montreal, Canada: Paper presented in the North East Universities Development Consortium.

Portes, A. (1998). Social capital: Its origins and applications in modern sociology. *Annual Review of Sociology, 24*, 1–24.

Prince, M., Livingston, G., & Katona, C. (2007). Mental health care for the elderly in low-income countries: A health systems approach. *World Psychiatry, 6*(1), 5–13.

Puri, J. (1999). *Woman, body, desire in post-colonial India: Narratives of gender and sexuality*. New York: Routledge.
Putnam, R. D. (1993). *Making democracy work: Civic traditions in modern Italy*. Princeton, NJ: Princeton University Press.
Putnam, R. D. (1995). Bowling alone: America's declining social capital. *Journal of Democracy, 6* (1), 65–78.
Putnam, R. D. (2000). *Bowling alone: The collapse and the revival of American community*. New York: Simon and Schuster.
Rajan, I. S. (2008). Home away from home. *Journal of Housing for the Elderly, 16*(1–2), 25–150.
Rajan, I. S., & Kumar, S. (2003). Living arrangements among Indian elderly: New evidence from the national family health survey. *Economic and Political Weekly, 38*, 75–80.
Rowe, J. W., & Kahn, R. L. (1997). Successful Aging. *The Gerontologist, 37*(4), 433–440.
Rumbaut, R.G. (1977). Ties that bind: Immigration and immigrant families in the United States. In A. Booth, A.C. Crouter, & N. Landale (Eds.), *Immigration and the Family: Research and Policy on US immigrants*. Mahwah, NJ: Erlbaum
Samanta, T., Chen, F., & Vanneman, R. (2015). Living arrangements and health of older adults in India. *Journals of Gerontology Series B: Psychological and Social Sciences., 70*(6), 937–947. doi:10.1093/geronb/gbu164
Samanta, T. (2014). Household context, social capital and subjective wellbeing of older adults in India. Building knowledge for population ageing in India. *Series II, Working Paper # 2*. New Delhi: United Nations Population Fund. Available at: http://countryoffice.unfpa.org/india/?publications=11981
Sathyanarayana, K. M., Kumar, S., & James, K. S. (2012). *Living arrangements of elderly in India: Policy and programmatic implications. Working paper: building knowledge for population ageing in India*. New Delhi: United Nations Population Fund.
Scientific American (2014). *Graying parent care falls to daughters not sons* by Dina Fine Maron. August 26. Accessed: http://www.scientificamerican.com/podcast/episode/graying-parent-care-falls-to-daughters-not-sons/
Silverman, D. (2006). *Interpreting qualitative data: Methods for analysing talk, text and interaction* (3rd ed.). Thousand Oaks, CA: Sage.
Waldinger, R. (1995). The 'other side' of embeddedness: A case-study of the interplay of economy and ethnicity. *Ethnic and racial studies, 18*(3), 555–580.
Times of India (2011). *Old age homes tripe in state*, December 5. Accessed: http://timesofindia.indiatimes.com/city/ahmedabad/Old-age-homes-triple-in-state/articleshow/10987962.cms
van Willingen, J., & Chadha, N. K. (2008). Social networks of old people in India. *Journal of Aging and Social Policy, 15*(2–3), 109–124.
Vera-Sanso, P. (2004). They don't need it, and I can't give it: Filial support in South India. In P. Kreager & E. Schroder-Butterfill (Eds.), *Ageing without children: European and Asian perspectives*. Oxford: Berghahn Books.
Vikram, K., Vanneman, R., & Desai, S. (2012). Linkages between maternal education and childhood immunization in India. *Social Science and Medicine, 75*, 331–339.
Vlassoff, C. (1990). The value of sons in an Indian village: How widows see it. *Population Studies, 44*(1), 5–20.
Yip, W., Subramanian, S. V., Mitchell, A. D., Lee, D. T., Wang, J., & Kawachi, I. (2007). Does social capital enhance health and well-being? Evidence from rural China. *Social science & medicine, 64*(1), 35–49.

Chapter 7
Engaging the Future as Ethnographic Object: Japan's Aging Society Crisis, Ontogenesis and Cybernetics

Ender Ricart

Abstract The aging society crisis in Japan represents a point of saturation, when the numbers of elderly in-need-of-care exceed the available resources, leading to economic, societal, and national collapse. The aging society crisis is an agent of change, a future potentiality that is inextricably linked with enactments of and approaches to aging, old age, and sociality in present-day Japan. The passing of Japan's national long-term care insurance in 2000 opened up new possibilities, new ground on which to compare and contrast, to carve out meaning, identity and particularities. I discuss a differentiation currently underway between the ordering of people, things, institutions, and ideas involved in "prevention" versus that of "care". Topologically, I argue, care and prevention are noticeably distinct orders of being. But, with an analysis of temporality, we discover that prevention is an action motivated by an anticipatory future that effects change in the present, while care is a direct evincing of the feared-for future. Both care and prevention are thus united in a shared becoming that is the aging society crisis. Engaging the future as an ethnographic object presents opportunities for innovative theoretical exploration and development. How do we go about analysing an object that lacks substantial existence, yet displays agency in the present by organizing networks of relations and motivating change? How might we theoretically frame and approach the future? This research analytically explores futurity as a force of change not unlike power, information and action, by applying the theory of ontogenesis and cybernetics.

Keywords Aging society crisis · Health promotion · Temporality · Ontogenesis and cybernetics · Japan

---

E. Ricart (✉)
Department of Health and Analytics, Battelle Memorial Institute, Seattle, Washington, USA
e-mail: ricart@battelle.org

T. Samanta (ed.), *Cross-Cultural and Cross-Disciplinary Perspectives in Social Gerontology*, DOI 10.1007/978-981-10-1654-7_7

7.1 Introduction and Theoretical Framework

In the last two decades, Japan has seen a dramatic decline in birth rates, a rise in life expectancy, and the retirement of the baby-boomer generation.[1] With a quarter of its population over the age of 65, Japan is one of the world's leading aged nations.[2] The problem-conscious of the aging society in Japan was born in the early 1970s with the tying together in statistical forecast models of economic projections and demographic trends. The crisis represents a point of saturation, when the numbers of elderly in-need-of-care exceed available financial resources and labour-power, leading to economic, societal and national collapse.

With their eyes on the coming crisis, the architects of Japanese healthcare policy have increasingly adopted an epidemiological approach to aging that regard certain conditions of old age, namely conditions that render persons dependent and in-need-of-care, as potentially preventable through health education, early detection and targeted prevention programmes and services. The goal is to limit the number of elderly in-need-of-care and enrolling in Japan's national and publicly financed Long-Term Care Insurance (LTCI) system.

Engaging the future as an ethnographic object presents opportunities for innovative theoretical exploration and development. How do we go about analysing an object that lacks substantial existence, yet displays agency in the present by organizing networks of relations and motivating change? How might we theoretically frame and approach the future? This research analytically explores futurity as a force of change not unlike power, information, and action, by applying the theory of ontogenesis (see Simondon 1980, 1992; Mackenzie 2002; Combes 2012; Lamarre 2012; Ricart 2013, 2014). If ontology is the study of being and worldliness, ontogenesis is the study of becoming and ordering. Ontogenesis follows the integrating principle, the force of change. It is a study of the needle as it threads assemblages of ideas and things so they "hang together". As scientists, politicians, local government employees and seniors rally around the problem of an aging society, this ethnographic analysis looks specifically at emerging ontologies of the self and life world and the particular formation of institutions of care.

The aging society crisis of which Japanese gerontologists, government officials, care providers, and the general public speak is positioned in the temporal future. In analysis, however, I do not approach the crisis as a temporal future, but a social imaginary to account for its agentive capacity in the present. Time, as it is generally understood as a sequencing of events, cannot be applied in our analysis of the

[1]In 2013, the birth-rate is estimated to have been 1.39 children born per woman. The average life expectancy was 80.85 years for men and 87.71 years for women (CIA World Factbook, 2013).

[2]According to a study conducted by the Ministry of Health, Labour, and Welfare, due to the number of lives lost in the March 11, 2011, Great East Japan Earthquake, for the last three years Japan has had to forfeit the title of "world's most aged nation," the first time it has done so in 27 years. The national average age of Japanese women is now second to Hong Kong, and Japanese men have taken up eighth place (Due to the great earthquake, 2012).

futurity of the aging society crisis, for here the future functions as both the end and the beginning. This simultaneity of end-as-beginning accounts for the agency of the future crisis as a force of change. The actions taken in the present by the government, scientists and people of Japan to ameliorate the future aging society crisis presuppose the end. This end cannot be fully prevented; the severity of the crisis can only be lessened. The actions being taken in the present are both in response to a specific social imaginary of the future crisis and, in turn, inform and confirm the reality of the future crisis.

The becoming of the aging society crisis ends in the actualization of the crisis, and this end is the presupposition for the action and changes in its parts in the present. The aging society crisis, then, is an encompassing whole that is both internal and external to the form and structure of its parts. Using the ethnographic method, I detail two parts of the aging society crisis in Japan—"care" (*kaigo*) as dependence on LTC, and "care-prevention" (*kaigo yobo*) as prevention of dependence on LTC. In the discourse of the aging society crisis, care and care-prevention are figured as two antithetical movements. With an analysis of the actors involved, networks of relations and spatial limits of movement, I demonstrate that care and care-prevention are discrete, occupying different configurations of institutions, knowledge and practice. However, they share an ontology of the aging process, and, temporally, both are oriented around and motivated by the future aging society crisis.

For aging studies specialists, government employees, caregivers and seniors in Japan, care is positioned as a direct window into a futurity of burden. It is a clustering of anxiety for the coming crisis. Thus, with each person who enrols in the publicly financed and universally available LTCI system, this future state-of-being is indexed, confirmed, and brought closer to the present reality. Care-prevention is discursively framed as care's antithesis. It is oriented around minimizing the need for care, specifically prevention of enrolment in LTCI, and holds the promise of subduing an epidemic of in-need-of-care seniors taxing the economy. Care-prevention is an anticipatory form of self-care, investing in the present to decrease the time spent in need of more costly care in the future.

This temporal loop of end-as-beginning characterizes the agency of the future crisis in the present. It is in the process of its own becoming, recursively formed and materially manifest in its parts: care and care-prevention. As the nation of Japan prepares for the aging society crisis, ontologies of the aging process and sociality have transformed. To be discussed further, in the discourse of care-prevention, each individual is depicted as possessing a well spring of energy, which, if properly maintained, continues to flow until the spring is tapped out, resulting in an ideal disease-free sudden death. Mental, physical and social activities are means of maintaining and stimulating the body's life force. When the flow of the life force is interrupted due to imbalanced activity, inactivity or genetic and environmental factors, illness and care-dependence result. Similarly, care is a matter of removing blockages and building canals so the spring of energy can continue to flow until its source reaches its pre-determined limit.

With introductions from my primary research site at a renowned gerontological research facility and geriatric hospital in Tokyo, Tokyo Metropolitan Institute of Gerontology (TMIG), over an 18 month period I was introduced to and interviewed national and local government employees, policymakers, community leaders in old age programmes, and resident seniors. I also participated and observed in care-institutions, geriatric hospitals, gerontological research lab and field research projects, local community initiatives, and other senior-oriented programmes in the Tokyo metropolitan area, surrounding suburbs and more removed rural towns. Despite the fact that scientists and the general population of seniors do not otherwise share knowledge cultures or practices of everyday life, I found there to be a high degree of overlap in their ontologizing of the human condition in old age. Scientists, government employees and seniors alike regarded the aging process as a matter of waning bodily energy that needs to be stimulated through activity, exchanged with the energy of social and natural living environments, and assisted through care-work to foster a higher quality of life and death.

This chapter concludes with the observation that Japan's aging society crisis is discursively framed in energetic terms, resembling the feedback systems of cybernetics theory (Pickering 2010a, b). I question whether or not ontogenesis and cybernetics theory might usefully be applied in the ethnographic study of other social imaginaries of the future, like global warming. While it is outside the purview of this chapter to explore further, other crises may be conceptually and temporally bounded wherein the future end-point, the actualization of the crisis, functions as the premise for taking action in the present to ameliorate the severity of the future condition. Other future imaginaries of crisis may also be bound topologically with specific configurations of actors, resources, institutions and habits of being designated, legitimized and naturalized as appropriate or inappropriate through policy and scientific research. Theoretically approaching these social imaginaries of the future as quasi-bounded "feedback systems" specific to particular socio-historical moments, allows us to attend the commonalities and connections between seemingly diverse sets of actors, disparate practices and distinct cultures (i.e. scientific and everyday). It also opens up the future to uncertainty, as changes in its' parts can lead to changes in the whole.

7.2 Diverging Sociality of Care and Care-Prevention

There is a division being drawn between the people, things and institutions involved in care-prevention and the various avenues through which it operates, and those involved in care. Care-prevention is situated at the level of the local-community and describes the actions of gerontologists, the local government, and residents. Care is situated at the national level, involving all institutions and services covered by Japan's long-term care insurance (LTCI) and the national government, specifically the Ministry of Health, Labour, and Welfare (MHLW). In what follows, I ethnographically detail the sociality for those seniors' in-need-of care and enrolled in

LTCI and still-independent seniors to demonstrate the diverging topological configuration of care and care-prevention.

For both care and care-prevention, the attending forms of sociality are always rhetorically framed in relation to a vanishing traditional three-generational household. In interviews and informal conversation with Japanese community residents, elderly, government employees, and gerontologists, the "traditional" three-generation Japanese family is described as being nearly extinct. In its place stands the nuclear family, visible not only in Japan's cities but also rural households, where most of the younger generations have fled to live in urban centres, leaving large family homes to waste away along with their aging occupants.

Prior to World War II, Japanese civil law stipulated that parental care responsibilities fell upon the wife of the first-born son (Arai and Ikegami 1998). The sentiments of this civil law were compounded by Confucian mandates to respect one's seniors and familial obligations. For a long time after the civil law was disbanded after WWII, the use of public services suggested a failure to fulfil such family obligations. However, this is waning. One of the explicit aims in instituting the LTCI system in 2000 was to counteract this socio-cultural stigma towards outside care, and transfer the burden of caregiver from family members to the nation as a whole.

The intervening years between 2000 and the LTCI revision of 2005 were geared towards public service announcements and regional promotions encouraging elderly and their families to welcome into their private lives the various care services offered by the new insurance system. Such public promotions tended to frame LTCI services positively, as relieving elderly, who are concerned with burdening family or friends, of a particular socio-historical stigmatized imposition of care. The socialization of care was additionally intended to free up younger generations of family caregivers so they could re-join the workforce and re-stimulate the sluggish Japanese economy.

When LTCI was enacted in 2000, the government's assumption was that enrolment would be low, giving them time to build-up the necessary care-infrastructure to support future growth. But, within the first few years, LTCI enrolment, particularly adults with low-level needs, was off the charts incurring higher than expected total spending (Ishibashi and Ikegami 2010, p. 2). There was a strong latent demand for this entitlement service, especially among seniors with light conditions. The implementation of national LTCI and its public promotion brought the family, old-age care, governance, and society into new relations. Care's sociality, as a private matter confined to collectives of the local, the family, the home, and the burdens of care being confined therein, has been dislodged, opening it to new social relations as part of an emerging order of care framed by LTCI policy.

Care is increasingly synonymous with "dependence on Long Term Care Insurance". The actors involved in care, as mentioned earlier, are situated on the national level and involve all LTCI services; national government-supported healthcare facilities, such as nursing homes, day cares and medical institutions; and all staff who are paid through government monies. The cost of care for elderly enrolled in LTCI is likewise paid for by the government, with only 10 % of total costs being out of pocket payments (funding for LTCI comes from a mandatory tax on working adults aged 40–65). The administration of LTCI requires a host of supportive staff to ensure its savvy navigation by elderly, family, services and institutions. With a few exceptions, these government workers, facilities, services, and tools cannot be used by anyone except for those classified as "in need of care". Thus, the topology of care is overwhelmingly dominated by collectives and networks affiliated with the national government and nationalized healthcare system (though family-based care administration is still common in Japan, it is increasingly supported and supplemented by LTCI services, institutions and care goods).

Due to physical or mental limitations that render them in-need-of-care, heavily dependent elderly are largely confined to spaces and places not unlike Goffman's total-institution, either in the home or at care facilities (1961). While in-need-of-care seniors' sociality is often regarded as shrinking, this is only in the traditional sense of person-centred interactions; their sociality grows in the number of human and nonhuman actors oriented around caregiving, forming what I call a "technical ensemble of care" (Simondon 1980). When speaking of technical ensembles, I am not speaking purely of tools or technological machineries. Technical ensembles of care refer to the coming together of a thick collective of actors and in-need-of-care elderly around a certain set of LTCI-identified needs. To provide an example of its topological reach, the ensemble of care includes an expansive collection of care workers, care facilities, care-managers, occupational therapists, speech therapists, nutritionists, geriatricians, medical specialists, nurses, assistive technology trainers, home visiting nurses and helpers, family members, and a variety of assistive technologies: hospital beds, walkers, wheelchairs, hand rails, etc.

The sociality of the in-need-of-care elderly has transformed. Collectives of care and the in-need-of-care aged individual have become a unit of being, and technical ensembles of care now define what age-related care is in Japan today. As mentioned briefly, elderly in-need-of-care are a direct manifestation of the feared-for crisis. The being of the in-need-of-care senior, this collective of human and non-human that forms the technical ensemble of care, requires significant man-power, technical-power, economic-power and institutional-power, which, with the anticipated levels of elderly in-need-of-care, would be cataclysmic for the nation's future. Thus, the sociality of care increasingly indexes a future crisis of present-day conditions that are magnified by the sheer quantity of in-need-of-care elderly and attending collectives of care. The in-need-of-care senior requires an army of caregivers, whereas the still healthy and independent senior is discussed by scientists, policymakers, and seniors as responsible for and capable of carrying out preventive care-work on their own.

With the erosion of the sympathetic network of the family, elderly people are, I have been told by gerontologists, government employees, seniors, and community leaders in old age programmes, liable to become shut-ins (*tojikomori*), grow unhealthy due to the shrinking of their social-world, become depressed, develop dementia (*ninchishō*), become frail and in-need-of-care, and eventually die a slow and miserable death. A growing trend in care-prevention discourse has been to encourage other forms of sociality, namely "community building" (*komyunite tsukuri*, equally referred to in Japanese as health building [*kenkō tsukuri*] or town-building [*machi tsukuri*]). The emphasized necessity for other forms of sociality to replace the loss of the three-generation family reinforces a specific conceptualization of human nature and the aging process: that social relation and exchange are a fundamental aspect of healthy aging.

Given the linkage made by gerontologists, local and national government employees, and seniors themselves between community and healthy aging, the community is a form of sociality reserved for independent and healthy seniors. In the first lesson of a three-month bi-weekly senior exercise class, the class instructor, a petite middle-aged woman with a small voice, stood in the front of the town-centre gymnasium leading a class of 30 seniors. Using the megaphone, she shouts, "We will begin with building your body strength, but our true goal is to make you strong enough to participants in community building!" Before social participation in the community can take place, one must be physically active and fit. Thus, in-need-of-care seniors do not partake in the social form of the community as they are not able to actively engage as independent agents.

Community, as described by gerontologists, local-government employees, community building leaders, and seniors, is a place of social exchange, commerce, and is in harmony with nature. It is described using organic terms, touching upon an atemporal, natural and innate drive lying dormant within the people of Japan that has been left to fallow over the last 60 years of the family's dissolution in the post-war period. Community building is the local responsibility of the regional government, and is a matter of establishing various community-wide programmes and services in an attempt to re-stimulate this "natural" and "spontaneous" (*shizen*) human drive for companionship and a shared-togetherness.

Gerontologists, community leaders, and local government officials tell me that once this latent human need is stimulated through community building projects, the community will no longer need to be actively built from the top down, but will begin to take on a life of its own. Mr. Katayama, vice president of a Tokyo-based senior community building project, tells me: "Little by little people's hearts are being moved. People involved in community building go out into the world and take this feeling of togetherness, this feeling of community with them and share it with others". For Mr. Katayama, the movement of people's hearts is pivotal for the action of community building to take place. As this latent need for community is rekindled, bottom-up communities will emerge to support a healthy and active lifestyle for its many aging residents.

Working towards the actualization of an intimate community that can fill the void left by the dissolution of the traditional family involves the topological mobilization of new sets of actors and networks of relations. These actors include local government officials, participants in community activities, NPOs, local business owners, medical institutions, aging studies scholars and gerontologists, and indirectly, the national government as granting agent. Community building is not a new subject for discussion, as it has been an ongoing trend for the last two decades, but community building for an aging society carries with it the extra distinction of merging epidemiological notions of prevention. Community building as a primary form of care-prevention strives to build an intimate and affective living environment, because this in turn is believed to affect positive change in the mental and physical health of its seniors.

As an aging studies scholar at Tokyo Metropolitan University emphasized in our conversations, healthy aging is a matter of maintaining independence well into old age, and critical to this is not just the individual's lifestyle and physical health, but also monitoring and transforming the living environment. This includes, for example, the architecture of one's home: Are there barriers to mobility? Is it climate controlled? Is it oriented towards the physical capabilities of an aging body? The safety and integration of the community and balance with the surrounding natural environment is also considered.

Researchers and local government employees with whom I have spoken, emphasized a natural synergy that must be cultivated between the aging individual and the surrounding living environment. It is not about the individual body of the aged, but the "aging society". That is, building a form of sustainable sociality, a community, within which generations of people can reside throughout the aging process. It literally is a society for all age groups whom, in turn, age together. In this community, there would be houses for young couples, young families, middle-aged couples or individuals, independent older adults, and dependent elderly. Each of these dwelling spaces would be particularly designed for the physicality and needs of each stage of life and placed in different areas of the town to promote integration, socialization, and ready access to those public institutions needed at each phase.

The example provided to me stems from one well-known aging studies researcher at Tokyo University who is currently involved in the construction of two aging society model towns. One of these model societies will feature an expansive recreation room where meals are also served. She envisions a young child going to the common room for breakfast in the morning as her mother goes off to teach at the nearby university and the father goes to the hospital to provide medical care for elderly residents. In the common room the parents know the child will not be alone, but safe with other community members. In these places and social venues, the elderly too will have a social role and function to fill. Intergenerational exchange would be a natural outcome. The gerontologists and local government workers with whom I have spoken about the preventative capacity of the living environment (spatially and socially) believe it is one important innovation that Japan, as the leading super aged society, can contribute to the science of aging.

In regard to aging and old age, community building is overwhelmingly situated as a mechanism of care-prevention. On the epidemiological scale the focus of tertiary prevention of aging concerns itself with matters of community building, and through community building, health building—hence the ready interchangeability of the terms "community building" (*machi tsukuri*) and "health building" (*kenkō tsukuri*) in various regional and national projects in Japan. The community is regarded as a positive, natural, and emergent kind of sociality and opens up space for hope, unlike the sociality of care that directly indexes the future crisis. The sociality of care is a thick collective of people and things focussed on maintaining the functional health of one individual. That individual and the collective are confined to such relationships, whereas within the community of care-prevention, we find able-bodied and healthy elderly who are, along with other community members, actively engaging, flowing and moving.

As the model "aging society", social events like exercise groups, classes and informational sessions, or festivals are all intended to stimulate and cement the social solidarity of the community. This community, because of its organic and natural ontology, would spontaneously generate, support and even encourage organizations and activities involving and tying together community residents (Social Welfare Research Centre [*shakai hoken kenkyujo*] 2013, pp. 37–39). In the sociality of the community, elderly are seen as satisfying basic human needs through the action of socialization, and therefore, live a longer and healthier life without the kind of support required when one becomes in-need-of-care under the LTCI scheme. As the Tokyo-based community building project vice president Mr. Katayama mentioned, these projects will rekindle awareness within each of us of our fundamental need for communion with others.

7.3 Shared Ontology of Aging Process in Care and Care-Prevention

Transitioning from our discussion of the social and topological difference between care and care-prevention, I now want to turn to their continuities. I ethnographically explore the shared ontological configuration of care and care-prevention through a discussion of "unhealthy aging" and "healthy aging". Age-related health issues have, in the field of gerontology, been termed "geriatric syndrome", and manifest in the form of delirium, dizziness, susceptibility to falling, incontinence, and, most commonly, "frailty" (*kyojyaku*). A widely accepted diagnostic system for frailty was derived by the American gerontologist Linda Fried and colleagues, who classified frailty using five characteristic dimensions: weight loss, weakness, slowness, decreased activity and exhaustion (Fried et al. 2001). One Japanese gerontologist at the TMIG argues that Fried's approach focuses only on the physicality of frailty, ignoring environmental aspects and how they too may delay or prevent its onset. His research team developed a questionnaire that includes, in addition to these core physical indicators, questions concerning social functioning, such as: "How often do you go out? Do you have hobbies? Do you see friends, family, or care workers?" The logic underlying this addition to Fried's questionnaire is that the more one goes out to socialize, the more opportunities one has to interact with people, requiring the regular use of the mind, mouth and body. In regular use, these functions are considered less likely to decline. Failure to use them results in functional decline, termed "disuse syndrome" (*haiyōshōkōgun*).

Many gerontologists, aging studies scholars, and elderly in Japan have explained to me the importance of the concept of "use it or lose it", reciting phrases such as:

"If you do not move, you will become unable to move" [*ugokanai to ugokenai yō ni naru*]; "you move in order to stimulate or revitalize" [*kasseika suru tame ni ugokasu*]; and, "put your body's functions to use so that they don't waste away" [*shintai no nōryoku otoroenai yō ni ikasu*].

Another gerontology concept similar to disuse syndrome but unique to Japan is "locomotive disease", colloquially called *rocomo byo*. Locomotive disease, as the name implies, refers to the declining functionality of the aspects of the muscular-skeletal system related to locomotion. In my interview with the Health Promotion and Exercise Department Chief at the National Institute of Health and Nutrition (NIHN), he detailed the importance of locomotion to both individual mental and physical health and the health of society. He explains that movement is a prerequisite for the formation of human ties (*tsunagari*), and that in turn, human ties necessitate movement, thereby circularly preserving both individual health and the health of society. He gives the example of two people walking along a road, crossing paths and exchanging greetings. With modern technology—cars, phones, computers, the Internet, and television—such exchanges decrease. He tells me that one can fulfil basic bodily needs remotely without leaving the comforts of one's home or interacting physically with other people. One may have a hundred friends online, but never meet them in person. This, he says, is not a true relationship because it lacks movement. Humans are "moving things." He draws out the kanji for "animal" in Japanese—*dōbutsu* (動物). The first character is the kanji for "motion" or "movement", and the second kanji means "thing". The kanji, for him, reveal the deeper importance of movement to human nature—that movement is necessary for our existence, for life as we know it. The movement and motion of locomotion is not only at the core of what it means to be a human animal but is also the backbone of social relations and community formation.

For the Japanese gerontologists, aging studies specialists, care-prevention programme staff, local and national government employees, and seniors I spoke with, physical, mental and social activity are both indices of human vitality and means of staving off the onset of stagnation and decline characteristic of unhealthy (and therefore costly) aging. In staving off, that is, preventing the onset of age-related physical and mental decline, one also prevents enrolment in Japan's LTCI system and decreases care-related costs, thereby sparing the economy and reducing the man-power and labour hours required to provide future care. It then follows that ideally one can remain active until the last day of one's life, upon which one dies suddenly and unexpectedly in a phenomenon popularly hoped for by elderly, *Pin Pin Korori* (often simply referred to as PPK). *Pin pin* means lively and active and *korori* means suddenly. Therefore: "lively and active, when suddenly—death". Its opposite, *Nen Nen Korori*, is a slow and gradual decline into bedridden dependence.

Such a discussion of the capacity of continued activity to prevent the slowing and eventual inactivity of old age, crystalized in the feared-for bedridden dependency of old age (*netakiri*), speaks to the ontology of the aging process. "Healthy aging" is characterized by continued activity well into old age whereupon one suddenly comes to a stop, whereas "unhealthy aging" is depicted as starting with inactivity and leads to a slow decline, necessitating assistance and care. As demonstrated in the "use it or lose it" approach and in the reading of the Japanese word for animal as "active-thing", activity is preventive because it maintains the continued flow of an internal and innate energetic life force. By stimulating this innate energy or life force through action, one promotes health. Community building was also discussed as emerging spontaneously from the wellspring of an innate, shared human instinct for social engagement with others to form an intimate and affective community. As the Department Chief of NCHN was quoted earlier, action is also the foundation of sociality, suggesting that the drive for affective communion with fellow humans is, at its base, the drive for activity.

Over time the body expends its life-energies and if unhindered by opposing forces, in old age the body will reach its "natural" conclusion wherein all the body's life-energy will have been equally and fully depleted—hence *Pin Pin Korori*, sudden death. Forces that can speed up or befuddle the mechanics of aging, leading to disease of the body, mind, and spirit, are often characterized as popular moral vices of modern civilization, such as overly processed and nutritionally depleted foods, alcohol, tobacco, inactivity or sedentary lifestyle, over-working, and social isolation from family, friends and society at large (*hikikomori, tojikomori, kodoku*). The emphasis placed on the natural community and promoting the safety and security of the living environment of the aged, often idealized as additionally being in harmony with nature, also reinforces the negative impact that present-day urbanization and metropolitan lifestyles have on healthy aging.

In care-prevention discourse, the body's life-energies are often referred to indirectly using phrases like *enerugi* (energy), *genki* (energy and health), *ikigai* (purpose in life, something to live for) and *iki iki* (lively, energetic). A recent Kanagawa Prefecture health initiative called *Mibyo Naosu*, however, goes one step further and offers direct discussion of the source of the body's energy. The concept of Mibyo Naosu is derived from Chinese medicine, or *kampo* in Japanese. It stresses a close association between health and diet. Mibyo is a type of illness similar to lifestyle diseases, as it builds up over a lifetime of poor lifestyle choices, and so, if we positively transform bad habits early enough, some major later life illnesses can be prevented entirely.

The author of the Mibyo Naosu healthcare initiative, Kanagawa Prefecture Governor Yuji Kuroiwa, explains that at its core is the concept of *ki* (気). *Ki* is the life force of all living things. The human body has its own *ki* that is part of a larger system of exchange and interaction with the living environment. Direct exchange of external and internal *ki* occurs most obviously through what you eat: The *ki* in food, for example, is broken down through the digestion process and is directly incorporated into one's own bodily *ki*. However, there is also a more atmospheric exchange of *ki* that takes place through laughing, in the admiration of beautiful things, relaxing in the bath, spending time with family or friends, and being in nature. The governor laments that though humans are of nature, we have recently fallen out of sync with nature and need to be guided in how to synchronize with it once again. This current prefecture-wide health project is about raising public awareness of the body's *ki*, a healthy and balanced lifestyle, and human's relationship to the surrounding living environment.

Despite its Confucian conceptual origins and association with Chinese medical practice in Japan and China, the concept of *ki* in Mibyo Naosu healthcare initiative gives direct ethnographic voice to the energetic principle of the aging process in the emerging discourse of care and care-prevention. Mibyo Noasu and other prevention services encourage certain kinds of activity deemed healthy and educate elderly about how to live and age healthily. Care-prevention programmes share and spread a life philosophy about healthy aging. The ideal lifestyle for an aging person, again, encourages regular exercise, adequate sleep, a "balanced" diet with plenty of protein, decreasing stress, pursuing hobbies that one will be able to continue well into old age, mental exercise (termed "mental training"), social activity, frequent health screenings, and avoiding tobacco or excess alcohol.

The aging body in-need-of-care is likewise ontologized as a matter of life-energy. However, the life-energy of in-need-of-care elderly have been unequally distributed either progressively from the very beginning (as in persons genetically predisposed to certain illnesses), through inactivity in lifestyle, or imbalance in environmental factors (vices of modern civilization), resulting in the need for outside assistance and stimulation. Much like the diseased society of Japan, which requires stimulation by the local government and volunteers through community building projects to draw out the latent natural potential for sociability, in the case of the in-need-of-care aging body, it requires a technical ensemble of care described earlier.

There is a measure of artificiality in this technical ensemble of care that differentiates its life-energy from the free flowing life-energy of the healthy independent body. The in-need-of-care aging body has been rendered artificial, as it cannot persist on its own outside of this technical ensemble of care. It requires intervention, the artificializing action of humans. It is important to note that artificialization can occur regardless of the object being intervened in—that is, it can be either a natural or fabricated object (Simondon 1980, pp. 46–47). Artificiality resides in the fact that it has been subject to sustained human intervention, without which it would not and could persist or even exist on its own (Simondon 1980, p. 46). The French philosopher of technology Gilbert Simondon presents the example of a greenhouse plant, which is unable to exist in its natural state of self-generativity and relies on the actions of man to aid in its reproduction. The greenhouse plant has been rendered artificial, as it can no longer exist outside of the greenhouse (Simondon 1980, p. 47). Like the greenhouse, networks of care-work with the aging body to stimulate and encourage the continued flow of bodily life-energy by minimizing the impact of already-present obstacles. Like the greenhouse plant, without immersion in a technical ensemble, here the ensemble of care, the in-need-of-care aging body would not be able to survive.

Care assists an encumbered flow around blockage or barrier—the neural plaque build-up of Alzheimer's disease or the stiffening of the arterial walls in atherosclerosis. When firmly set, this blockage or barrier cannot be removed and requires a technical ensemble of caregivers to shoulder the burden of moving around the barriers. The aging body's once free flowing life-energy becomes artificialized; it must be encouraged and stimulated to continue. To conclude, in both care-prevention and care the ontology of the aging process and old age is a matter of stimulating through activity or through technical assistance the continued flow of life-energy.

7.4 Part and Whole: Recursivity of the Future

Care and care-prevention are divided into different networks of relations, involving discrete configurations of institutions, actors and knowledge systems. To demonstrate this, I described the diverging formation of sociality in care as technical ensembles of care and the community building projects of care-prevention. Care is the business of the national government, financing, managing and licensing care providers and institutions. Care-prevention is the responsibility of the local government, organizing community building projects and educating still independent seniors to care for themselves.[3]

Ontologically, the body and aging process are similarly conceptualized in the discourse and practice of care and care-prevention, suggesting they rest on a background of commonality that forms a larger and unifying whole. The aging body in-need-of-care is dependent on a technical ensemble of care. It indexes a future aging society crisis in which there are insufficient resources to provide care for the baby boomers. Care-prevention is an investment of time and energy in the present by the local government and still-independent baby boomers with the idea that it will pay off in the future by decreasing care-needs.

Pulling from cybernetics (Pickering 2010a, b), we can conceptually approach the aging society crisis as an energetic system in a quasi-open feedback loop. In the problem-solutions of gerontologists, government employees, and policymakers, the aging society crisis is described to possess a predetermined set of resources: economic resources, labour-power dedicated to providing care, the care systems in place and promised to senior citizens, scientific knowledge for intervening in unhealthy aging, and the elderly themselves. The emergence of the crisis cannot be stopped, but the balance of energies can be diverted or ordered in a more favourable way. For example, using energetic terms, care-prevention is framed by gerontologists, geriatric specialists, policymakers, and elderly as a means of diffusing the otherwise unbalanced and concentrated expenditure of time and energy in care by investing that same quantity of energy over longer periods of time through

[3]In the most recent LTCI reform of April 2014, the Ministry of Health, Labour and Welfare has removed two levels – Support levels 1 and 2 – that had care-prevention components from LTCI coverage and administration (Ministry of Health, Labour, and Welfare, 2014). Support levels 1 and 2 are now included in a "New Care-Prevention System" (*shin kaigo yobo seido)*, becoming the responsibility of the regional, as opposed to the national, government to design, promote, and run. With this 2014 reform, care and care-prevention have become completely topologically differentiated, split into different institutions and administrations. When Support levels 1 and 2 were included in LTCI, the national government was obligated to provide a standard of service and support to prevent continued functional decline, but now prevention has separated from care to form its own system alongside other prevention services and programmes.

preventive and maintenance activities. A similar logic is repeated in the likeminded ontologizing of the body as a micro-energetic system. If the life-energy of aging individuals is regulated and maintained through self-care work, then the need for sudden and concentrated expenditure of energy in care-work by ensembles of care will be minimized.

My turn to cybernetics here stems from three sources: First, from the particular language used to describe the aging society crisis and the underlying ontology of the aging process shared by gerontologists, government employees, community leaders and seniors in Japan. It is possible that the science of cybernetics popularized in America in the 1950s and 1960s likewise influenced broader conceptualizations of self and life world in Japan as well. Certainly, cybernetics left an imprint on other sciences including statistics, biological sciences, population studies and political science. Second, others have noted that the science of cybernetics itself has conceptual overlaps with "Eastern" philosophy and spirituality, particularly concerning a connection between part and whole and decentring of selves (Pickering 2010b, p. 396). Finally, the analytic employed here, ontogenesis, has thematic and conceptual overlaps with cybernetic theory (Simondon 1980; Mackenzie 2002). Because of my analytical orientation, I likely brought to attention those patterns and forces in my field of analysis that resonate with cybernetics and ontogenesis (Ricart 2014).

Regardless, when we approach the aging society crisis as an energetic system in a quasi-open feedback loop we open up the future to the possibility of change. Cybernetic theory allows us to approach the future as both a force of change, imposing form on the present, but also transformable. In the aging society crisis of Japan, the practice and concept of care-prevention attempts to shift the balance of energy to ameliorate the conditions of possibility—an overwhelming number of seniors in-need-of-care—by contributing to and altering the present-being. The aging society crisis, then, recursively feeds back into itself, inciting change in its parts that in turn can bring about change in the whole.

How might ontogenesis and cybernetics be usefully applied to the analysis of other social imaginaries of the future? Namely imaginaries that are a product of statistical forecast modelling, legitimized by science, and materialized through public policy, healthcare initiatives, and public health programmes? Ecological crises, overpopulation, epidemics, and other such crises have similar temporal organization wherein the end is also the source of change in the present, mobilizing the formation of new networks of association, configurations of institutions, and involving new actors towards the shared goal of ameliorating this future state of being.

The particular material and conceptual make-up of crises are distinct to the socio-historical moment and cultural context in which they emerge. The historical conjuncture of present-day Japan adds an important ethical dimension to the future reality of the aging society crisis and the actions taken in the present.

The baby boomers are arguably the last generation of seniors promised care by the national government in exchange for a lifetime of hard work. LTCI was the final piece of social welfare legislation, following the 1961 passing of the National Pension Act and modern National Health Insurance system. The people of Japan are described as being disillusioned with national ideologies, questioning previously taken-for-granted social roles and value systems (Harootunian and Yoda 2006). What youth remain in Japan (children of the 1970s, 1980s, 1990s, and new millennium) are, with each generation, increasingly regarded as not truly Japanese—said to be disconnected from so-called traditional beliefs and practices, Westernized and apathetic about politics and future wellbeing of their own nation (Arai 2006; Ivy 2006). With the death of the baby boomers comes a more profound death of the Japanese citizen-state.

The aging baby boomers are the last generation that looks to the state for paternalistic guidance. The community-based care-prevention initiatives currently forming are not grassroots efforts, but policy-directed and government-funded programmes. The older generations do not question the government's directives, as younger generations do. Their bodies are marked by the post-war, post-industrial era. They worked hard and sacrificed their health and family for the greater economic good of the nation. They grew up not receiving much, and still do not expect much. Products of their environments, the liver and lung cancers of older men and the rounded backs and arthritic hands of old women are the diseases of the post-industrial body. So is their marked willingness to participate in scientific research and care-prevention programme development and to alter their bodily habits to help the nation lessen the severity of the crisis that they have come to feel responsible for.

Given the sacrifices made by seniors during their prime years for the betterment of the nation, I expected a sense of entitlement to the care the nation promised them in old age. However, when I spoke with seniors about recent trends to reduce LTCI services, they often expressed support. Perhaps it is because those I spoke with are still healthy and active, making it easier to agree with the MHLW's withdrawal of entitlements. Nonetheless, seniors often explained to me that it was his or her responsibility to take preventive measures into their own hands so that they will not burden the nation with care-needs in the future, a future imagined to be unable to support them. One man in his late 70s and active in senior social clubs tells me:

> K: The message the Ministry of Health, Labour, and Welfare is sending those of us in our seventies is, "die quickly and quietly".
>
> Me: That's upsetting, isn't it?
>
> K: No. This is a good thing. Without a will of one's own people will get sick easier and grow more dependent on others. We need to do things for ourselves, take responsibility for our own health and happiness, rather than expect someone else to do it for us.

References

Arai, A. (2006). The 'Wild Child' of 1990s Japan. In T. Yoda & H. Harootunian (Eds.), *Japan after Japan: Social and cultural life from the recessionary 1990s to the present*, selected volumes (pp. 216–238). Duke University.

Arai, Y., & Ikegami, N. (1998). Health care systems in transition: An overview of Japanese health care systems. *Journal of Public Health Medicine, 20*(1), 29–33.

CIA World Factbook. (2013) Retrieved October 30, 2013, from https://www.cia.gov/library/publications/theworldfactbook/geos/ja.html

Combes, M. (2012). *Gilbert Simondon and the philosophy of the transindividual* (T. Lamarre, Trans.). Cambridge, MA: The MIT Press.

Due to the great earthquake, the ranking of Japanese women's lifespan fell from first. [Nihonjin jyosei no jumyo, sekai ichi kara tenraku, daishinsai ga eikyou]. (July 27, 2012). http://www.asahi.com/national/update/0726/TKY201207260647.html.

Fried, L. P., Tangen, C. M., Walston, J., Newman, A. B., Hirsch, C., Gottdiener, J., et al. (2001). Frailty in older adults: Evidence for a phenotype. *The Journals of Gerontology Series A, Biological Sciences and Medical Sciences, 56*(3), M146–M156.

Goffman, E. (1961) *Asylum: Essays on the social situation of mental patients and other inmates*. Anchor Press.

Harootunian, H. (2006). Japan's long postwar: The trick of memory and the ruse of history. In T. Yoda & H. Harootunian (Eds.), *Japan after Japan: Social and cultural life from the recessionary 1990s to the present*, selected volumes (pp. 98–121). Duke University.

Ishibashi, T., & Ikegami, N. (2010). Should the provision of home help services be contained? Validation of the new preventive care policy in Japan. *BMC Health Services Research, 10* (224), 1–7. Retrieved from http://www.biomedcentral.com/1472-6963/10/224

Ivy, M. (2006). Revenge and recapitation in recessionary Japan. In T. Yoda & H. Harootunian (Eds.), *Japan after Japan: Social and cultural life from the recessionary 1990s to the present*, selected volumes (pp. 195–215). Duke University.

Lamarre, T. (2012). Afterword. Humans and machines. In Combes, M. (Ed.), (T. Lamarre, Trans.), *Gilbert Simondon and the philosophy of the transindividual* (pp. 79–108). Cambridge, MA: The MIT Press.

Mackenzie, A. (2002). *Transduction: Bodies and machines at speed.* New York, NY: Continuum.

Ministry of Health, Labor, and Welfare (MHLW). (2014). *Rōjin Hoken Ka Kankei* [Division of Senior Health Care]. In *Zenkoku kaigo hoken – kōreisha hoken fukushi tantōkachō kaigi shiryō ni tuite* [*National health insurance, sectional manager's papers concerning health and welfare of elderly*] (pp. 455–496). Retrieved on June 14, 2014, from http://www.mhlw.go.jp/file/05-Shingikai-12301000-Roukenkyoku-Soumuka/0000038304.pdf

Pickering, A. (2010a). Cybernetics as nomad science. In C. B. Jensen & K. Rodjel (Eds.), *Deleuzian intersections: Science, technology, anthropology* (pp. 55–162). New York, NY: Berghahn Books.

Pickering, A. (2010b). *The cybernetic brain: Sketches of another future*. Chicago, IL: University of Chicago Press.

Ricart, E. (2013). From being to ontogenetic becoming: Commentary on analytics of the aging body. *Anthropology and Aging Quarterly, 34*(3), 52–60. (Theme Issue, The Body).

Ricart, E. (2014). Field of difference: Limitations of the political in ontological anthropology. In *Fieldsights, Theorizing the Contemporary*. "*The Politics of Ontology*", *Cultural Anthropology*. Retrieved April 22, 2014 from http://culanth.org/fieldsights/524-field-of-difference-limitations-of-the-political-in-ontological-anthropology

Simondon, G. (1980 [1958]). On the mode of existence of technical objects: Part 1. (M. Ninian, Trans.). Ontario, Canada: University of Ontario.

Simondon, G. (1992 [1964]). The genesis of the individual. In J. Crary & S. Kwinter (Eds), *Zone 6: Incorporations* (pp. 297–317). (M. Cohen & S. Kwinter, Trans). Zone Books.

Social Welfare Research Center [*shakai hoken kenkyujyo*]. (2013). Explanation of the long term care insurance system [*kaigo hoken seido no kaisetsu*].

Part III
Health and Well Being

Chapter 8
Antecedents of Subjective Wellbeing Among Older Adults in Kerala

S. Irudaya Rajan, Anusmita Devi, Tannistha Samanta and S. Sunitha

Abstract Several attempts have been made to assess the quality of life of older people through objective measurement of various factors such as health, income, economic and social status. However, the self-perception of wellbeing has emerged as a critical axis to understand the experience of older people and their overall wellbeing. This emphasis on subjectivity and experience coincides with the cultural turn that the field of gerontology has recently encountered. This study locates the conceptual framework within this epistemological turn and uses the subjective wellbeing inventory (SUBI) from the *Kerala Aging Survey 2013*. The goal of this study is straightforward. *First*, it reviews theoretical (particularly, the "convoy model of family relations", Antonucci in Social support: theory, research and applications. Springer, The Netherlands, pp 21–37, 1985) and empirical work from an interdisciplinary standpoint and sifts through diverse cultural contexts to highlight the complex interactions between family structure, social membership, (physical) health and subjective wellbeing. *Second*, it offers a descriptive analysis to determine the socio-cultural antecedents of wellbeing among older adults in Kerala. The analysis suggests that age, marital status, gender, household size, education, grand parenting roles and normative expectations are key determinants of subjective wellbeing. Not surprisingly, subjective wellbeing is shown to become worse as one grows older, though this association is amplified for older women. Analysis also suggests that loneliness experienced by the respondents is often a product of a perceived sense of lag between expectations and their fulfilment. Overall, the study provided a nuanced understanding of the critical links between family structure, cultural expectations, socio-demographic factors and subjective wellbeing of older Indians in a changing demographic context.

S. Irudaya Rajan (✉) · S. Sunitha
Centre for Development Studies, Thiruvananthapuram, Kerala, India
e-mail: rajancds@gmail.com

A. Devi · T. Samanta
Indian Institute of Technology, Gandhinagar, India
e-mail: anusmita.devi@iitgn.ac.in

T. Samanta
e-mail: tannistha@iitgn.ac.in

T. Samanta (ed.), *Cross-Cultural and Cross-Disciplinary Perspectives in Social Gerontology*, DOI 10.1007/978-981-10-1654-7_8

Keywords Subjective wellbeing · Gender · Health · Loneliness · Aging · Kerala

8.1 Background

Though objective measures to assess wellbeing of older adults have dominated the gerontological literature for a significant period, recent advances in examining self-perception of wellbeing has emerged as a critical axis to understand the experience of older people and their overall wellbeing. This study borrows the concept of subjective wellbeing from the Kerala Aging Survey 2013, which measures subjective wellbeing on the basis of a SUBI and a health questionnaire and examines the factors that influence subjective wellbeing among older adults in Kerala. The objective of this study is twofold. *First*, it would review theoretical and empirical work from diverse disciplines and cultural contexts to highlight the complex interactions between family structure, cultural scripts, economic location, social networks, health and subjective wellbeing. *Second*, it would also provide a descriptive empirical analysis to determine the socio-cultural antecedents influencing wellbeing of older adults in Kerala. Though descriptive studies on older adult health outcomes are not entirely new in India, a systematic study of factors affecting subjective wellbeing is missing from the gerontological literature.

The southern state of Kerala provides an interesting site for this study. Demographically, Kerala has a significant proportion of older persons (60 years and above) in the total population. In particular, while the national average of older population in India stood at 8 % according to the 2011 Census, Kerala reported a staggering 12.6 %. Demographers have projected a further increase in the older population by 166 % within a period of 50 years from 2001 to 2051 (Rajan, et al. *Kerala Migration Report*, *2008*). Additionally, Kerala also reports exponentially high rates of emigration of working adults. According to the 2011 Census, 2.8 million working-age adults had migrated out of Kerala, which has critical implications on the caregiving and support framework of older persons in a country where multigenerational living is common and filial obligation is expected and maintained. Interestingly, the state of Kerala also ranks the highest in the Human Development Index suggesting gains in overall health, education and income, even after adjusting for inequality (UNDP 2011). These contrasting social and economic forces make Kerala a fertile intellectual site to examine factors influencing subjective wellbeing of older persons in the state.

8.2 Subjective Wellbeing: Conceptual Underpinnings

There is a considerable body of empirical research on happiness and subjective wellbeing of older people identifying the objective, personal and social factors favouring or hindering the achievement of high levels of happiness. Much of the conceptual clarity about the idea of "subjective wellbeing" was introduced by Gergen (1971) through his seminal contribution, *The Concept of Self*. Gergen defined subjective wellbeing as being an individual's sum of performance and satisfaction in his/her endeavours. Bradburn (1969), another significant contributor to the understanding of subjective wellbeing, defined subjective wellbeing in terms of two components, positive affects (emotions such as joy, happiness, etc.) and negative effects (emotions including anger, sadness, etc.). Later, Diener et al. (1999) added a third component, life satisfaction, in understanding the concept. He defined subjective wellbeing as "a broad category of phenomenon that includes people's emotional responses, domain satisfaction and global judgments of life satisfaction" (Diener et al. 1999). This perspective of looking at subjective wellbeing has become one of the most widely accepted in social gerontology. The subjective wellbeing inventory (SUBI henceforth) and the General Health Questionnaire of the *Kerala Aging Survey* have incorporated both positive and negative wellbeing dimensions, as suggested in the conceptual literature of subjective wellbeing.

8.3 Antecedents of Subjective Wellbeing: Sifting Through Empirical Work

8.3.1 Socio-cultural Factors

Empirical work on subjective wellbeing has been both methodologically and culturally varied. For instance Ku et al. (2007) in their study of older people in Taiwan pointed out that there have been many instruments designed to map subjective wellbeing but most of these are either designed with samples from the western countries or are not age-specific. Ku referred to the existing literature pointing out the difference in experiencing old age according to the societal structure, especially in societies where family traditions are contrasting. Ku also pointed out that people in individualistic societies such as America and Western Europe were more likely to focus on their own life conditions while the people in collectivist societies such as China, Japan, Korea, India, and Taiwan might link their own wellbeing to the larger familial picture and the associated expectation of obligation and filial piety. In order to understand the impact of socio-cultural factors influencing the perception of wellbeing many cross-national studies have been carried out. Interestingly, studies conducted in the developed world have reported mixed findings. Fagerstrom et al. (2007) in their comparative analysis of elderly people in The Netherlands, Luxembourg, Italy, Austria, United Kingdom and Sweden found that though most of the participants reported being satisfied with their lives, the Italians showed a relatively lower level of life-satisfaction. Similarly, Katz's (2009) comparative analysis of the elderly in the five countries of Norway, Germany, England, Spain and Israel revealed that Israelis reported lower level of subjective wellbeing owing to difference in normative expectations of a 'good old-age'.

8.3.2 Health and Demography

Physical health is indisputably an important factor in shaping up the sense of wellbeing in an individual (Glifford and Golde 1978). Though health and functional capabilities decline with age, certain studies argue that life satisfaction does not necessarily decline owing to the beneficial cognitive adaptation to changing life situations (Felce and Perry 1997). Many authors have even described old age as a wholly positive stage in life where an individual can acquire new knowledge, seek pleasant experiences, realize their potential and have higher subjective wellbeing (Gergen and Gergen 2003). Contrary to these findings, certain studies (Cambell et al. 1976) showed an inverted U-shaped (Easterlin 2006; Easterlin and Sawangfa 2007) relation between age and wellbeing, whereby subjective wellbeing is reported to regress with the advent of age. Deaton (2008) concluded that the U-shaped relation (Clark and Oswald 1994; Frey and Stutzer 2002) of age and wellbeing is present only in rich English-speaking countries where older people are relatively happy with their lives.

Several research supports the important link of gender and subjective wellbeing. In particular, gender becomes crucial in understanding the differential life-experience and self-perception not only during adult lifetime but also during advanced stages in life. For example, Cheung and Ngan (2011) found that older women in Hong Kong report higher life-satisfaction than men while Meggiolaro (2013) reported men in Italy ranking higher on the SWB scale than women. Shyam and Yadev (2006) in their comparative analysis of subjective wellbeing of older adults in institutionalized and non-institutionalized settings in certain parts of India, found higher level of life satisfaction among men. They pointed out that this conclusion is inconsistent with other similar studies (Braun 1977; Cameron 1975; Spreitzer and Snyder 1974) reporting lack of or marginal difference in levels of life satisfaction between men and women. The gender roles prescribed for the traditional Indian men and the realization of these expectations (which men have better chances of realizing than women in India), according to them, could probably explain this reported discrepancy.

8.3.3 Family Structure and Social Networks

Household level factors such as living arrangement, or more generally, family structures are prominent factors influencing the perception of wellbeing among older people, particularly in the context of Asian and east Asian countries (Martin 1989; Hashimoto 1991; Amin 1998; Rajan and Kumar 2003). Rajan and Kumar (2003) in their study of living arrangement among the elderly in India based on the National Family Health Survey, reported, 80 % of the older adults continue living with their adult children. Living alone, as suggested by several empirical accounts, is believed to be negatively associated with general wellbeing. Cracolici et al. (2014) in their study based on the Italian Survey on Income and Living Conditions found that one-person households were more fragile and sensitive to income stress than couples. Interestingly, the study also suggested that larger number of children decreased the probability of reaching a higher level of subjective economic wellbeing, which in turn was related to the overall wellbeing of the individual. Certain other studies (Gierveld 1997; Chan 2011; Samanta 2014), on the contrary, reported a positive relation between living arrangement, household structure and intra-familial relationships. Chan et al. (2011) in their study of community dwelling older men and women in Singapore reported older adults living alone or with just a single child showed more depressive symptoms than people living with spouse or more number of members in their family. In congruence with these findings, Samanta's (2014) analysis of the relation between living arrangement, social capital and wellbeing among the older adults in seven Indian states, also suggested the existence of positive association between living arrangement and perceived wellbeing among the older adults. The study further highlighted the role of marital status, proximity of adult children and social ties as other mediating factors within the larger household structure significantly influencing the self-perception of wellbeing.

8.3.4 Intergenerational Relationships and Social Contract

Another important factor gaining currency in gerontological literature on subjective wellbeing is intergenerational relationships and social exchange. Many studies have established that older parents often live near or with at least one child, interact with their children, and often exchange aid with their children (Rajan and Kumar 2003). Parents and older children come to play important roles in each other's lives. Katz (2009) examined the impact of family relations(solidarity, conflict and ambivalence) on the subjective wellbeing of elderly in five countries(Norway, Germany, England, Spain and Israel) and found family solidarity to be an important factor influencing all the three components of subjective wellbeing (Positive effects, Negative effects and life-satisfaction).

Curiously, other studies show no or negative association between intergenerational exchange and subjective wellbeing (Blau 1973; Campbell et al. 1976). For instance Lee and Ellithorpe (1982), in their analysis of frequency of exchange of economic and instrumental support between adult children and older parents in the United States found no relation between "morale" (notion of wellbeing) and intergenerational exchanges. Shyam and Yadev (2006) in their study of older adults in institutionalized and non-institutionalized settings in North India found the older males in institutionalized setting reporting lesser depression than their peers in non-institutionalized setting. This depression among the non-institutionalized older adults could be attributed to the gap between normative expectations of filial obligation from their adult children and the actual realization of those duties, which would be lower in the case of institutionalized older adults owing to cognitive adaptation. The study further suggested that, though the non-institutionalized older adults had more social support available yet the level of satisfaction from those relations might have been lower leading to higher rates of depression.

Within the context of intergenerational exchange, grandparenting roles hold significant value, especially in the light of numerous changes in the morphology of family, ushered in often by the rapid processes of globalization and urbanization in many countries. Luo and Zhan's (2012) analysis of left-behind families of migrants in rural China suggested that older parents who look after their grandchildren had a positive evaluation of filial piety than those who did not. A similar study of the impact of caregiving role on the health of older adults in Taiwan by Ku et al. (2013) found that long-term multigenerational caregivers reported better self-rated health, higher life satisfaction and fewer depressive symptoms as compared to non-caregivers with living arrangement and duration of caregiving acting as other mediating factors in the overall subjective wellbeing of grandparents. Though one might expect such role obligations to be a part of only developing collectivist Asian societies, Forsyth's (1994) examination of age and familial roles in the level of SWB of grandparents in America also indicated positive association between the subjective wellbeing of grandparents and their relationship with their grandchildren. Furthermore, this role became more important with advancing age. However, these finding are contrary to other studies (Goudy and Goudeau 1981; Kutner 1962) which claimed that people with lesser contact with children and grandchildren have better opportunities to gain a feeling of autonomy and self-reliance.

8.4 Theoretical Considerations

This study will heavily draw from the *Convoy Model* of social relationships forwarded by Toni Antonucci (Antonucci 1985; Antonucci et al. 2014). The term 'convoy' was first employed by Plath (1980) to refer to a cohort of people or "consociates" who help in shaping up the personalities of the Japanese kids that he was working with. This concept was further elaborated by Antonucci later to explain how close relationships (especially within the family structure) for the older people help in shaping their sense of wellbeing by instilling in them a sense of self-worth and efficacy. This in turn allowed them to cope better with life circumstances and crises. The model also takes into consideration the scope for negative emotions, as often evident in very close relations, and describes how 'over-care' can translate into irritation and later bitterness in such relationships. This model is useful because it highlights the double-edged and ambivalent nature of relationships that may characterize multigenerational families. Silverstein and Bengston's (1994) theory on intergenerational solidarity and conflict echoes similar concerns. They advocate the presence of not only familial piety and obligation between the multiple generations within the family structure but also the scope for ambivalence and conflict in performing roles and meeting these role-expectations.

Furthermore, this ambivalence could lead to a gap between conventional role-expectation and pragmatic role performance and might result in loneliness among the elderly. The *Cognitivist Model* of loneliness (Peplau et al. 1982) is based on this perceived sense of deficit in need and actual realization of those expectations. Cognition is seen as a mediator between a perceived deficit and the experience of loneliness. The experience of loneliness is viewed as a personal perception and is expected to be dependent on levels of need, meeting of those needs and the social norms that structure those expectations.

This particular study would attempt to analyse the notions of subjective wellbeing among the older people in Kerala within these three broad theoretical frameworks.

8.5 Data and Method

The study is primarily based on a comparative analysis of the data drawn from the *Kerala Aging Surveys (KAS) 2004 and 2013*. The sample for KAS is based on the second Kerala Migration Survey (2003) conducted by the Centre for Development Studies (CDS), Thiruvananthapuram and coordinated by Dr. S. Irudaya Rajan in collaboration with the Government of Kerala (Ref. Zachariah and Rajan 2005, 2009). The first survey fielded in 2004 had covered 5013 elderly households (that is, households where there is at least one member aged 60 and above) throughout Kerala in all the fourteen districts. Since then every 3 years, the survey has been conducted to assess the health status of the elderly. In 2007, the same households were canvassed and this survey identified 3697 elderly persons as "alive". Of the remaining elderly persons, i.e., 11.5 % were reported as 'died' and 14.7 % as "untraced". The same set of elderly who were alive in 2007 was taken for study in 2010. Subsequently again in 2013 survey it was found that the number of elderly interviewed came down to 2030 and death marked as 336.

The surveys collected detailed information on living arrangements, socio-demographic and economic status of households, in addition to health conditions and health utilization of older people in Kerala. The survey refers to two sets of questionnaire—the SUBI and The General Health Questionnaire (GHQ) to measure the level of wellbeing among older adults, who were the primary respondents. The SUBI included Likert scale questions such as *Compared to the past, do you feel your present life is "very happy", "quite happy" or "not so happy"*; GHQ questions included general health questions that also tapped into the older person's own perception of health, such as *Have you recently felt capable about making decisions about things*? (Response options: *more than usual, same as usual, less capable than usual, much less capable*).

For the purpose of this study, a subjective wellbeing inventory (SUBI) was constructed with nine variables, each having three response codes-. An additive index (0–18) was developed by adding the nine variables where, higher values indicate higher levels of subjective wellbeing. The mean of the index was calculated at 8.6 with a standard deviation of 3.9. Reliability coefficient Cronbach's Alpha was reported as 0.895, suggesting robust internal consistency of index items. See Fig. 8.1.

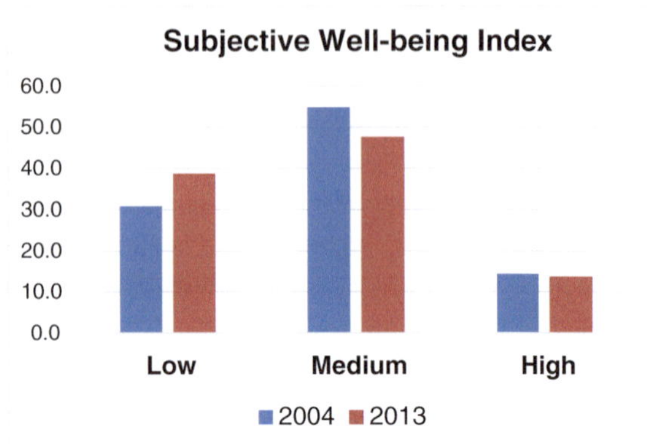

Fig. 8.1 Subjective wellbeing index for 2004 and 2013. *Source* Kerala Aging Surveys (2004, 2013)

8.6 Findings

Drawing from the theoretical and empirical literature, we have provided a descriptive table that maps the association of selected indicators with subjective wellbeing of older adults in Kerala. In the next few sections, we attempt to make sense of the data driven observations.

Age of older adults and subjective wellbeing

The role of age in the sense perception of wellbeing has been debated about at length, as discussed in the review of literature. While some scholars believe in the positive, liberating effects of age pushing a person towards the pinnacle of happiness, others look at the social and morphological degeneration of the human body with age. Table 8.1 demonstrates the levels of wellbeing reported by older adults in this study, across three age cohorts in the years 2004 and 2013. The study found that increasingly larger percentage of people reported lower levels of wellbeing with increasing age in both the years. Also, lesser percentage of people reported medium level of wellbeing with increasing age. Thus, as expected, age, here, clearly has a dampening effect on the perception of wellbeing among older adults. Since the likelihood of physical ailments are heightened in older ages, this finding is not surprising. In fact, compared to other Asian countries, older adults in India are projected to experience higher levels of disability and disease burdens (Chatterji et al. 2008).

Gender and subjective wellbeing

In a context where health and healthcare utilization are bounded by gender and social hierarchies (Lamb 1997; Lee et al. 2014), differences in subjective wellbeing among older men and women are not surprising. In congruence with the findings of an earlier study of older adults in India by Shyam and Yadev (2006) which showed men reporting higher levels of wellbeing, this study too found that a higher percentage of females reporting lower levels of wellbeing as compared to males in both the years. Also, the rise in the percentage of women reporting lower level of wellbeing as compared to the rise in percentage of males reporting lower wellbeing over the years was significantly higher. Furthermore, the decrease in percentage of

Table 8.1 Selected descriptive statistics of older adults by levels of subjective wellbeing index (% reported)

Factors	Low level of SWB		Medium level of SWB		High level of SWB	
	2004	2013	2004	2013	2004	2013
Age						
60–69	27.3	33.0	56.6	53.0	16.1	14.0
70–79	32.1	37.9	54.6	47.9	13.2	14.3
80+	41.7	42.7	48.5	45.0	9.8	12.3
Sex						
Male	26.9	32.0	54.8	18.3	18.3	17.8
Female	34.2	43.6	54.9	10.9	10.9	10.7
Education						
Illiterate	43.0	53.1	54.8	38.0	2.2	8.9
Literate without school education	40.6	53.9	52.3	38.3	7.1	7.8
Primary not completed	37.7	47.5	55.0	36.6	7.3	15.9
Primary	30.5	36.5	57.7	53.9	11.8	9.6
Upper primary up to secondary	24.9	34.0	56.3	52.5	18.8	13.5
Secondary passed but have no degree	17.0	26.6	55.0	56.9	28.0	16.5
Degree and above	10.2	15.0	43.0	52.9	46.9	32.1
Marital status						
Never married	33.0	42.5	56.4	52.5	10.6	5.0
Married	26.4	35.1	56.5	49.2	17.1	15.6
Widowed	38.7	44.4	51.8	44.4	9.5	11.2
Divorced	48.5	54.5	48.5	45.5	3.0	0.0
Separated	50.0	50.0	45.8	50.0	4.2	0.0
Total	30.9	38.7	54.8	14.3	14.3	13.7

Source Kerala Aging Surveys (2004, 2013)

respondents reporting medium level of wellbeing across the years is also steeper in case of the female respondents as compared to the males; while respondents reporting medium level of wellbeing saw a drop of 4.6 % among the male older adults, there was a drop of 9.2 % among the female respondents across the years (see Table 8.1; Fig. 8.2).

Education level of older adults and subjective wellbeing

The study found that the percentage of people reporting lower levels of wellbeing decreased as one moved higher on the level of educational attained. Interestingly, though no significant difference in the percentage of people reporting medium level of wellbeing across the various education levels was found in 2004; 11 years later, in 2013, significantly lesser percentage of people with education below primary level reported medium level of wellbeing as compared to people with higher levels of education (see Table 8.1). Thus, education becomes an important factor in the perception of wellbeing levels, more so as one advances on the age trajectory.

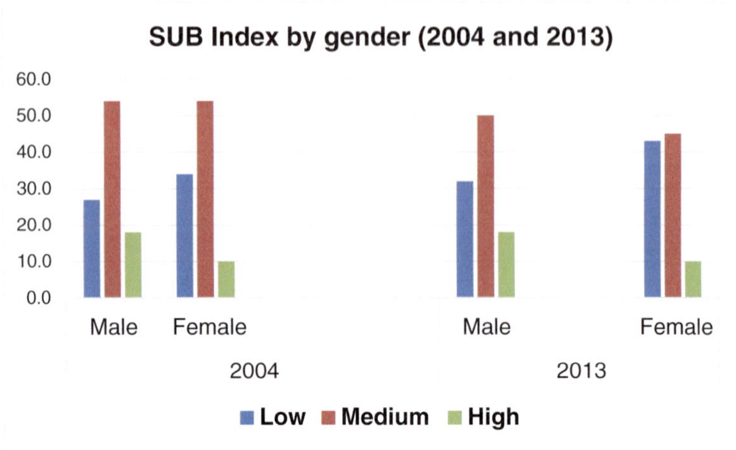

Fig. 8.2 Subjective wellbeing Index by gender (2004 and 2013 compared). *Source* Kerala Aging Surveys (2004, 2013)

Marital status and subjective wellbeing

Data revealed that higher percentage of people who have divorced or separated from their spouses reported lower levels of wellbeing than people who had never married or were widowed. Also, in both the years, fewer percentage of married older adults reported low wellbeing levels as compared to people with other marital status. This finding is consistent with the demographic literature on marriage and happiness elsewhere where married people are reported being happier, healthier and better off (Waite and Linda 1995; Waite and Gallagher 2002). No significant difference was seen in the percentage of people reporting medium level of wellbeing across the different categories of marital status (see Table 8.1).

Physical health perceptions and subjective wellbeing

Physical health is established as an important factor determining the state of happiness of a person (Glifford and Golde 1978). In congruence with such empirical studies, this study too reports a negative link between the physical state of health and the level of subjective wellbeing of older adults. As shown in Fig. 8.3a, b, higher percentage of people reporting good health conditions ranked higher on the wellbeing scale indicating higher levels of wellbeing. People with poor health conditions, on the other hand, reported significantly lower level of wellbeing.

Living arrangement and subjective wellbeing

Household size is often believed to be a significant indicator for determining the extent of isolation or sociality of the older adults in that household (Mahmood et al. 2008). Mahmood et al. in their study of older adults in Bangladesh, further elaborate that a larger household offers more social interaction and more support for the older adults. Table 8.2 here, demonstrates the interaction between household size and the levels of subjective wellbeing of the older adults in Kerala in the years 2004

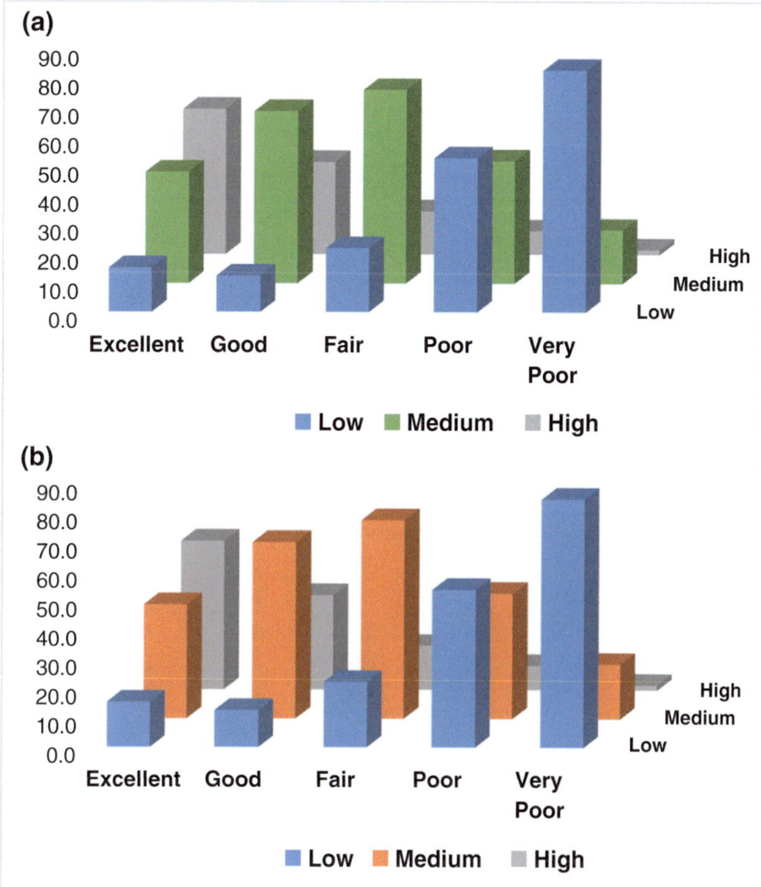

Fig. 8.3 **a**, **b** Subjective wellbeing and perception of physical health (2004 and 2013 compared). *Source* Kerala Aging Surveys (2004, 2013)

and 2013. It was found that older adults living alone reported lower levels of wellbeing as compared to older adults living with one or more resident(s) in the household in both the years. Further, the percentage of people reporting medium level of subjective wellbeing in both the years was significantly higher for households with more than one member as compared to single-member households. However, as can be seen from Table 8.2, the influence of household size on the wellbeing levels dampen with increasing age of the respondent. Increasingly higher percentage of people reported lower levels of wellbeing in 2013 even when living in households with more than two members.

Grandparenting role, as pointed out in the literature review, is another important component in shaping the sense perception of wellbeing among the older adults, especially in a country like India where filial obligations and expectations are the

Table 8.2 SUBI by household size of the older adults

# members in household	Levels of SWB					
	Low		Medium		High	
	2004	2013	2004	2013	2004	2013
1	52.6	43.0	36.3	41.4	11.1	15.6
2	32.2	34.0	47.1	51.1	20.7	14.9
3	30.5	36.6	52.2	51.1	17.3	12.3
4	29.5	31.4	56.8	57.0	13.7	11.6
5	29.2	34.1	56.9	53.3	14.0	12.6
6	30.2	33.6	57.1	57.5	12.7	8.8
7+	30.6	41.3	57.1	48.6	12.3	10.1
Total	30.9	38.7	54.8	47.6	14.3	13.7

Source Kerala Aging Surveys (2004, 2013)

Table 8.3 SUBI by older adults living with grandchildren

	Level of subjective wellbeing					
	Low		Medium		High	
	2004	2013	2004	2013	2004	2013
No child	33.3	38.1	50.7	48.4	16.0	13.5
One	26.7	35.8	60.4	49.5	12.9	14.6
Two	29.8	40.7	56.7	47.3	13.5	12.0
Three	31.8	37.4	57.1	50.3	11.1	12.3
Four and above	29.5	47.8	56.6	31.9	13.9	20.4
Total	30.9	38.7	54.8	47.6	14.3	13.7

Source Kerala Aging Surveys (2004, 2013)

normative order of the family structure. About 80 % of the respondents who felt taking care of the grandchildren was a responsibility of their grandparents reported higher levels of wellbeing. Also, 75 % of the respondents who were meaningfully engaged in providing care to their grandchildren pragmatically, reported higher levels of wellbeing. The study further, found that greater percentage of people reported lower levels of wellbeing when living without grandchildren. However, increasingly more number of older adults reported lower levels of wellbeing in 2013 irrespective of the number of grandchildren they were residing with. Also, very little difference could be seen in percentage of people reporting medium levels of wellbeing depending on the number of grandchildren they resided with, in 2013, as compared to 2004 (Table 8.3). However, in 2013 a disproportionately higher percentage (47.8 %) of older adults with four or more grandchildren reported lower level of wellbeing. This might be attributed to the failing physical health as one ages and the added responsibility of looking after many grand children could have had negative association with the levels of wellbeing of the older adults.

Perceived normative obligations

Finally, we examined if normative expectations and perceptions of what constitutes "good" old age affected people's subjective wellbeing. In a culture where filial obligations are common and intergenerational reciprocity is traditionally cultivated and maintained (Vera-Sanso 2004; Croll 2008) it is important to investigate if the perception of these cultural practices are disrupted, sustained or remain unchanged. About 75 % older adults who reported looking after older parents as a responsibility of the sons, showed higher levels of wellbeing; whereas, only about 66 % of people reporting caregiving to older parents as a responsibility of daughters felt higher levels of happiness. About 85 % of respondents who felt they had enough money reported higher levels of wellbeing. Further, 80 % respondents who had bank account reported higher levels of wellbeing. Also, the sense perception of monetary sufficiency for a range of items such as food, clothing, housing, health care and medicines, reflected a hierarchical pattern. Interestingly, while 59 % of people experiencing shortage of money for food reported low level of wellbeing, only about 40 % respondents reporting insufficient funds for health care and medicines reported a corresponding low level of wellbeing. About 55 % respondents with the belief that parents should provide for their male children till they reach adulthood (18–20 years) reported low wellbeing. Whereas, higher proportion of respondents who believed that parents should be the source of financial support till their male children enter the work force or are married, reported medium to high levels of wellbeing. Not so incidentally, a staggeringly higher percentage (47.6) of respondents who felt the parents were responsible for providing financial support to their female children till they find a job reported low level of wellbeing. On the other hand, a comparatively lower percentage of respondents (29.8) believing that financial responsibility of female child should be borne by parents only till her marriage experienced low level of wellbeing.

8.7 Concluding Thoughts

The two-year time point (2004 and 2013) enabled us to reflect on the indicators that seemed to have affected wellbeing of older adults in Kerala. We observe an increased reporting of lower levels of wellbeing among the older adults over the years. In particular, as shown in Fig. 7.1, older adults reporting low levels of subjective wellbeing in the year 2004 was 31 % while it was close to 40 % in the year 2013.

Consistent with the theoretical and empirical literature, this study found age, marital status, gender, household size, education, grand parenting roles and normative expectations as key determinants of subjective wellbeing. Specifically, we observed a decrease in level of wellbeing of older adults with an advancement in age. Also, in contrast to other studies reporting either higher level of wellbeing among older women or showing insignificant relationship between gender and wellbeing levels (Cheung and Ngan 2011; Braun 1977), this study found that an increasingly higher percentage of women reported lower levels of wellbeing as one grew older. Furthermore, educational level of the older person seemed to have an increasing influence on the level of satisfaction as one aged. For example, people with no education or education below primary school, reported lower levels of wellbeing over the years. Household size and grandparenting roles were also seen to be positively associated with the levels of wellbeing reported by people; though age seemed to have mediated both the antecedents.

In our attempt to interpret these observations, we come back to our discussion on the utility of theoretical frameworks in guiding empirical findings. Specifically, we adopt (previously discussed) Antonucci's "Convoy model of family relations" to understand the influence of living arrangement and grandparenting roles on the levels of wellbeing of older adults. In particular, the model suggests how close relationships (especially within the family structure) for the older people help in shaping their sense of wellbeing by instilling in them a sense of self-worth and efficacy. Our bivariate examination both with household size (a proxy for living arrangement) and grand parenting with subjective wellbeing confirm this. Further, Silverstein and Bengtson's theory on intergenerational conflict and solidarity helps in understanding the relation between the perceived normative structure and the level of subjective wellbeing. According to this theory the family structure is guided by the expectation of intergenerational obligation and filial piety. As suggested by the findings of the study, respondents who believed that they had a greater role to play in the lives of their children and grandchildren and could realize these normative obligations, reported higher levels of satisfaction. As a corollary to this, the failure of realization of these normative filial obligations, suggests lower level of satisfaction among the respondents. This lack of wellbeing resulting from a lag between perceived sense of expected normative behavior and their actual realization can be illustrated with the help of the cognitivist model of loneliness which suggests loneliness is a product of a perceived sense of lag between expectations and their fulfilment. The model suggests, experience of loneliness is dependent on the levels of need, meeting of those needs and the social norms that structure those expectations.

A few limitations of this study may be noted. Though we have been able to identify the possible antecedents of wellbeing through bivariate relationships, we are unable to confirm the likelihood of such occurrences owing to the lack of a multivariable model. Since, studies on subjective wellbeing of older persons are rare in India, the goals of this study were modest. It aimed at providing a broad brush overview of the factors that might influence wellbeing. In the process, the study hoped to demonstrate how theory guides empirical interpretations. As next steps, we hope researchers will take this intellectual agenda forward and provide a more nuanced understanding of the critical links between family structure, cultural expectations, socio-demographic factors and subjective wellbeing of older Indians.

References

Amin, S. (1998). Family structure and change in rural Bangladesh. *Population Studies, 52*(2), 201–213.

Antonucci, T. C. (1985). Social support: Theoretical advances, recent findings and pressing issues. *Social support: Theory, research and applications* (pp. 21–37). Netherlands: Springer.

Antonucci, T., Arouch, J. K., & Birditt, K. (2014). The convoy model: Explaining social relations from a multidisciplinary perspective. *The Gerontologist, 54*(1), 82–92.

Blau, Z. S. (1973). *Old age in changing society*. New York: New Viewpoints.

Bradburn, N. M. (1969). *The structure of psychological well-being*. Oxford, England: Aldine.

Braun, P. M. W. (1977). Psychological well-being and location in the social structure (Doctoral dissertation, University of Southern California, 1976). *Dissertation Abstracts International, 38,* 2351A

Campbell, A., Converse, P., & Rogers, W. (1976). *The quality of American life*. New York: Russell Sage Foundation.

Chan, A., Malhotra, C., Malhotra, R., & Ostbye, T. (2011). Living arrangements, social networks and depressive symptoms among older men and women in Singapore. *International Journal of Geriatric Psychiatry, 26*(6), 630–639.

Chatterji, S., Kowal, P., Mathers, C., Naiddo, N., Verdes, E., Smith, J. P., & Suzman, R. (2008). The health of aging population in China and India. *Health Affairs, 27*(4), 1052–1063.

Cheung, C., & Ngan, R.M. (2011). Filtered life satisfaction and its socioeconomic determinants in Hong Kong. *Social Indicators Research, 109*(2), 223–242.

Clark, A. E., & Oswald, A. J. (1994). Unhappiness and unemployment. *The Economic Journal*, 648–659.

Cracolici, M. F., Giambona, F., & Cuffaro, M. (2014). Family structure and subjective economic well-being: Some new evidence. *Social Indicators Research, 118*(1), 433–456.

Croll, E. J. (2008). The intergenerational contract in the changing Asian family. In R. Goodman & S. Harper (Eds.), *Ageing in Asia*. London, UK: Routledge

Deaton, A. (2008). Income, health and wellbeing around the world: Evidence from the Gallup World Poll. *The Journal of Economic Perspectives: A Journal of the American Economic Association, 22*(2), 53.

Diener, E., Suh, E. M., Lucas, R. E., & Smith, H. L. (1999). Subjective well-being: Three decades of progress. *Psychological Bulletin, 125*(2), 276.

Easterlin, R. A. (2006). Life cycle happiness and its sources: Intersections of psychology, economics, and demography. *Journal of Economic Psychology, 27*(4), 463–482.

Fagerström, C., Borg, C., Balducci, C., Burholt, V., Wenger, C. G., Ferring, D., & Hallberg, I. R. (2007). Life satisfaction and associated factors among people aged 60 years and above in six European countries. *Applied Research in Quality of Life, 2*(1), 33–50.

Felce, D., & Perry, J. (1997). Quality of life: The scope of the term and its breadth of measurement. In Roy I. Brown (Ed.), *Quality of Life for People with Disabilities: Models, Research and Practice*. London: Stanley Thornes.

Fernandez-Ballesteros, R., Kruse, A., Zamarron, M. D., & Caprara, M. (2007). Quality of life, life satisfaction, and positive aging. In R. Fernandez-Ballesteros (Eds.), *Geropsychology: European perspectives for an aging world*. Ashland: Hogrefe & Huber Publishers.

Forsth, C. J. (1994). An examination of age, familial roles and sex differences in levels of subjective well-being of grandparents. *International Journal of Sociology of the Family, 24*(2), 69–84.

Frey, B., & Stutzer, A. (2002). *Happiness and economics: How the economy and institutions affect well-being*. New Jersey: Princeton University Press.

Gergen, K. J. (1971). *The concept of self*. Oxford, England: Holt, Rinehart & Winston.

Gergen, K. J., & Gergen, M. M. (2003). Positive aging: Living well is the best revenge. *Ways Of Aging*. 203–224.

Gifford, A., & Golde, P. (1978). Self-Esteem in an aging population. *Journal of Gerontological Social Work, 4*, 35–50.
Goudy, W. J., & Goudeau, J. F. (1981). Social ties and the life satisfaction of older persons. *Journal of Gerontological Social Work, 4*, 35–50.
Hashimoto, A., (1991). Living arrangements of the aged in seven developing countries: A preliminary analysis. *Journal of Cross-Cultural Gerontology, 6*, 359–381.
Katz, R. (2009). Intergenerational family relations and subjective well-being in old age: A cross-national study. *European Journal of Ageing, 6*(2), 79–90.
Ku, P., McKenna, J., & Fox, K. R. (2007). Dimensions of subjective well-being and effects of physical activity in Chinese older adults. *Journal of Ageing, 15*(4), 382–397.
Ku, L. J. E., Stearns, S. C., Van Houtven, C. H., Lee, S. Y. D., Dilworth-Anderson, P., & Konrad, T. R. (2013). Impact of caring for grandchildren on the health of grandparents in Taiwan. *The Journals of Gerontology Series B: Psychological Sciences and Social Sciences, 68*(6), 1009–1021.
Kutner, B. (1962). The social nature of aging. *The Gerontologist, 2*, 5–8.
Lamb, S. (1997). The making and unmaking of persons: Notes on aging and gender in North India. *Ethos, 25*(3), 279–302.
Lee, G. R., & Ellithorpe, E. (1982). Intergenerational exchange and subjective well-being among the elderly. *Journal of Marriage and the Family*, 217–224.
Lee, J., Shih, R., Feeney, K., & Langa, K. M. (2014). Gender disparity in late-life cognitive functioning in India: Findings from the longitudinal aging study in India. *Journals of Gerontology, Series B: Psychological and Social Sciences, 69*(4), 603–611.
Luo, B., & Zhan, H. (2012). Filial piety and functional support: Understanding intergenerational solidarity among families with migrated children in rural China. *Ageing International, 37*(1), 69–92.
Martin, L. (1989). Living arrangements of the elderly in Fiji, Korea, Malaysia and the Philippines. *Demography, 26*, 627–643.
Mahmood, R. A., Begum, S., & Shamsul Islam, A. B. M. (2008). *Bangladesh*. In Rajan, S. I. (Eds.), *Social Security for the Elderly: Experiences from South Asia,* (pp. 163–224). Routledge: India, New Delhi.
Plath, D. W. (1980). *Long engagements: Maturity in modern Japan (Vol. 75*). Stanford University Press.
Peplau, L. A., Miceli, M., & Morash, B. (1982). Loneliness and selfevaluation. In L. A. Peplau & D. Perlman (Eds.), *Loneliness: A sourcebook of current theory, research, and therapy* (pp. 131–151). New York: Wiley.
Rajan, S. I., & Kumar, S. (2003). Living arrangement among Indian elderly: Evidence from national family health survey. *Economic and Political Weekly, 38*(1), 75–80.
Shyam, R., & Yadev, S. (2006). Indices of well-being of older adults: A study amongst institutionalized and non-institutionalized elderly. *Pakistan Journal of Psychology Research, 21*(3–4), 79–94.
Silverstein, M., & Bengtson, V. L. (1994). Does intergenerational social support influence the psychological well-being of older parents? The contingencies of declining health and widowhood. *Social Science and Medicine, 38*(7), 943–957.
Vera-Sanso, P. (2004). They don't need it, and I can't give it: Filial support in South India. In P. Kreager & E. Schroder-Butterfill (Eds.), *Ageing without children: European and Asian Perspectives*. Oxford: Berghahn Books.
Waite, & Linda, J. (1995). Does marriage matter? *Demography, 32*(4), 483–507.
Waite, L., & Gallagher, M. (2002). *The case for marriage: Why married people are happier, healthier and better off financially*. Broadway Books.
Zachariah, K. C., & Rajan, S. I. (2005). Unemployment in Kerala at the turn of the century: Insights from CDS Gulf migration studies.
Zachariah, K. C., & Rajan, S. I. (2009). *Migration and development: The Kerala experience*. New Delhi: Daanish Publishers.

Chapter 9
Early Life Critical Transitions in the Relationship Between Current Life Stressors and Depressive Symptoms Among Community-Residing Older Lebanese Adults

May H. Aydin, Abla Mehio Sibai and Anthony Rizk

Abstract This study uses the life course and life stress paradigms to better understand the role of earlier life transitions and current late life stressors on depressive symptomatology among a subset of older Lebanese. Occurrence, timing and life stage of five critical transitions and their impact on stressful events and current depression levels were investigated. Results show elevated depressive symptoms among older Lebanese who experienced health decline transitions during the 16-year Lebanese war. For those experiencing health decline off-time, or every early in life, increased depressive symptoms were observed when exposed to an increased number of current life stressors. Current depressive symptoms were exacerbated among older Lebanese with no earlier history of a health decline but who had a recent health event compared to those with experiences of health decline at younger ages. The cross-sectional nature of the data warrants caution in interpreting the study's results; however, findings highlight the importance of the life course perspective, and, in particular, the appropriateness of understanding early life experiences in shaping trajectories of late life psychological wellbeing and coping in research and practice.

Keywords Life course · Transitions · Stressors · Depression · Coping · Trauma

---

M.H. Aydin (✉)
National Center for Science and Engineering Statistics, National Science Foundation, Arlington, VA, USA
e-mail: maydin@nsf.gov

A.M. Sibai · A. Rizk
Faculty of Health Sciences, American University of Beirut, Beirut, Lebanon
e-mail: am00@aub.edu.lb

A. Rizk
e-mail: ar47@aub.edu.lb

T. Samanta (ed.), *Cross-Cultural and Cross-Disciplinary Perspectives in Social Gerontology*, DOI 10.1007/978-981-10-1654-7_9

9.1 Introduction

Mixed findings in the aging literature suggest that researchers cannot fully understand the impact of current life events on health outcomes without taking into consideration stressors that were encountered earlier in the life course (Wykle et al. 1992; Krause 1998; Seabrook and Avison 2012; Ferraro and Wilkinson 2013; Luecken and Roubinov 2012; Maschi et al. 2013; Pearlin and Skaff 1996; Ross and Mirowsky 2008). This posits that theoretical perspectives which suggest the connection between distress and health (e.g. *life stress paradigm*, Ensel and Lin 1991), as well as the resources that moderate those relationships, merit investigation within a life course lens that takes into account culture, history and the individual experience. More specifically, it becomes imperative to examine whether responses to old age stressors are the result of immediate and current constraints, and whether these responses are, at least in part, shaped by early adversity and associated psychological resources accumulated across the life course (George 2003). These relationships may also be significantly amplified if the earlier stressors have occurred within the context of non-normative traumatic environments, for example, during disasters and wars (Strauss et al. 2011; Ogle et al. 2013).

9.2 Theoretical Rationale

The life course perspective has been used by some to understand the complexities of the human experience of aging and old age. Although the study of the life course can be subjective due to the associated diversity in meaning, in general, it assumes that aging is a lifelong process and that successful understanding of current wellbeing can only occur by looking at change and development through time. One of the challenges inherent in the life course perspective is the measurement of individual life experiences within the context of sociocultural meaning. Likewise, the challenges in using longitudinal data to measure life course experiences must take into account age as well as the intersection of time, individual experience and culture. The rationale for using the life course perspective in this study stems from a need to understand how recent war history, as it implicitly exists in the memories of individuals subjected to these events, shapes life choices and trajectories and exposes individuals to differential risks of illnesses (Fry 2003; George 2003).

This study uses the life course paradigm to investigate how choices made earlier in life can shape the trajectories of wellbeing later in life among a sample of community-residing older Lebanese. We utilize this framework to explore how *occurrence*, *timing* and *life stage* of five critical transitions, namely marriage, divorce, work, retirement, and health decline, as experienced in younger age and during the backdrop of the Lebanese Civil War (1975–1991), impact current depressive symptomatology. We also investigate whether these early life critical transitions moderate the relationship between current stressful events and depression for older Lebanese.

9.3 Literature Review

The general assumptions of the *life course perspective* suggest that trauma is likely to have its most powerful health-damaging effects during critical stages of development (Kessler 1997), and that early traumas may lead to a life trajectory in which impact of eventful and chronic stressors is elevated (Turner and Lloyd 1995; Schafer et al. 2013; Feng et al. 2014). It is also argued that the degree to which current stressors impact psychological wellbeing may also depend on what point in the life course one is exposed to earlier life stressors. This study investigates both the direct and moderating effects of occurrence, timing and life course stage of five critical life course transitions during the Lebanese Civil War on depression symptomatology related to current stressful life events.

Occurrence: Some empirical work supports the theoretical life course perspective by identifying the effect that early life experiences have on later life health outcomes. Krause (1998) and later other researchers (Alastalo et al. 2013a, b) found that older men separated from their parents in childhood were physically more vulnerable and had lower physical and psychosocial functioning and higher blood pressure in late adulthood compared to their non-separated cohorts. Schafer et al. (2013) also show that an accumulated number of early adverse events in childhood were associated with decreased quality of current life evaluation. Based on these studies and the current knowledge base, it is hypothesized that early life course experiences play a significant role in modifying the current stress-depression relationship.

Timing: Normative Versus Off-Time Events: The impact of early stressors may also depend on the timing of transitions within critical life course stages. Although normal life course stage changes prompted by transition events may not be necessarily harmful to wellbeing, the timing of these events may lead to adverse effects (*Social Clock Theory*, Neugarten and Hagestad 1976). In particular, *off-time* events may be especially detrimental, since these are unscheduled transitions that generally do not fit with the normative experiences of cohorts within a specific life stage, for example, entering the workforce at a later life stage or retirement in middle-age. Elder et al. (1994) found that men who were mobilized for military duty at relatively older ages during World War II were at more risk for physical health problems in the future than men mobilized at younger ages. The authors argue that social life course disruption in family and career caused by older men's military service negatively influenced access to and utilization of informal and formal resources associated with positive health outcomes in older age. The timing of these transitions within a traumatic environment, such as a civil war context, may also be reflected in the stress-depression relationship. Rook et al. (1989), however, found limited empirical support for the social clock theory, suggesting that the impact on psychological wellbeing of the timing of a major life event depends upon the desirability of the event involved, rather than the actual timing (i.e. on-time/off-time) of the event. It is hypothesized that critical off-time transitions occurring earlier in the life course will have both a direct and indirect negative impact on current psychological wellbeing for the study sample of older Lebanese.

Life Stage: A life course stage can be thought of as a significant period that marks a specific physical, social and historical context for the individual (Pearlin and Skaff 1996). At any given life stage, an individual has an accumulated level and amount of knowledge, resources and coping skills that can be used to deal with particular stressful situations. For example, Ogle et al. (2013) show that older adults who report their most traumatic event occurring after transition into adulthood exhibited less severe symptoms of PTSD than those who experienced their most traumatic event in childhood. It may be that the successful ability to employ the social and psychological resources needed for coping with a traumatic event is a factor of when in the life stage the traumatic event occurred. The *when* may be defined as chronological age at the time of occurrence or the experience of other concomitant stressors within the same time frame. Timing disruptions to normative or "on-time" role transitions in early life may, in turn, impact response to late life stressors through shifts in normative life trajectories (George 2003). During the course of the Lebanese Civil War, younger civilians may have had competing transitional roles, such as singlehood to married or healthy to health decline, which influenced their ability to cope effectively with the experienced war trauma. Conversely, younger civilians exposed to wartime stressors may have not been able to effectively manage disruptions to these critical life transitions. It is argued that critical life transitions occurring at particular developmental life stages (regardless of normative timing expectations) may differentially impact the current stress-depression relationship, as measured by age of transition experience, and that the timing of these life transitions within the context of trauma (i.e. war), as measured by whether the transition occurred during the Lebanese war time frame (1975–1991) directly and indirectly impact depression levels related to current life stressors.

9.4 Methods

9.4.1 Setting: Older Lebanese in the Context of the Civil War

The Lebanese Civil War, which extended over 16 years from 1975 to 1991, provided a unique opportunity to investigate the effect of life transitions and life stressors on the psychological morbidity of current cohorts of older adults. In the 1970s, the proportion of Lebanese older adults did not exceed 5 % of the total population. Currently, this percentage measures 8.4 % and is expected to increase past 15 % by 2030 (World Population Prospects 2012 Revision, n.d.). This demographic shift has been accelerated by economic development, especially in the health sector, contributing towards a decrease in mortality rates from 7.2 per 1000 population in 1980 to 4.4 per thousand in 2010 and a concomitant increase in life expectancy from 68.4 in 1980 to 79.8 in 2010 (World Population Prospects 2012 Revision, n.d.). Population aging in Lebanon coincides with significant changes in family structure, with rising age of first marriage, a shift from multigenerational to nuclear families, increasing female workforce participation and waves of migration of younger cohorts because of the protracted crisis situation and uncertainties, all of which may contribute to exacerbating vulnerability in later life (Tohme et al. 2011). Our current cohorts of the older population were the young adults of the Civil War era, the casualties of which are estimated at 150,000 deaths and up to 360,000 wounded. The war also concomitantly induced a profound deterioration of economic, social and health systems and massive internal displacement and forced migration, the toll of which on adults at the prime of their productive years cannot be underestimated (Sibai et al. 2001, 2007).

9.4.2 Data Source and Study Population

Data are obtained from the Older Adult Component of a larger project, the Urban Health Study (UHS) (CRPH 2004). The UHS was administered by the Center for Research on Population Health at the American University of Beirut, Lebanon in 2003 and aimed to investigate the health consequences of population change and health needs of individuals residing within three under-privileged communities in Greater Beirut area, namely Hey el Selloum, Nabaa, and Burj al Barajneh. Further details on the methods are presented elsewhere (Khawaja and Mowafi 2006; Jawad et al. 2009; Aydin and Sibai 2015). Briefly, the UHS followed a two-stage sampling design, where a sample of 3300 households was initially selected using a probability sampling proportional to population size. All older adults aged 60 years and over were subsequently approached to participate in the Older Adult Component and individual interviews were conducted within subjects' place of residency using a structured multi-dimensional health-interview questionnaire specifically developed for the objectives of the study.

The study sample was limited to Lebanese citizens aged 60 years and older (n = 490; average age 68.2 years) regardless of religion or legal status. Institutionalized individuals, migrant workers, and those residing within abandoned buildings were not included in the original survey; and because of their tumultuous social and political histories that are very distinct from those of the Lebanese, Palestinian refugees were excluded from the study analysis.

9.4.3 Variables

Dependent Variable: The Arabic version of the 15-item shortened Geriatric Depression Scale (GDS-15) served as the dependent variable and was used to measure depressive symptomatology (Yesavage et al. 1983). The scale has been tested in the Lebanese context and was found to be a valid instrument for measuring geriatric depression in older persons without dementia in both community and primary care settings (Chaaya et al. 2008). The mean level of depressive symptoms in our sample was estimated to be 6.7 (SD = 3.8, Cronbach's alpha = 0.83, indicating relatively strong reliability among the construct items).

Independent Variables: The five specific early life transitions that are addressed in this study included (1) singlehood to first marriage, (2) unemployed to career or job, (3) married to divorced or widowed, (4) healthy to health decline, and (5) working to retirement. Each individual was assigned 1 if the specific life transition occurred and 0 if it did not occur, regardless of when that transition occurred during one's lifetime. Total number of critical life transitions was assessed by summing the total experiences across the life course (regardless of timing or life stage; range 0–5 transitions).

To address normative timing, a dummy variable was created that measures whether a transition occurred normatively (i.e. on-time) or non-normatively (i.e. off-time). Transitions were considered normative given social expectations, as well as the political and environmental climate of current Lebanese older cohorts. Ages for off-time transitions were assigned as follows: (1) 30+ years for singlehood to married; (2) any age for married to divorced and younger than 45 years for married to widowhood; (3) younger than 35 years for incident comorbidity (i.e. first chronic sickness or disability); (4) younger than 50 years for working to retirement. The career transition was not included as an off-time/on-time variable because shifts into the career role are considered normative (and often necessary) regardless of what stage in the life course this occurs. All transitions were coded 1 if they occurred off-time and 0 if they occurred on-time, or normatively.

To address developmental life stage, number of years since occurrence of the critical life transition was calculated (data collection year (2003)—age of the specific transition). Data (not shown) indicate that the average number of years (SD) since the occurrence of the specific transition was: 44.5 (10.0) for singlehood to married; 37.9 (14.5) for not working to career; 15.0 (13.2) for married to divorced or widowed; 11.1 (11.5) for healthy to health decline; and 15.3 (13.9) for working to retirement. Additionally, the age at which the transition occurred was compared against the time period of the Lebanese Civil War (1975–1991), and a dummy variable was created to assign those which were experienced within the war time frame (1 = 1975–1991; 0 = other). Table 9.1 provides the age distribution of the study sample in relation to civil war exposure and date of the survey.

A total of nine current life stressors included the experience of six conceptually distinct event domains occurring in the year prior to the study: loss of loved ones (death), work-related and financial problems (pressing loans, retirement of main

Table 9.1 Age distribution of the study sample across time (1975–2003)

Age in 1975 (start of war)	Age in 1991 (end of war)	Age in (2003) elder health survey	*n*	%
32–36	48–52	60–64	173	35.2
37–41	53–57	65–69	138	28.1
42–46	58–62	70–74	94	19.1
47+	63+	75+	85	17.6

breadwinner), familial conflicts (divorce, separation of children), exposure to violence (assaults, verbal abuse), health-related incidents (diagnosis of a new disease), and serious accidents (falls). A cumulative score was assigned to each respondent based on the summation of the total number of life events experienced in the past year (range 0–9, mean = 2.9, SD = 1.9).

Socio-demographic covariates included gender, current age, literacy, working status (never worked, used to work and currently working), self-perceived religiosity (1 = very religious; 0 = otherwise), and area of residence. Health status was measured as a composite score of limitations with activities of daily living (ADL) (Katz et al. 1963) and instrumental activities of daily living (IADL) (Lawton and Brody 1969). Although socioeconomic status has been identified as an important indicator of late life wellbeing (Haas et al. 2012), the present study sample of older Lebanese comprised an economically deprived, homogeneous group; therefore, using socioeconomic status as a control would not be meaningful.

9.4.4 Data Analysis

Frequencies and means (SD) were calculated to describe characteristics of the study sample. Item non-response error was minimal given the random nature of the missing data on most independent variables. Random missing data on the depressive symptom score were imputed through mean substitution methods. The main effects were examined using a series of Ordinary Least Squares (OLS) regression models since the outcome variable, depressive symptomatology, was normally distributed and ordinal. Additionally, interaction terms were created to test for the potential moderating roles of occurrence, timing, and life stage of transitions in the stress-depression relationship. Unstandardized and adjusted (beta) coefficients controlling for the covariates and significance levels were reported. All models were adjusted controlling for gender, age, literacy, work status, residency area, ADL/IADL difficulties and religiosity. Multicollinearity was addressed using variance inflation factor tests for all variables included in the regression models. Using relative weights to adjust for sample design, all analyses were conducted using SAS (version 8). For theoretical variables of interest, *p*-value thresholds are based on one-tailed tests of significance.

9.5 Results

9.5.1 Descriptive

The sample included 490 Lebanese with a mean age of 68.2 years (SD = 6.4). Over half (58.2 %) were females and approximately 41 % were able to read and write (Table 9.2). Three-quarters of the subjects lived in the Beirut area of Nabaa, and over one-third considered themselves to be very religious (34.9 %). The average

Table 9.2 Descriptive characteristics of the study sample (n = 490)

Variable	%
Age (range 60–95)	
Mean ± SD	68.2 ± 6.4
Gender (% female)	58.2
Having a spouse (%)	64.7
Literate (% able to read and write)	41.3
Religiosity (% very religious)	34.9
Employment status (%)	
Never worked	38.5
Used to work	46.3
Currently working	15.2
Residency area	
Hey al Selloum/Burj al Barajneh	24.9
Nabaa	75.1
Total ADL/IADL (range 0–17)	
Mean ± SD	2.7 ± 4.1
Type of critical life transition	
Singlehood to married	96.0
Unemployed to working	49.5
Married to divorce/widowed	31.3
Healthy to health decline	70.0
Working to retirement	46.1
Total number of critical life transitions (range 0–5)	
Mean ± SD	2.9 ± 1.2
Type of current life event stressors	
Loss of loved ones	66.0
Work/Financial problems	34.8
Familial conflicts	19.9
Incident health decline	28.8
Exposure to violence	20.0
Serious accident	32.8
Total number of current life event stressors (range 0–9)	
Mean ± SD	2.9 ± 1.9
Depressive symptoms (GDS[a]) (range 0–15)	
Mean ± SD	6.7 ± 3.8

[a]Geriatric depression scale

score for functional limitations in activities and IADL scale was 2.7 (SD = 4.1). The overwhelming majority (96 %) of the sample experienced marriage sometime in their life, but only one-third experienced a transition into divorce or widowhood (31.3 %). About half experienced a transition from unemployment to working/career (49.5 %) as well as from working to retirement (46.1 %) sometime in their life, and nearly two-thirds experienced a health decline (70 %). The summative score for the total critical life transitions was 2.9 (SD = 1.2) and that of total current life events was 2.9 (SD = 1.9). Depressive symptoms based on the GDS was estimated at a mean of 6.7 (SD = 3.8).

9.5.2 *Main Effects*

Results for the main effects models suggest that an increase in the total number of critical life transitions experienced over the life course exacerbated current levels of depressive symptoms for the older Lebanese adult ($b = 0.49$, $p < 0.001$) (Table 9.3). Results for the *occurrence* of specific life transitions suggest an inverse statistically significant relationship between depression scores with shift from

Table 9.3 Ordinary least square regression main effects for occurrence of critical life course transitions on depressive symptomatology

Variables	Unadjusted model		Adjusted model	
	Beta (SE)	ß	Beta (SE)	ß
Number of critical life transitions	0.49 (0.14)***	0.15	0.31 (0.23)	0.09
Type of life transition				
Singlehood to married	−0.19 (0.88)	−0.01	0.50 (0.84)	0.02
Unemployed to working/career	−1.69 (0.96)*	−0.21	0.48 (0.94)	0.06
Married to widowhood/divorce	0.51 (0.37)	0.06	−0.49 (0.40)	−0.05
Healthy to health decline	1.44 (0.37)***	0.17	0.94 (0.35)**	0.11
Working to retirement	2.2 (0.97)**	0.28	0.94 (0.91)	0.12
Model adequacy				
F-value	5.35*–12.39***		10.38***–12.16***	
R^2	0.02–0.05		0.18–0.19	
Adjusted R^2	0.02–0.04		0.17	

Notes Unstandardized (beta) and standardized (ß) coefficients are reported for parameter estimates. Standard errors are in parentheses. Total number and type of life transition variables were entered into separate models

The adjusted model controlled for gender, age, literacy, current work status, residency, number of ADL/IADL difficulties and religiosity

*$p < 0.05$, **$p < 0.01$, ***$p < 0.001$

Table 9.4 Ordinary least square regression main effects for timing and developmental life stage of critical life course transitions on depressive symptomatology

Variables	Unadjusted model		Adjusted model	
	Beta (SE)	ß	Beta (SE)	ß
Critical life transition: off-time[a]				
Married	−0.71 (0.48)	−0.06	0.15 (0.47)	0.01
Widowhood/Divorce	−0.10 (0.74)	−0.00	0.03 (0.74)	0.00
Health decline	0.42 (0.56)	0.04	0.30 (0.54)	0.02
Retirement	−0.05 (0.54)	−0.00	−1.12 (0.55)*	−0.13
Critical life transitions: 1975–1991				
Married	−0.74 (0.85)	−0.03	0.37 (0.80)	0.02
Career[b]	0.73 (0.54)	0.06	0.80 (0.53)	0.06
Widowhood/Divorce	0.61 (0.56)	0.04	−0.16 (0.54)	−0.01
Health decline	0.90 (0.43)*	0.09	0.53 (0.40)	0.05
Retirement	0.02 (0.46)	0.00	−0.49 (0.48)	−0.04
Model adequacy				
F-value	0.01–1.65		4.00***–11.48***	
R^2	0.000–0.01		0.17–0.20	
Adjusted R^2	0.000–0.01		0.14–0.17	

Notes Unstandardized (beta) and standardized (ß) coefficients are reported for parameter estimates. Standard errors are in parentheses. Main independent variables were entered into separate models The adjusted model includes variables that control for gender, age, literacy, current work status, residency, number of ADL/IADL difficulties and religiosity

*$p < 0.05$, **$p < 0.01$, ***$p < 0.001$

[a]Career transition is not considered in this analysis as it is considered an 'always' on-time normal transition in the life course of any individual

[b]The 'Current work status' variable was omitted from the list of covariates in the adjusted model for the 'career' critical life transition owing to the multicollinearity with the net effect of years since start of work/career on depressive symptomatology

unemployed to working status ($b = -1.69$, $p < 0.05$) and a positive relationship with health decline ($b = 1.44$, $p < 0.001$) and retirement ($b = 2.2$, $p < 0.01$). After introducing controls for various socio-demographic variables, health decline was the only transition that remained statistically significant ($b = 0.94$, $p < 0.01$).

Findings from the adjusted models for the *timing* of the life transition suggest that over a one point significant decrease in current depression symptoms was observed for those experiencing retirement earlier than expected (i.e. 50 years or younger) ($b = -1.12$, $p < 0.05$) (Table 9.4). No other off-time transitions were observed to have statistically significant main effects on depression symptoms. Similarly, findings did not support a direct impact of the *stage* of the life transition occurring during the 16 year wartime (1975–1991) on depressive symptomatology among older Lebanese.

9.5.3 Moderating Effects

Interaction terms were created to assess the buffering role of occurrence, timing and life stage of the critical life transitions experienced on the relationship between recent life stressors and depressive symptomatology. Only occurrence and off-time transitions played a significant role in moderating the relationship for both total number and type of current life stressor and these are presented in Tables 9.5, 9.6 and 9.7 respectively.

Some empirical support was provided for the moderating role that off-time critical life transitions have in the relationship between total number of recently experienced stressful life events and current depressive symptomatology (Table 9.5). In the adjusted models, the positive effect of total number of current life stressors on depressive symptoms significantly increased for persons experiencing a health decline transition off-time (i.e. 35 years and younger) compared to those who had undergone this transition at normative ages ($b = 0.79$, $p < 0.01$). No other off-time transitions were observed to moderate the relationship between number of current stressors and depression levels of older Lebanese.

Results of the influence of the type of lifetime transition experienced on the relationship between types of recent stressful events and depression are presented in Tables 9.6 and 9.7. Owing to the large number of hypotheses tested in this analysis, only selected findings whereby the p-value was less than 0.01 were considered

Table 9.5 Interaction effects for off-time critical life transitions and total number of life event stressors (TLE) on depressive symptomatology

Variables	TLE[a]	Adjusted models[b]	R^2	Adjusted R^2
	Beta (SE)	Beta (SE)		
Married (off-time)	0.38 (0.10)***	−0.20 (0.89)		
Widowhood/Divorce (off-time)	0.52 (0.20)*	1.08 (1.41)		
Health decline (off-time)	0.22 (0.12)	−1.31 (0.94)		
Retirement (off-time)	0.34 (0.18)	−0.57 (0.98)		
Married (off-time) × TLE		0.12 (0.26)	0.21	0.21
Widowhood-divorce (off-time) × TLE		−0.48 (0.49)	0.23	0.23
Health decline (off-time) × TLE		0.79 (0.32)**	0.21	0.21
Retirement (off-time) × TLE		−0.22 (0.30)	0.18	0.18

Notes Unstandardized (beta) coefficients are reported for parameter estimates in all models. Standard errors are in parentheses

Main independent variables were entered into separate models

*$p < 0.05$, **$p < 0.01$, ***$p < 0.001$

[a]Total number of recent life events

[b]The adjusted models controlled for gender, age, literacy, current work status, residency, number of ADL difficulties and religiosity

Table 9.6 Interaction effects of critical life transition and type of recent life event stressor on depressive symptomatology

Critical life transition	Recent life event stressors					
	Loss of loved ones		Work/Financial problems		Familial conflicts	
	Beta (SE)		Beta (SE)		Beta (SE)	
	Yes	No	Yes	No	Yes	No
Singlehood to married						
Yes	0.79 (1.37)	0.91 (1.69)	0.42 (1.09)	0.01 (1.07)	0.90 (0.99)	0.17 (0.94)
No	0.46 (1.39)	Ref	0.15 (1.66)	Ref	1.03 (1.97)	Ref
Unemployed to career						
Yes	1.25 (1.06)	0.73 (1.10)	1.35 (1.08)	1.39 (1.04)	1.81 (1.10)	0.87 (1.03)
No	0.18 (0.46)	Ref	0.82 (0.48)	Ref	0.55 (0.57)	Ref
Married to widowed or divorced						
Yes	−0.01 (0.53)	−0.35 (0.62)	0.52 (0.60)	−0.79 (0.46)	−0.68 (0.85)	−0.08 (0.42)
No	0.35 (0.41)	Ref	0.02 (0.42)	Ref	1.10 (0.47)*	Ref
Healthy to health decline						
Yes	0.93 (0.58)	0.30 (0.62)	1.14 (0.48)*	0.73 (0.43)	1.41 (0.52)*	0.89 (0.39)*
No	−0.23 (0.63)	Ref	0.05 (0.64)	Ref	0.94 (0.80)	Ref
Working to retirement						
Yes	1.63 (0.48)**	1.11 (0.58)	1.60 (0.52)**	1.56 (0.43)**	2.34 (0.63)**	1.25 (0.39)**
No	0.18 (0.45)	Ref	0.60 (0.46)	Ref	0.38 (0.55)	Ref

Notes Unstandardized (beta) coefficients are reported for parameter estimates in all models. Standard errors are in parentheses

Main independent variables were entered into separate models

$*p < 0.01$, $**p < 0.001$

[a]All models are adjusted and include variables that control for gender, age, literacy, current work status, residency, number of ADL difficulties and religiosity

Ref = Reference group is no critical life transition and no recent life event stressor

significant and thus reported here. For those experiencing a transition into retirement sometime in the past, the impact of a recent death ($b = 1.63$, $p < 0.001$), work-related or financial troubles ($b = 1.60$, $p < 0.001$), familial conflict ($b = 2.34$, $p < 0.001$), exposure to violence ($b = 2.30$, $p < 0.001$), and a serious accident ($b = 2.03$, $p < 0.001$), on depressive symptoms was significantly exacerbated. Noteworthy were also the effects between incident health decline and depressive symptoms, whereby this positive relationship was strongest among those with no health decline transition ($b = 1.88$, $p < 0.01$). Interestingly, depression scores decreased for those who experienced a health decline sometime in their life, but did not report a recent familial conflict ($b = 0.89$, $p < 0.01$). Experiencing a health decline worsened the positive relationship between recent exposure to violence ($b = 1.66$, $p < 0.001$) or serious accident ($b = 1.87$, $p < 0.001$) on depressive symptoms.

Table 9.7 Interaction effects of critical life transition and type of recent life event stressor on depressive symptomatology

Critical life transition	Recent life event stressors					
	Incident health decline		Exposure to violence		Serious accident	
	Beta (SE)		Beta (SE)		Beta (SE)	
	Yes	No	Yes	No	Yes	No
Singlehood to married						
Yes	0.33 (0.97)	−0.65 (0.94)	0.59 (1.07)	−0.11 (1.02)	0.61 (0.96)	−0.36 (0.93)
No	−2.33 (2.05)	Ref	−0.27 (1.71)	Ref	−0.91 (2.06)	Ref
Unemployed to career						
Yes	1.28 (1.07)	1.37 (1.00)	1.85 (1.08)	0.70 (1.01)	1.70 (1.04)	1.53 (1.02)
No	1.80 (0.50)**	Ref	0.01 (0.60)	Ref	1.80 (0.49)**	Ref
Married to widowed or divorced						
Yes	0.52 (0.61)	−0.42 (0.46)	1.13 (0.75)	−0.42 (0.46)	0.69 (0.57)	−0.60 (0.46)
No	0.85 (0.44)	Ref	0.21 (0.47)	Ref	0.78 (0.41)	Ref
Healthy to health decline						
Yes	1.54 (0.48)**	1.08 (0.40)*	1.66 (0.56)**	0.84 (0.40)	1.87 (0.49)**	1.02 (0.43)*
No	1.88 (0.78)*	Ref	0.45 (0.70)	Ref	1.18 (0.61)	Ref
Working to retirement						
Yes	1.82 (0.53)**	1.88 (0.41)**	2.30 (0.58)**	1.10 (0.39)*	2.03 (0.50)**	1.76 (0.42)**
No	1.75 (0.49)**	Ref	0.03 (0.57)	Ref	1.55 (0.47)**	Ref

Notes Unstandardized (beta) coefficients are reported for parameter estimates in all models. Standard errors are in parentheses

Main independent variables were entered into separate models

*$p < 0.01$, **$p < 0.001$

[a]All models are adjusted and include variables that control for gender, age, literacy, current work status, residency, number of ADL difficulties and religiosity

Ref = Reference group is no critical life transition and no recent life event stressor

9.6 Discussion

This study adds to the body of knowledge in the aging literature on the relationship between distress and health with a life course lens that takes into account the context of the one's culture and history. Findings from this study show that the total cumulative number of critical life transitions experienced in adulthood, and in particular life transitions related to health decline, are independent and statistically significant predictors of higher depressive symptomatology in old age. This relationship was notable when exposure to incident comorbidity had occurred during the 16-year Lebanese war, corroborating findings from the West among elders who experienced World War 2 during adolescence (Strauss et al. 2011). Thus, for this sample of older Lebanese, health obstacles during wartime are salient elements leading to long-lasting effects on current psychological wellbeing. Within the context of limited access to health care resources and health coverage, and even during times of peace, the consequences of early health decline may linger and lead to psychological distress in later life as evidenced by more depressive symptoms.

The finding of a negative relationship between off-time retirement (i.e., retirement occurring at young non-normative ages) and current depressive symptomatology merits some consideration. The association may have been confounded by some further characterization of the event of early retirement, and more information is needed on the nature of the retirement, whether it was voluntary and/or desired or forced during the civil war. It may also be the case that the timing of the event within the Lebanon context of wars and violence was inconsequential; this corroborates suggestions by critics of social clock theory (Rook et al. 1989; Bell and Lee 2006). Better proxies for timing, desirability and one's own perception of specific transitions are needed to better understand these effects.

Depending on the timing and type of transition experienced early in the life course of older persons, depression levels were moderated following exposure to recently occurring life stressors. For example, an additional number of current life stressors increased depressive symptoms for older Lebanese experiencing a health decline transition off-time (i.e. 35 years and younger) compared to those that underwent this transition at normative ages. Thus, older Lebanese persons experiencing a health decline very early in life were found to be psychologically more vulnerable to stressful situations occurring later in the life course compared to their counterparts. Turner and Lloyd (1995) propose that earlier traumas may lead to life trajectories in which the risk for additional life stressors is elevated, and older adults are particularly vulnerable to the effects of accumulated trauma (Ogle et al. 2014; Krause et al. 2004; Cohen et al. 2012). Ensel and Lin (1991) suggest a resource-deterioration model where older persons with off-time health transitions have stressors that are perceived as more eventful and and/or lack the necessary resources and skills to ward off the effects of additional life stressors; however, this has been contested in literature that assesses resources available to older adults in non-Western countries (Chokkanathan 2009). Moreover, Feng and colleagues (2014) suggest that, where resource deterioration in later life occurs, disability plays a central role in impacting social support available to older adults.

Notable was the modifying impact of a health decline transition in the relationship between recent health-related events and depression. Results indicate that for older individuals with no earlier history of a health decline, the magnitude of the association between a recent health event and depression was heightened. Lack of experience associated with a worsening in health status earlier in the life course may actually exacerbate one's ability to effectively cope with a current health stressor. This latter finding supports a perspective suggesting that the accumulation of past episodes may buffer the negative impact of recently occurring stressors (Kendler et al. 2000; Slavich et al. 2011). However, the extent to which this is accurate may be dependent upon the nature of the life transition as well as the specific current stressor.

The occurrence and timing of these transitions within the life course support the notion that current depression levels are, in fact, shaped by the experience of early life experiences. However, caution should be used in interpreting these results given the cross-sectional nature of the data used in the current study. Panel data employed within a longitudinal research design is necessary to establish with confidence whether changes in current depressive symptomatology can be attributable to early life transitions and/or their moderating role in the stress-depression relationship. Nevertheless, the study's findings not only underscore the importance of early life experiences on late life psychological wellbeing and coping in research and practice, it also brings to light the significance of the life course perspective in developing age-related theory that can be used as the basis to understanding observed social and psychological phenomena among adults at older ages.

References

Alastalo, H., von Bonsdorff, M. B., Räikkönen, K., Pesonen, A. K., Osmond, C., Barker, D. J., et al. (2013a). Early life stress and physical and psychosocial functioning in late adulthood. *PLoS ONE 8*(7), e69011. doi:10.1371/journal.pone.0069011

Alastalo, H., Räikkönen, K., Pesonen, A. K., Osmond, C., Barker, D. J., Heinonen, K., et al. (2013b). Early life stress and blood pressure levels in late adulthood. *Journal of Human Hypertension, 27*, 90–94.

Aydin, M. H., & Sibai, A. M. (2015). Life stressors and psychological well-being: Does access to health care help the older Lebanese? *Lebanese Medical Journal, 63*(1), 34–41.

Bell, S., & Lee, C. (2006). Does timing and sequencing of transitions to adulthood make a difference? Stress, smoking, and physical activity among young Australian women. *International Journal of Behavioral Medicine, 13*(3), 265–274.

Center for Research on Population and Health (CRPH). (2004). *The urban health study: Design and methodology* (unpublished work). Center for Research on Population and Health, American University of Beirut.

Chaaya, M., Sibai, A. M., El Roueiheb, Z., Chemaitelly, H., Chahine, L. M., Al-Amin, H., et al. (2008). Validation of the Arabic version of the short geriatric depression scale (GDS-15). *International Psychogeriatrics, 20*(3), 571–581.

Chokkanathan, S. (2009). Resources, stressors and psychological distress among older adults in Chennai, India. *Social Science & Medicine, 68*(2), 243–250.

Cohen, C. I., Palekar, N., Barker, J., & Ramirez, P. M. (2012). The relationship between trauma and clinical outcome variables among older adults with schizophrenia spectrum disorders. *The American Journal of Geriatric Psychiatry, 20*(5), 408–415.

Elder, G. H, Jr., Shanahan, M. J., & Clipp, E. C. (1994). When war comes to men's lives: Life course patterns in family, work, and health. *Psychology and Aging, 9*(1), 5–16.

Ensel, W. M., & Lin, N. (1991). The life stress paradigm and psychological distress. *Journal of Health and Social Behavior, 32*, 321–341.

Feng, D., Ji, L., & Xu, L. (2014). Mediating effect of social support on the association between functional disability and psychological distress in older adults in rural China: Does age make a difference? *PLoS ONE, 9*(6), e100945.

Ferraro, K. F., & Wilkinson, L. R. (2013). Age, aging and mental health. In *Handbook of the sociology of mental health* (pp. 183–203). Netherlands: Springer.

Fry, C. L. (2003). The life course as a cultural construct. In A. Settersten Jr. (Ed.), *Invitation to the life course: Towards new understands of later life* (pp. 269–294). Amityville, New York: Baywood Publishing Company Inc.

George, L. K. (2003). What life-course perspectives offer the study of aging and health. In A. Settersten Jr. (Ed.), *Invitation to the life course: Towards New understands of later life* (pp. 161–188). Amityville, New York: Baywood Publishing Company Inc.

Haas, S. A., Krueger, P. M., & Rohlfsen, L. (2012). Race/Ethnic and Nativity Disparities in Later Life Physical Performance: The Role of Health and Socioeconomic Status Over the Life

Course. *The Journals of Gerontology Series B: Psychological Sciences and Social Sciences, 67B*(2), 238–248.
Jawad, M. H., Sibai, A. M., & Chaaya, M. (2009). Stressful life events and depressive symptoms in a post-war context: Which informal support makes a difference? *Journal of Cross-Cultural Gerontology, 24*(1), 19–32.
Katz, S., Ford, A. B., Moskovitz, R. W., Jackson, B. A., & Jaffe, M. W. (1963). Studies of illness in the aged. The index of ADL: A standardized measure of biological and psychological function. *JAMA, 185*, 914–919.
Kendler, K. S., Thornton, L. M., & Gardner, C. O. (2000). Stressful life events and previous episodes in the etiology of major depression in women: An evaluation of the "kindling" hypothesis. *American Journal of Psychiatry, 157*(8), 1243–1251.
Kessler, R. C. (1997). The effects of stressful life events on depression. *Annual Review of Psychology, 48*, 191–214.
Khawaja, M., & Mowafi, M. (2006). Cultural capital and self-rated health in low income women: Evidence from the Urban Health Study, Beirut, Lebanon. *Journal of Urban Health, 83*(3), 444–458.
Krause, N., Shaw, B. A., & Cairney, J. (2004). A descriptive epidemiology of lifetime trauma and the physical health status of older adults. *Psychology and Aging, 19*(4), 637.
Krause, N. (1998). Early parental loss, recent life events, and changes in health among older adults. *Journal of Aging and Health, 10*(4), 395–421.
Lawton, M. P., & Brody, E. M. (1969). Assessment of older people: Self-maintaining and instrumental activities of daily living. *Gerontologist, 9*, 179–186.
Luecken, L. J., & Roubinov, D. S. (2012). Pathways to lifespan health following childhood parental death. *Social and Personality Psychology Compass, 6*(3), 243–257. doi:10.1111/j.1751-9004.2011.00422.x
Maschi, T., Viola, D., Morgen, K., & Koskinen, L. (2013). Trauma, stress, grief, loss, and separation among older adults in prison: The role of coping resources on physical and mental well-being. *Journal of Crime and Justice*,. doi:10.1080/0735648X.2013.808853
Neugarten, B. L., & Hagestad, G. O. (1976). Age and the life course. In R. E. Binstock & E. Shanas (Eds.), *Handbook of aging and the social sciences* (pp. 35–55). New York: Van Nostrand Reinhold.
Ogle, C. M., Rubin, D. C., & Siegler, I. C. (2013). The impact of developmental timing of trauma exposure on PTSD symptoms and psychosocial functioning among older adults. *Development Psychology, 49*(11), 2191–2200.
Ogle, C. M., Rubin, D. C., & Siegler, I. C. (2014). Cumulative exposure to traumatic events in older adults. *Aging & Mental Health, 18*(3), 316–325.
Pearlin, L. I., & Skaff, M. M. (1996). Stress and the life course: A paradigmatic alliance. *The Gerontologist, 36*(2), 239–247.
Rook, K. S., Catano, R., & Dooley, D. (1989). The timing of major life events: Effects of departing from the social clock. *American Journal of Community Psychology, 17*, 233–358.
Ross, C. E., & Mirowsky, J. (2008). Age and the balance of emotions. *Social Science and Medicine, 66*, 2391–2400.
Sibai, A. M., Fletcher, A., & Armenian, H. K. (2001). Variation in the impact of long-time wartime stressors on the mortality among the middle-aged and older population in Beirut, Lebanon, 1983-1993. *American Journal of Epidemiology, 154,* 128–37.
Sibai, A., Kanaan, M., Chaaya, M., & Campbell, O. (2007). Mortality among married older adults in the suburbs of Beirut: estimates from offspring data. *Bulletin of the World Health Organization, 85*(6), 482–486.
Schafer, M. H., Ferraro, K. F., & Mustillo, S. A. (2013). Children of misfortune: Early adversity and cumulative inequality in perceived life trajectories. *American Journal of Sociology, 116*(4), 1053–1091.
Seabrook, J. A., & Avison, W. R. (2012). Socioeconomic status and cumulative disadvantage processes across the life course: implications for health outcomes. *Canadian Review of Sociology, 49*(1), 50–68.

Slavich, G. M., Monroe, S. M., & Gotlib, I. H. (2011). Early parental loss and depression history: Associations with recent life stress in major depressive disorder. *Journal of Psychiatric Research, 45*(9), 1146–1152.

Strauss, K., Dapp, U., Anders, J., Renterln-Kruse, W., & Schmidt, S. (2011). Range and specificity of war-related trauma to posttraumatic stress; depression and general health perception: Displaced former World War 2 children in late life. *Journal of Affective Disorders, 128*, 267–276.

Tohme, R. A., Yount, K. M., Yassine, S., Shideed, O., & Sibai, A. M. (2011). Socioeconomic resources and living arrangements of older adults in Lebanon: Who chooses to live alone? *Ageing and Society, 31*(1), 1.

Turner, R. J., & Lloyd, D. A. (1995). Lifetime traumas and mental health: The significance of cumulative adversity. *Journal of Health and Social Behavior, 36*, 360–376.

World Population Prospects (WPP). (n.d.). 2012 Revision. United Nations Department of Economic and Social Affairs (UNDESA). Accessed at http://esa.un.org/wpp/

Wykle, M. L., Kahana, E., & Kowal, J. (1992). *Stress and health among the elderly*. New York: Springer.

Yesavage, J. A., Brink, T. L., Rose, T. L., Lum, O., Huang, V., Adey, M., et al. (1983). Development and validation of a geriatric depression screening scale: A preliminary report. *Journal of Psychiatric Research, 17*, 37–49.

Chapter 10
Analysis of Perceived Health Status Among Elderly in India: Gender and Positional Objectivity

Zakir Husain and Dona Ghosh

Abstract Self-reported health status is often used as a simple index of healthy aging by researchers. However, respondents' perceptions about his/her own health may be conditioned by individual and familial (micro-level) characteristics and, at the macro-level level, by social conditions. In this paper, using data from National Sample Survey Office's surveys on "Morbidity and Health Care" (1995 and 2004), we examine how self-reported health status and the pattern of inconsistencies varies across gender. Given the observed level of gender discrimination in South Asian societies we would expect that *actual* health status of males will be better than that of females. Social conditioning can, however, mould expectations of women and lower their expectations about desirable health status. This may lead to a situation when *perceived* health status is better off for women. This hypothesis is tested using bivariate analysis across different socio-economic correlates like education, socio-religious identity, living arrangement and engagement in economic activity, etc. We find that women consistently report better health status than males. Econometric analysis based on a logit model, on the other hand, fails to find any statistically significant difference in self-reported health status across gender. In the next step of our analysis, we examine possible inconsistencies in actual and perceived health status. The former is measured by mobility of respondent and whether he/she is suffering from any ailment. Inconsistencies may arise when (a) the respondent reports any of these problems but perceives him/herself to be in good health, and (b) the respondent reports poor health despite not reporting any of the above indicators. Variations in the inconsistencies across gender are examined and found to vary significantly across gender.

Keywords Aging · Gender discrimination · Health status · Perceptions · India

Z. Husain (✉) · D. Ghosh
Humanities and Social Sciences Department,
Indian Institute of Technology Kharagpur, Kharagpur, India
e-mail: dzhusain@gmail.com

T. Samanta (ed.), *Cross-Cultural and Cross-Disciplinary Perspectives in Social Gerontology*, DOI 10.1007/978-981-10-1654-7_10

10.1 Introduction

Biologically, "aging" refers to "a persistent decline in the age-specific fitness components of an organism due to internal physiological degeneration" (Rose 1991). Operationally, most of the countries in the world define "elderly" as a person aged 65 years and above. This cut-off, however, does not fit in well with socio-economic conditions in Afro-Asian countries. In such countries, a more economically and socially relevant, though somewhat arbitrary definition, may be in terms of the age at which one can begin to receive pension benefits. At the moment, there is no United Nations (UN) standard numerical criterion, but the UN agreed cut-off is 60 years and above to refer to the older population.[1]

Aging becomes a problem when the elderly population comprise a large proportion of a country's population, requiring a diversion of resources from productive areas to the provisioning of social security of the aged section of the population. Although the phenomenon of aging first appeared in developed countries of Europe and North America—where aged comprises 21 % of the population in 2009—the decline in fertility, coupled with the increase in life expectancies, in developing countries of Africa and Asia implies that aging will soon become a major policy issue in these countries. It has been projected that in 2050, 80 % of total elderly population of world will be in developing countries, with 55 % residing in Asia.

India is no exception to this trend. Based on 2001 Census data, a study reports that 6.9 % of total population is elderly in India; it also projects that this percentage will increase to 12.4 and 20 % in 2026 and 2050, respectively (Registrar General and Census Commissioner, India 2006). The process of aging in India has several features (Raju 2011; Rajan and Mishra 2011):

(a) Unprecedented pace of industrialization, urbanization and globalization has ruptured traditional family values, isolating them from their young members altering the living arrangements for elderly.
(b) The dominance of the unorganized sector in the Indian economy, limited employment opportunities for elderly and low level of savings implies that the aged are often in a financially precarious position and close to, if not below, the poverty line.
(c) Domestic and cross-border migration increases isolation of young members from their residual family members by creating physical distance between generations.
(d) The state has not been successful in ensuring social security the elderly, especially, for those who are in a vulnerable position.
(e) Lower mortality rates among women resulting in feminization of the aged population. A high proportion of the female population will be widows, lacking financial security.

[1]Retrieved from http://www.who.int/healthinfo/survey/ageingdefnolder/en on 1 July 2014.

Hence, population aging has become a major area of concern for Indian society generating policy challenges in domains like, economic security, social security, housing, productive aging, welfare, multigenerational bonding and health care. We shall emphasize on the last issue in this paper.

Aging aggravates several health problems as elderly people are sensitive to chronic and multiple illnesses—captured by morbidity (World Bank 1993; Rajan et al. 1999; Alam 2006; Kumar 2003; Rajan 2006), chronic ailments (Alam and Mukherjee 2005; Albert et al. 2005) and functional autonomy (Alam and Mukherjee 2005; Albert et al. 2005; Ghosh and Husain 2010). Different geriatric studies have identified micro-level determinants of health status in old age. These may be broadly divided into two groups:

(a) *Pecuniary status*: Economic conditions (Gupta and Sankar 2003), old age poverty (Alam and Mukherjee 2005; Albert et al. 2005), economic independence (Gupta and Sankar 2003; Alam 2008; Mini 2009; Ghosh and Husain 2010) and per capita expenditure of household (Chou and Chi 2002; Gupta and Sankar 2003; Rajan 2006; Alam 2008; Mini 2009; Ghosh and Husain 2010; Husain and Ghosh 2010); and,
(b) *Social status*: Living arrangements (Rajan and Kumar 2003; Gupta and Sankar 2003; Ghosh and Husain 2010), intergenerational residence (Gupta and Sankar 2003; Sen and Noon 2007), offspring support (Sobieszczyk et al. 2002; Arifin 2006), caste (Alam 2008; Husain and Ghosh 2010), widowhood (Alam 2008), social security aspects (Kumar 2003; Subrahmanya 2005) and educational attainments (Chou and Chi 2002; Gupta and Sankar 2003; Rajan 2006; Alam 2008; Mini 2009; Ghosh and Husain 2010; Husain and Ghosh 2011).

In addition, macro-level factors like familial care (Rajan and Kumar 2003), healthcare expenses (Nyce and Schieber 2005) and healthcare facilities (Kumar 2003; Rajan 2006; Alam 2006, 2008) also affect health status.

All the earlier-mentioned studies have used self-reported health status. This is a measure that has been criticized for its subjective nature (Gupta and Sankar 2003). For instance, feminist gerontologists point out, old men and women do not exist in isolation from their racial, ethnic and class-based locations. Thus, the experience of aging and its associated challenges (like health), are shaped not just by gender differences—there are other hierarchies, operating at the macro-level, that influence both aging and gender. This implies that perceptions of health status may be socially constructed and the product of macro-level factors in which the agent is embedded (Sen 1993). As researchers on aging try to generate improved indicators to capture healthy aging, it is necessary to re-examine the ability of perceived health status to reflect healthy aging. This calls for examination of the interaction between micro-level factors identified earlier and societal forces that mould perceptions about own health, driving a wedge between self-reported and actual health status. In particular, given levels of gender discrimination prevailing in South Asian societies, it is worth investigating how the gendered nature of society interacts with individual circumstances to determine perceived health status.

This paper examines gender differences in self-reported health among elderly population. Hence we have computed the difference in self-reported heath of elderly among men and women in India, after controlling for individual and household factors (like, marital status, educational attainment, economic independence, living arrangements, age groups and geographical place of residence), operating at the micro-level. Such differences have been estimated for each of the two NSSO rounds and by rural and urban areas. Furthermore, we have compared the reported health with the actual health status of the elderly population, using different proxy variables like, condition of mobility, record of hospitalization and recent ailment. This enables us to identify possible inconsistencies between the two different health statuses. The study uses unit level data from National Sample Survey Office (NSSO) 52nd (Survey on Health Care) and 60th (Morbidity and Health Care) rounds conducted in 1995–1996 and 2004, respectively.

The structure of the paper is as follows: The conceptual framework of the study is presented in Sect. 10.2. This is followed by a discussion of the empirical methodology used and a description of the data set and its generation. Findings are presented and discussed in Sect. 10.4. A concluding Sect. 10.5 sums up the findings and discusses their implications.

10.2 Conceptual Framework

10.2.1 Theoretical Development in Social Gerontology

In India, aging gerontology has emerged as an important area of research in recent years. A surprisingly vast body of literature has emerged in the past two decades examining different facets of aging in India—work force participation of aged, health status of elderly, residential arrangements, social network between aged, effectiveness of policy measures targeting elderly, etc. Such works are based on what Bengston et al. (1997) calls the positivist framework. This paradigm, a common approach existing since the early nineteenth century in Western research, involves certain stages (Achenbaum and Bengston 1994; Schrag 1967):

(a) Observation and description of data—generally from large scale surveys (which may sometimes be of dubious quality) and the decadal Census;
(b) Classification of observed data into categories reflecting similarities and differences;
(c) Explanation of the difference observed, and
(d) Prediction.

This is a process "involving informational feedback, whereby hypotheses defined on the basis of previous findings and theory are judged by current empirical results, and where researchers are continually looking for confirming or contradictory evidence by which to refine or dismiss theory" (Bengston et al. 1997: S74).

The process of theory development may lead to a further step—designing appropriate intervention strategies to alter and improve the welfare of the group being studied.

This approach overlooks the importance of understanding the data and its results and placing them in a broader social context. In particular, the nuances of the gender dimension is not fully appreciated, as it is treated as a category not dissimilar to other analytical categories defined in terms of caste, religion, occupation, education, economic status, etc. All such categories are treated as primarily descriptive categories, to be incorporated in econometric models either as dummy variables or as continuous variables, without delving into the broader social meaning and implications of such categorization. This has led the Indian literature on gerontology into a cul-de-sac with researchers restricting themselves to quantitative analysis of greater econometric sophistication, but with little effort to capture social realities and how they impinge on the experience of aging. The studies on perceived health status of aged are an example. While earlier studies were content to identify socio-economic determinants of health using a binary logit model (Mini 2009), Ghosh and Husain (2010) introduced an ordered logit, followed by a generalized ordered logit (Husain and Ghosh 2011). In all three studies, however, the impact of social determinants on perceived health was studied without any attempt to deconstruct the formation of perceptions and/or study how macro realities shape the interaction between micro-level realities, social determinants and perceived health.

In particular, the importance of gender has remained an unexplored phenomenon. Although gender is an organizing principle of social life across the life span (Rossi 1985), there has been little effort to examine what differentiates between male and female aged and how gender alters the experience of aging (Ginn and Arber 1995). It is also necessary to recognize the presence of intersecting inequalities. Just as gender shapes the aging process so do other hierarchies influence both gender and aging (Calasanti 2004; Calasanti and Slevin 2001; Connidis 2001; McMullin 2000). Old men and women do not exist apart from their racial and ethnic, sexual and class-based locations. These categories must be used to "theorize old age … as a social location, part of a system of age relations that intersects with other forms of inequality" (Calasanti 2009: 474).

To sum up, it is necessary to redefine the approach adopted by the majority of researchers on social gerontology and turn to a deeper (and richer) understanding of gender differences (as well as variations across other analytical categories). The statistical sophistication of the Indian studies should be integrated with an attempt to situate the results in their social context and analyse the interaction between macro and micro-level factors. In the succeeding sections, we apply such an approach in the context of gender differences in perceived health status.

10.2.2 Gender Differences in Actual Health Status

The starting point of our analysis is the hypothesis that actual health status of males is better than that of females. The possible reasons are summarized in Fig. 10.1.

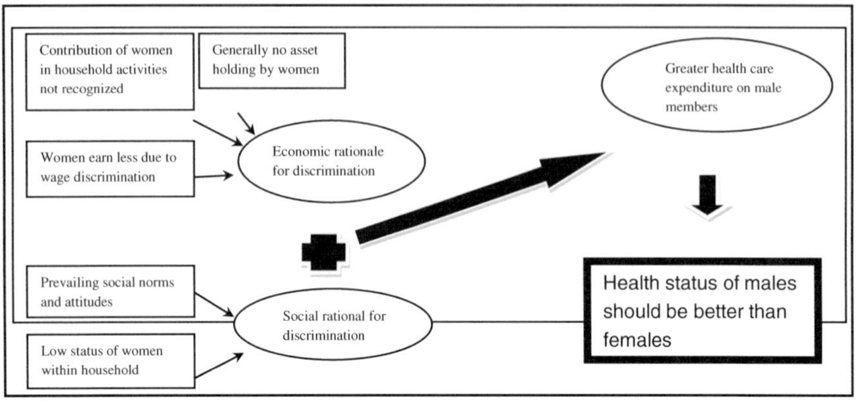

Fig. 10.1 Explaining gender differences in actual health status

In Indian society, status of women within household is very low as asset (particularly land) holding of women is not very common. This creates financial dependence on male relatives, lowering their fall-back position and hence their bargaining capacity (Sen 1990; Agarwal 1996). The participation of women in economic activities is poor or, even if they participate in such activities, they generally face wage discrimination and earn less than that male workers (Rustagi 2005; Haspels and Majurin 2008, World Bank 2012). Moreover, their contributions in household activities are seldom recognized even by themselves (Agarwal 1996). Prevailing socio cultural laws and attitudes are also biased against women, placing men on a higher pedestal than females. As a result, households treat male members as the most important members of the households and, following the principles of rationality, spend more on the health care expenses of male members. This results in males enjoying a better health status. There has been some evidence in support of this proposition (Das Gupta 1987; Morduch and Stern 1997; Nayyar 2009).

10.2.3 Formation of Perceived Health Status

In this paper, however, we are studying not actual but perceived health status. In case of self-reported health status, contrary to what we argued in Sect. 10.2.2, it may be possible to observe women reporting better (perceived) health status compared to males. This may be explained in terms of Sen's theory of positional objectivity (Sen 1993).

Sen starts with the proposition that "What we decide to believe is influenced by what we observe. How we decide to act relates to our beliefs" (Sen 1993: 126). However, what we observe and what we choose to believe depends upon our position. As a result, positionally dependent observations, beliefs and actions become central to our knowledge and practical reason. The nature of objectivity in observations, beliefs and actions will be determined by the position of the observer and certain parametric influences on this agent. The influences that define the objectivity of the agent's reasoning are, in part, locational. This refers to what we may call as the micro-level factors shaping the world of the agent in the form of family and local community level forces. Parametric influences on the process of analysing and interpreting information are, however, also important. Such macro-level factors comprises of broader society or country level factors. Together these forces lead to positional variability between agents that leads, in turn, to differences in:

(a) the set of information available;
(b) the ability to process information; and
(c) the ability to interpret the processed information.

These bounds on rationality affect the objectivity of agents' decisions about beliefs and actions.

Positional objectivity can also lead to what Marxian philosophers call "objective illusion" (mentioned in, for instance, in Marx's *Capital*, volume I and *Theories of Value*; see also Cohen 1978). This refers to a positionally objective belief—based on a limited class of positional observations, or in the absence of access to other means of positional scrutiny—that is, in fact, mistaken. Sen provides two instances of such cultural illusions.

Kerala has one of the highest life expectancy among Indian states and has successfully attained a transition in the health sector. In contrast, the states of Bihar and Uttar Pradesh lie at the other end of the spectrum. Both states have under-developed public health sector and low life expectancy. The reported self-morbidity, surprisingly, is much higher in Kerala, compared to the states of Bihar and Uttar Pradesh. The reason for this paradox is that people are, in general, more literate and aware of ailments in Kerala than in Bihar and Uttar Pradesh. The resultant positional variability generates a cultural illusion in the form of a factually wrong perception about morbidity status in these states.

The second example also relates to self-reported morbidity status. Although females have survival disadvantages vis-à-vis males and have higher mortality rates, the self-perceived morbidity rates of females are similar to, if not lower than that of their male counterparts. Sen (1985, 1993) cites a study by Lall and Seal (1949) which observed that, in post-famine Bengal in 1944, widows hardly reported being in "indifferent health", whereas widowers complained massively about that. Such behaviour may be explained in terms of positional differences caused by tradition of women's deprivation resulting in the social tendency to normalize gender inequality as part of the prevailing mode of living. Social influences may reduce expectations and keep a person content in situations that appear unsatisfactory to others: "…the malleability of mental attitudes … may tend to hide and muffle the extent of deprivation in many cases," (Sen 2006: 387).

10.2.4 Hypothesis

The theory of positional objectivity as applied to gender relations provides the inspiration for our study. This paper is an attempt to verify whether the long prevailing gendered relations and deprivation of females have conditioned women into accepting their survival disadvantages and poor health status as normal. We note, however, that apart from such macro-level conditioning through a patriarchal society and the socio-religious group of which the agent is part of, individual factors—like age, economic status, educational level, living arrangements, economic activity status, economic dependence, whether any other person in the family is ill, etc.—may also determine their positional objectivity and shape beliefs about their own health status.

10.3 Database and Methods

10.3.1 Database

Information on morbidity and health care is been captured by NSSO in two rounds—52nd round (July, 1995–June, 1996) and 60th round (January–June, 2004). Both rounds covered the morbidity and utilization of medical services and problems of aged persons. A two-stage stratified sample design was followed to collect data from sample households by interview method using a close-end structured questionnaire. First Stage Units (FSUs) comprised selection of rural stratum and urban stratum from census villages and NSSO Urban Frame Survey (UFS) blocks, respectively. At Second Stage Units (SSUs), 10 households were surveyed in each selected FSU. In this study, datasets from 52nd round and 60th round are merged together to obtain data on 68,852 respondents aged 60 years or more (49.4 % from 52nd round and 50.6 % from 60th round) along with the information on their marital status, educational attainment, economic independence, living arrangements, age groups and geographical zone and expenditure groups. Perceived health status is evaluated from respondent's perception about his/her health, based on a three point Likert type scale where the response variable is coded as "poor", "fair/good" and "very good/excellent". The percentages of elderly in these classes are 22.2, 70.1 and 7.7 % for poor, good and very good, respectively. Since, the proportion of sample reporting excellent health is relatively small, we have, for analytical convenience, merged the latter two categories to form a binary grouping—poor health and good health.

10.3.2 Sample Profile

Appendix Tables 10.7 and 10.8 briefly summarize the sample characteristics by place of residence. We find that, in both the areas, more than 75 % of elderly males are married. In contrast, more than half of the females are widows, reflecting the greater life expectancy of females. Most of the male elderly (above 60 %) live with their spouse, while the majority of females live with their children as a large proportion are widows. Further, a substantial proportion of females are completely financially dependent on others. Analysis of education profile reveals that the majority of the elderly are illiterate, irrespective of place of residence. It has been found that, the majority of elderly belongs to General caste category[2] and are within 60–69 years of age. In rural areas, the highest proportion of aged is found in Central India; in urban areas, on the other hand, South India has a relatively higher proportion of aged.

10.3.3 Methodology

Our analysis starts as a bivariate analysis of gender differences the perceived health status between males and females, across the study period and place of residence. We compare the percentage of males reporting themselves to be in good health with the corresponding proportion for females, across different socio-economic and demographic correlates like marital status, age group, geographical zone of residence, level of economic independence and expenditure group.

This analysis, however, neither captures the statistical significance of observed differences nor does it control for micro and macro-level factors that determine self-reported health. Hence, we follow up the bivariate analysis with an econometric analysis of the determinants of self-reported health. Since, reported health is bivariate in nature (good versus bad), we use a logistic regression model (estimated for rural and urban areas for each of the two survey years). The logit model regresses perceived health status on a female dummy and the following control variables:

[2]Scheduled castes (SCs) are Hindus belonging by birth to the lowest of the four castes. They were formerly untouchables and, even now, are often economically and socially depressed. Scheduled tribes (STs), on the other hand, are members of economically and socially depressed tribes (which may be non-Hindu also) who were also treated as untouchables. In post-Independence India, Articles 341 and 342 of the Constitution provide a list of all SCs and STs under the Constitution (Scheduled Castes) Order, 1950, and the Constitution (Scheduled Tribes) Order, 1950, respectively, to facilitate affirmative action targeting such social groups. In the 1991, the Government of India introduced another list of castes, apart from SC and STs, who are socially and economically backward. These castes are called Other Backward Castes (OBCs).

(a) Socio-religious categories: HSC (Hindu Scheduled Castes), HST (Hindu Scheduled Tribes); Reference category is HGEN (Other Hindu Castes)
(b) Marital status categories: Currently Married; Reference category is the residual, comprising of Never Married, Currently Married, Widowed and Divorced/Separated
(c) Educational status categories: Illiterate, Just literate, Below Primary, Middle, Secondary, Higher Secondary and Other Educational Qualification; Reference category is Primary
(d) Economic independence status categories: Not Dependent and Partially Dependent; Reference category is Fully Dependent
(e) Living Arrangement's categories: Alone and Inmate of Old, Not Inmate of Old Age, With Spouse, Without Spouse and with Children, Other Relation and Other Non-Relation; Reference category is With Spouse and Others
(f) Usual activity status: Employed, Employed in extra-SNA (System of National Accounts) activities, Unemployed; Reference category is residual, comprising of rentiers, pensioners, beggars and prostitutes
(g) Geographical zonal categories: North, Central, Northeast, East, West and South; Reference category is Central[3]
(h) Social-group categories: ST (Schedule Tribe), SC (Schedule Caste) and General; Reference category is General
(i) Age group categories: Young Old (age 60–69 years), Middle Old (age 70–79 years) and Older Old (80 years and above); Reference category is Middle Old.

The sign and statistical significance of the female dummy indicates whether there is a substantial difference in perceived health status across gender.

Finally, we examine the consistency of reported health status with some symptomatic proxies of actual health. These proxies are mobility of the respondent and whether he/she was suffering from some ailment at the time of the survey. Assuming that respondents who report good (poor) health will be mobile (immobile), not have been hospitalized (been hospitalized), and had not suffered from any ailment (had suffered from any ailment) in the day/fortnight preceding the survey we find out the proportion of inconsistent responses and examine whether there is any variation in these proportions across gender.

[3]Zones are defined as follows: North comprises Jammu and Kashmir, Himachal Pradesh, Uttarakhand, Rajasthan, Delhi, Punjab, Haryana and Chandigarh; East comprises West Bengal, Assam, Tripura and Orissa; Northeast comprises of Sikkim, Manipur, Meghalaya, Arunachal Pradesh, Nagaland and Mizoram; Central comprises Bihar, Jharkhand, Uttar Pradesh, Chhattisgarh and Madhya Pradesh; West comprises Gujarat, Maharashtra, Goa, Daman and Diu, and Dadra and Nagar Haveli; South comprises Karnataka, Andhra Pradesh, Telangana, Tamil Nadu, Kerala, Pondicherry and Andaman and Nicobar Islands.

10.4 Findings and Discussion

10.4.1 Bivariate Analysis

In Table 10.1, we report the percentage of elderly population who reported themselves to be in good health across different socio-economic correlates, by gender, place of residence and year. We also report the difference between percentages of males and females who reported themselves to be in good health. We found that, out of 39 groups in each place of residence and year, there are negative values in almost all cases.[4] This implies that a higher proportion of women tend to report themselves to be in good health, relative to men. While this may appear surprising, we should keep in mind that the health status being analysed is not actual health status, but reported health status. The findings are, therefore, in line with our hypothesis that women will tend to report good health due to conditioning at the macro and micro-level.

Although the bivariate analysis provides support for our hypothesis regarding the impact of conditioning on variations in reported health status over gender, the evidence has to be tested using statistical methods. Although tests of sample proportions can indicate whether the differences observed in Table 10.1 are statistically significant, such tests do not control for the presence of other socio-economic variables. Hence, in the next section, we estimate a logit model.

10.4.2 Econometric Analysis

Table 10.2 reports the results of a logit model of reported health status (good versus bad health) upon a female dummy and control variables. The presence of gender difference would be revealed in a statistically significant coefficient for the female dummy, while a positive sign of the coefficient would indicate the presence of conditioning. Although the coefficients of the female dummy are positive in all four models, only the coefficient in the model for urban areas in 2004 is significant. This would imply that, in all other cases, there is no gender difference in reported health status. While even the absence of any gender differential is surprising in highly gendered societies like India, the absence of statistically significant results is surprising and needs to be explored further.

One possible reason may be social differences across geographical reasons. In South India, for instance, women are more empowered than North Indian women (Dyson and Moore 1983; Basu 1992; Jejeebhoy 2001). Several reasons have been

[4]In 1995–96 there are negative values in 35 groups in rural areas and 34 groups in urban groups. The corresponding figures for 2004 are 36 and 37, respectively.

Table 10.1 Percentage of people in good health across socio-economic correlates—by gender, place of residence and time

Correlate	1995–96			2004		
	Male	Female	Difference	Male	Female	Difference
Rural areas						
Marital status						
Never married	87.13	92.13	−5.00	90.91	100.00	−9.09
Currently married	89.17	91.47	−2.30	93.33	96.13	−2.80
Widowed	92.96	94.35	−1.39	95.41	96.91	−1.50
Divorced/separated	91.18	86.67	4.51	91.89	90.38	1.51
Educational status						
Illiterate	90.59	93.40	−2.81	94.89	96.69	−1.80
Just literate	89.28	87.20	2.08	92.07	95.56	−3.49
Below primary	90.44	92.08	−1.64	94.35	96.92	−2.57
Primary	89.17	89.29	−0.12	92.13	94.76	−2.63
Middle	87.35	90.20	−2.85	91.33	95.30	−3.97
Secondary	85.46	100.00	−14.54	89.34	97.73	−8.39
Higher secondary	92.00	92.31	−0.31	90.48	88.89	1.59
Other educational qualification	72.88	92.00	−19.12	86.08	90.48	−4.40
Economic independence						
Not dependent	85.77	88.78	−3.01	90.60	93.96	−3.36
Partially dependent	91.03	91.59	−0.56	94.95	95.49	−0.54
Fully dependent	95.76	94.03	1.73	98.05	97.24	0.81
Living arrangements						
Alone and inmate of old	88.10	91.11	−3.01	81.58	96.83	−15.25
Not inmate of old age	87.22	92.52	−5.30	94.05	96.00	−1.95
With spouse	91.50	92.86	−1.36	94.79	97.62	−2.83
With spouse and others	88.84	91.30	−2.46	92.87	95.89	−3.02
Without spouse and with children	92.76	94.43	−1.67	95.63	96.88	−1.25
Other relation	92.00	93.94	−1.94	95.37	97.13	−1.76
Other non-relation	87.18	77.78	9.40	88.89	95.92	−7.03
Geographical zones						
North	90.80	95.63	−4.83	93.20	97.01	−3.81
Central	89.63	92.85	−3.22	93.36	96.07	−2.71
Northeast	81.04	84.07	−3.03	86.12	94.36	−8.24
East	91.24	94.45	−3.21	95.87	98.55	−2.68
West	89.68	93.92	−4.24	92.86	96.35	−3.49
South	92.17	93.04	−0.87	95.86	96.70	−0.84
Social groups						
ST	87.34	91.45	−4.11	90.63	95.55	−4.92
SC	91.25	94.80	−3.55	95.13	96.81	−1.68
General	90.11	92.96	−2.85	93.83	96.67	−2.84

(continued)

Table 10.1 (continued)

Correlate	1995–96			2004		
	Male	Female	Difference	Male	Female	Difference
Age groups						
60–69 years	87.43	91.69	−4.26	91.96	95.79	−3.83
70–79 years	93.23	95.33	−2.10	96.68	98.08	−1.40
80 years and above	96.05	96.67	−0.62	97.45	98.38	−0.93
Expenditure groups						
Poorest	90.67	93.64	−2.97	95.07	97.46	−2.39
Poor	91.02	93.66	−2.64	94.71	96.20	−1.49
Middle	89.86	94.29	−4.43	94.31	96.39	−2.08
Rich	90.29	92.78	−2.49	93.17	96.85	−3.68
Richest	88.06	91.19	−3.13	91.28	95.96	−4.68
Urban areas						
Marital status						
Never married	81.40	87.16	−5.76	91.67	95.59	−3.92
Currently married	85.73	87.57	−1.84	91.09	94.03	−2.94
Widowed	89.48	91.77	−2.29	94.31	95.94	−1.63
Divorced/separated	90.32	93.94	−3.62	95.00	96.15	−1.15
Educational status						
Illiterate	91.33	91.73	−0.40	93.55	95.83	−2.28
Just literate	92.42	92.02	0.40	97.79	97.52	0.27
Below primary	86.75	88.09	−1.34	96.03	96.05	−0.02
Primary	88.22	89.30	−1.08	92.80	94.07	−1.27
Middle	84.07	85.84	−1.77	90.85	93.61	−2.76
Secondary	83.15	86.61	−3.46	90.97	92.49	−1.52
Higher secondary	83.44	73.02	10.42	85.48	96.04	−10.56
Other educational qualification	79.48	81.31	−1.83	87.03	91.21	−4.18
Economic independence						
Not dependent	82.22	83.67	−1.45	88.78	94.02	−5.24
Partially dependent	87.14	88.61	−1.47	92.30	95.49	−3.19
Fully dependent	92.34	90.99	1.35	96.86	95.44	1.42
Living arrangements						
Alone and inmate of old	93.75	86.67	7.08	85.71	100.00	−14.29
Not inmate of old age	83.93	91.16	−7.23	91.30	97.21	−5.91
With spouse	86.16	85.71	0.45	90.58	94.43	−3.85
With spouse and others	85.65	87.77	−2.12	91.07	94.06	−2.99
Without spouse and with children	90.08	91.58	−1.50	95.14	96.09	−0.95
Other relation	80.42	92.58	−12.16	93.21	93.58	−0.37
Other non-relation	80.00	82.14	−2.14	88.24	96.30	−8.06

(continued)

Table 10.1 (continued)

Correlate	1995–96			2004		
	Male	Female	Difference	Male	Female	Difference
Geographical zones						
North	86.13	89.90	−3.77	90.53	95.02	−4.49
Central	87.88	91.47	−3.59	91.82	95.57	−3.75
Northeast	73.93	77.67	−3.74	86.38	89.69	−3.31
East	89.60	92.06	−2.46	94.85	97.56	−2.71
West	81.89	88.49	−6.60	89.38	93.65	−4.27
South	88.47	91.29	−2.82	92.74	95.84	−3.10
Social groups						
ST	84.27	86.56	−2.29	89.38	89.5	−0.12
SC	90.82	92.64	−1.82	94.53	95.38	−0.85
General	85.85	89.97	−4.12	91.39	95.46	−4.07
Age groups						
60–69 years	84.29	88.68	−4.39	89.22	93.94	−4.72
70–79 years	88.74	91.98	−3.24	95.6	97.2	−1.60
80 years and above	91.51	93.6	−2.09	96.18	97.85	−1.67
Expenditure groups						
Poorest	89.49	91.45	−1.96	93.72	95.62	−1.90
Poor	88.28	91.87	−3.59	92.89	96.27	−3.38
Middle	86.19	91.1	−4.91	92.34	95.93	−3.59
Rich	84.51	88.63	−4.12	90.32	93.93	−3.61
Richest	83.23	87.23	−4.00	88.72	93.64	−4.92

Table 10.2 Logistic regression results for reported health status—by place of residence

Indicators	Rural		Urban	
	(1995–96)	2004	(1995–96)	2004
Female	1.03	1.07	0.89	1.37***
Household size	0.98***	0.97***	0.97***	1
Reference category: general caste				
ST	1.06	0.8**	1.48***	0.73*
SC	1.12	1.03	1.25**	1.12
Log of per capita income	0.8***	0.72***	0.91	0.9
Reference category: age 70–79 years				
Age 60–69 years	0.66***	0.54***	0.83**	0.48***
Age 80 years and above	1.29*	1.02*	1.05	0.99
Reference category: other than currently married (includes never married, widowed, divorced/separated)				
Currently married	0.89	1.06	0.9	1.12*

(continued)

Table 10.2 (continued)

Indicators	Rural		Urban	
	(1995–96)	2004	(1995–96)	2004
Reference category: primary education				
Illiterate	1.18*	1.46***	1.21**	1.13
Just literate	1.16	1.09	1.4***	2.99***
Below primary education	1.18	1.4**	0.89	1.68***
Middle education	0.93	1.06	0.75**	0.91
Secondary education	0.89	0.88	0.71***	0.87
H.S. education	1.66	1.04	0.6***	0.61***
Other educational qualification	0.48**	0.68	0.56***	0.65***
Reference category: others				
Engaged in economic activities	0.62**	0.56***	0.65***	0.64***
Unemployed	1.18	–	1.31	0.34
Engaged in extra-SNA activities	0.63***	0.73***	0.71***	0.64***
Reference category: fully dependent				
Not dependent	0.56***	0.46***	0.62***	0.65***
Partially dependent	0.85**	0.77	0.85*	0.86
Reference category: with spouse and others				
With spouse	1.25	1.41***	1	1.07
With children and others	1.08**	1.02	1.05	1.28
Other living arrangements	0.9	1.05	0.87	1.12
Reference category: central				
North	1.33***	1.13	0.88	0.9
Northeast	0.45***	0.59***	0.32***	0.58***
East	1.13	1.76***	1.16	1.74***
West	1.08	0.99	0.69***	0.72***
South	1.19**	1.33***	1.02	1.1
Number of observations	20,384	21,043	12,729	11,953
Chi square	658.04	632.35	473.99	386.41
Pseudo R^2	0.06	0.08	0.05	0.07

Note ***denotes $p < 0.01$, **denotes $p < 0.05$, *denotes $p < 0.1$

suggested for such geographical variations (Bardhan 1974, Miller 1981). However, what is relevant to our analysis is that, differences in empowerment level imply that the impact of the conditioning forces may be weaker in some geographical zones. To test this possibility, we have estimated the logit models for each geographical zone separately. The results, reported in Table 10.3, however, belie our expectations. In rural areas, the female dummy has a significant and positive coefficient in North India, and a significantly negative coefficient in South India. This matches our hypothesis—less empowered North Indian women are more susceptible to conditioning and report good health status. In South India, on the other hand,

Table 10.3 Logit regression results for females of good health in different zones in India

Geographical zone	Rural sector				Urban sector			
	1995–96		2004		1995–96		2004	
	Female	N	Female	N	Female	N	Female	N
North	2.07***	2560	1.68	2108	0.62	1433	1.82*	1384
Central	1.00	7319	0.99	7980	0.94	3101	1.03	3112
Northeast	1.25	1139	1.54	1250	0.64	593	2.16*	512
East	0.88	2789	1.28	2915	1.09	1503	1.32	1379
West	1.29	2355	1.07	2226	0.9	2442	1.09	1970
South	0.60***	4178	0.70	4523	1.03	3615	1.44	3489

Note ***denotes $p < 0.01$, **denotes $p < 0.05$, *denotes $p < 0.1$

empowerment reduces the potential impact of conditioning and women are able to report their poor health status. However, all the coefficients of the female dummy are statistically insignificant in urban areas (1995–96) and rural areas (2004). In 2004, in urban areas, the coefficient is again positive and significant for North India, but not for South India. Instead the female dummy takes a value of 2.16 in Northeast India. This is somewhat unexpected as several states in Northeast India have matriarchal societies, where women may be expected to be more empowered and less susceptible to any forms of conditioning.

It is of course possible that the statistical tests are not "lying", but reflect the genuine absence of health differences across gender. While this may appear unlikely in highly gendered South Asian societies, we should keep in mind an important fact observed by Das Gupta (1987). After reporting discrimination in health care seeking behaviour for girls, along with high levels of abortion and infanticide, in the district of Khanna, Punjab, Das Gupta (1987) noted that the surviving girls were robust and had, inter alia, high levels of education. Based on a study of elderly in Thailand, Sobieszczyk et al. (2002) report that elderly men, compared to women, face disadvantages like worse survivorship, a lower likelihood of receiving money from adult children and greater probability of debt and other financial problems. Further, men are more likely than women to engage in activities that are harmful to health (WHO 2001).

In the present context, it is possible that, over the process of aging, the relatively weaker and less healthy women may have succumbed due to gender disparity in health care seeking behaviour. This would leave behind only "robust" survivors—who are healthier than the average Indian female. This would result in an absence of gender differential in health status. One possible way of checking this hypothesis is to look at the age-wise sex ratio (number of females per 1000 males) in India. Data from the 2011 Census has been plotted in Fig. 10.2. It may be seen that sex ratio starts to increase from roughly the mid-50s in both rural and urban areas, so that—by the late 50s (in urban areas) and mid-60s (in rural areas)—the sex ratio crosses 1000. This implies that there are more surviving females than males beyond this point—which implies that older women may, genuinely, be in better health than their male counterparts.

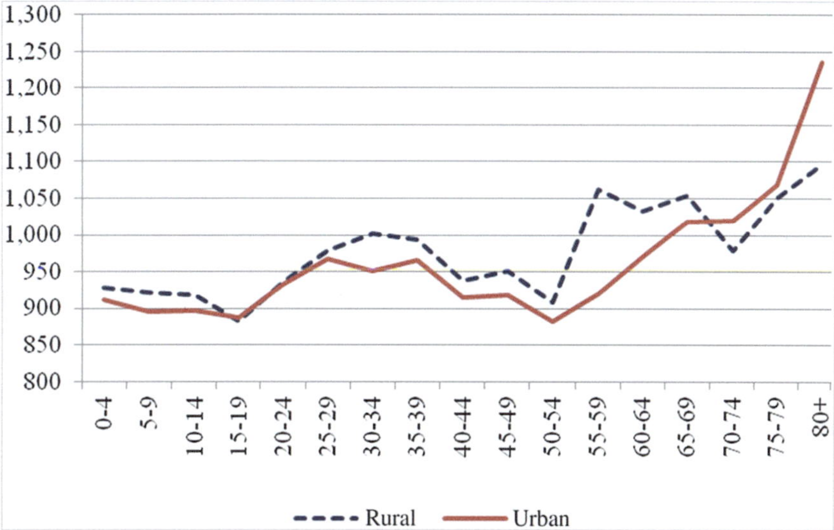

Fig. 10.2 Age specific sex ratio-census 2011. *Source* Office of the Registrar General and Census Commissioner of India (2011). Five year age group data (Table C14), accessed at http://www.censusindia.gov.in/2011census/C-series/C-14.html on 17 March 2016

The analysis of age-specific sex ratio does not, however, rule out conditioning. It may be possible that both factors—the conditioning effect *and* the robustness of elderly women—may have resulted in the absence of gender differences in perceived health status. This calls for a comparison of actual and reported health status to check the level of (in)consistencies in reported health status.

10.4.3 Inconsistency in Actual and Perceived Health Status

In this section, we examine gender variations in consistency of actual and reported health status. NSSO provides information on six indicators of actual health status:

(a) Whether the respondent is mobile,
(b) Whether the respondent had been hospitalized in the year preceding the survey,
(c) Whether the respondent had sought treatment as an out-patient in the fortnight preceding the survey,
(d) Whether the respondent had sought treatment as an out-patient in the day preceding the survey,
(e) Whether the respondent is suffering from any ailment, and
(f) Expenditure on health care.

Table 10.4 Inconsistency between actual and perceived health status

Perceived health	Health problem	
	Present	Absent
Poor	Consistent	Positive extreme response bias
Good	Negative response bias	Consistent

Since data on expenditure on health care has been elicited for only one member in each family, it will not be wise to use this measure. Hospitalization is also an unreliable measure as it may have cured the patient. Similarly, the reference period for inpatient treatment is too short (1 day, 15 days) for measures (c) and (d) to be used. We have, therefore used information on only two symptomatic indicators—mobility and ailments to prepare the matrix given in Table 10.4 for each of these indices individually. Consistent responses occur when either of the following occurs—healthy (mobile or without ailment) respondents report good health or sick (having mobility-related problems or suffering from ailments) respondents report poor health. The northeast and southwest cells, on the other hand, record inconsistent responses—healthy respondents report poor health or sick respondents report good health. Such inconsistencies may be referred to as negative response bias and positive response bias, respectively.[5]

A comparison of the relative frequencies in these two cells between matrices estimated for male and female elderly respondents is interesting and may reveal the presence of conditioning. A priori, the hypothesized influence of macro and micro-level factors on positional objectivity would result in higher negative response bias among males and higher positive response bias among females. The matrices are estimated for only the 60th round and for only those respondents who have reported about their health status themselves.

The results, presented in Table 10.5, do not provide support for our hypothesis. We find that a higher proportion of females displays negative response bias, while a higher proportion of males displays positive response bias (with the exception of urban respondents suffering from mobility-related problems). Nor do statistical tests of gender differences in sample proportions (results reported in Appendix Table 10.9) confirm our hypothesis.

Before rushing to any conclusions, we should first check the robustness of the results using statistical methods. We regress different types of bias against a female dummy and control variables. The control variables are

> Age of respondent, household size, socio-religious identity, log of per capita monthly household expenditure, dummies for educational level, living arrangement, economic independence and economic activity status, geographical zone of residence and whether any other aged member in family is ill.

[5]Extreme response bias is defined as the "tendency to endorse the most extreme response categories in spite of item content" (Baumgartner and Steenkamp 2006).

Table 10.5 Gender differences in consistency of responses by place of residence

Residence	Gender	Perceived health status	Not ill	Suffering from ailment
Whether suffering from any ailment				
Rural	Male	Poor self-reported health	10.92*	42.80
		Good self-reported health	89.08	57.20[a]
	Female	Poor self-reported health	13.55*	45.69
		Good self-reported health	86.45	54.31[a]
Urban	Male	Poor self-reported health	7.99*	33.78
		Good self-reported health	92.01	66.22[a]
	Female	Poor self-reported health	11.25*	34.82
		Good self-reported health	88.75	65.18[a]
Residence	Gender	Perceived health status	Mobile	Immobile
Whether suffering from mobility problem				
Rural	Male	Poor self-reported health	20.01*	74.06
		Good self-reported health	79.99	25.94[a]
	Female	Poor self-reported health	23.15*	74.47
		Good self-reported health	76.85	25.53[a]
Urban	Male	Poor self-reported health	16.19*	69.77
		Good self-reported health	83.81	30.23[a]
	Female	Poor self-reported health	18.87*	65.47
		Good self-reported health	81.13	34.53[a]

Note *denotes negative response bias while
[a]Denotes positive response bias

We estimate four logit models. The dependent variables in these models take the values as defined below,

(i) Value of 1 if respondent suffering from ailment claim to be in good health and 0 if they claim to be in bad health (positive response bias),
(ii) Value of 1 if respondent not suffering from any ailment claim to be in bad health and 0 if they claim to be in bad health (negative response bias),
(iii) Value of 1 if respondent suffering from mobility-related problem claim to be in good health and 0 if they claim to be in bad health (positive response bias), and
(iv) Value of 1 if respondent not suffering from mobility-related problem claim to be in bad health and 0 if they claim to be in bad health (negative response bias).

Our objective is to test whether a relatively higher proportion of females (males) exhibit positive (negative) response bias. If this is true then the coefficient of the female dummy should be statistical significant and positive for regression models of positive response bias; for regression models of negative response bias, on the other hand, the coefficient should be statistically significant and negative.

Results, reported in Table 10.6, provide some support for our hypothesis. As hypothesized, the female dummy has a positive dummy for all models with positive

Table 10.6 Results of logit models of response bias on female dummy

Indicators of actual health	Items	Positive response bias		Negative response bias	
		Rural	Urban	Rural	Urban
Ailment	Female dummy (β)	0.23	0.43	−0.25	0.03
	Z	2.97***	3.98***	−2.53***	0.21
	LR χ^2	562.03***	310.3***	491.16***	194.09***
	Pseudo R^2	0.08	0.08	0.09	0.09
	N	5171	3110	7294	3621
Mobility	Female dummy (β)	0.05	0.57	−0.32	−0.35
	Z	0.28	2.11	−4.82***	−3.59***
	LR χ^2	92.95***	60.11***	1738.89***	699.74***
	Pseudo R^2	0.09	0.10	0.15	0.12
	N	965	501	11,448	6212

Note ***denotes $p < 0.01$. Control variables mentioned in text have been included

response bias as the dependent variable, although the coefficients are significant only when the indicator of actual health status is ailment. Women are active within the house and its immediate surroundings; if their mobility is affected they may feel themselves to be worthless and unable to contribute to the family. This may reduce their tendency, created by a patriarchal society, to report themselves to be in good health. This may explain why the female dummy is insignificant for persons suffering from mobility-related problems but reporting themselves to be in good health. On the other hand, the female dummy has a negative coefficient for persons not suffering from ailments or from mobility-related problems but reporting themselves to be in good health—with the exception of persons suffering from ailments in urban areas. This may be explained by greater awareness among females in urban areas.

10.5 Conclusion

With the increased aging of the world's population, provisioning of social security and health care services to the elderly becomes a major challenge before policy-makers. Active aging has become the catch word with the Madrid Action Plan's call to build a society for all age groups (UN 2002). Ensuring healthy aging is a major component in the plan to make sure that the elderly participates in all spheres of life with dignity. This issue has created the need to define and measure health of the aged. In the past, perceived health status has often been taken as an index of the health status of the aged on the grounds that it is readily obtained and generally consistent with actual health status. This paper contests this claim by arguing that the literature on active aging often misses out on the gender perspective. We point out that attitudes and perceptions of women are conditioned by their position within a matrix defined by, on one hand, their socio-economic and cultural reality and, on the other hand, their immediate familial characteristics. These factors may colour their perceptions about what is good health and lead to inconsistencies between their perceived health status and more objective indicators of actual health.

Our analysis shows that, in the highly discriminatory society of India, gender differences in perceived health status are insignificant. This may partly be explained by the actual disadvantages faced by elderly males, vis-à-vis elderly females, leading to a reversal of age-specific mortality. Simultaneously, we have shown that the gendered nature of society and the immediate realities of the family influence supposedly objective assessments of health status by respondents. This leads to inconsistencies between symptomatic measures of health and self-reported health status. Specifically, we observe that a higher proportion of male aged tend to report poor health even if they do not suffer from any symptomatic health problem, while even females reporting symptomatic health problems tend to state that they are in good health.

This casts aspersions on the reliability of self-reported health status. In fact, even the dependability of symptomatic indicators becomes doubtful as positional objectivity may influence the process of information on symptoms. On a broader context, the definition of active aging—"process of optimizing opportunities for health, participation and security in order to enhance quality of life as people age" (WHO 2002: 12)—itself can become suspect. Quality of life is defined as "an individual's perception of his or her position in life" (WHO 1994). We have seen, however, positional dependence can mould observations, beliefs and actions so that what is an acceptable quality of life can become highly context and individual-specific concepts.

Does this imply that "objective" measures of health cannot be derived? We conclude by suggesting that two changes are required in our current approach towards healthy aging. First, it is necessary to deconstruct the concept of active aging and engender it. Second, there must be a shift to anthropometric or medical measures, rather than rely on self reporting—even of symptoms.

Acknowledgments The authors are grateful to Sugata Senroy, Dipankar Coondoo and Bandana Das for their suggestions on the statistical methodology. Tannistha Samanta also provided some useful comments. We are also grateful to participants of the workshop, particularly U.S. Mishra, for their comments on an early draft. Responsibility for any remaining errors lies with the authors.

Appendix

See Tables 10.7, 10.8 and 10.9.

Table 10.7 Sample profile of elderly population in rural areas

Correlate	Rural male			Rural female		
	1995–96	2004	Change	1995–96	2004	Change
Marital status						
Never married	1.88	1.61	−0.27	1.24	0.69	−0.55
Currently married	76.08	78.51	2.43	40.98	40.99	0.01
Widowed	21.72	19.52	−2.20	57.48	57.80	0.32
Divorced/separated	0.32	0.36	0.04	0.29	0.52	0.23
Educational status						
Illiterate	63.95	57.53	−6.42	90.60	87.98	−2.62
Just literate	4.00	2.65	−1.35	1.22	1.37	0.15
Below primary	12.28	12.98	0.70	3.58	4.45	0.87
Primary	10.83	12.12	1.29	3.01	3.98	0.97
Middle	5.30	7.58	2.28	1.00	1.44	0.44
Secondary	2.63	4.35	1.72	0.22	0.46	0.24
Higher secondary	0.47	1.34	0.87	0.13	0.10	−0.03
Other educational qualification	0.55	1.46	0.91	0.24	0.22	−0.02
Economic independence						
Not dependent	48.88	51.69	2.81	11.55	13.41	1.86
Partially dependent	19.95	16.19	−3.76	15.65	13.31	−2.34
Fully dependent	31.17	32.12	0.95	72.81	73.28	0.47
Living arrangements						
Alone and inmate of old	0.79	0.36	−0.43	0.89	0.62	−0.27
Not inmate of old age	1.69	2.3	0.61	3.97	6.31	2.34
With spouse	10.61	15.57	4.96	6.80	9.14	2.34
With spouse and others	63.52	61.47	−2.05	33.96	30.31	−3.65
Without spouse and with children	19.97	17.46	−2.51	49.11	48.39	−0.72
Other relation	3.05	2.52	−0.53	4.91	4.77	−0.14
Other non-relation	0.37	0.32	−0.05	0.36	0.47	0.11
Geographical zones						
North	12.85	9.72	−3.13	12.31	9.48	−2.83
Central	35.83	38.81	2.98	36.23	37.86	1.63
Northeast	6.43	7.57	1.14	5.10	6.31	1.21
East	14.03	14.54	0.51	13.04	13.90	0.86
West	11.00	9.66	−1.34	12.24	10.63	−1.61
South	19.86	19.70	−0.16	21.09	21.83	0.74
Social groups						
ST	12.43	12.30	−0.13	10.99	11.02	0.03
SC	17.23	17.75	0.52	17.70	17.12	−0.58
General	70.34	69.95	−0.39	71.32	71.86	0.54
Age groups						
60–69 years	61.02	64.73	3.71	63.90	66.80	2.90
70–79 years	29.30	26.60	−2.70	27.00	25.28	−1.72
80 years and above	9.68	8.67	−1.01	9.11	7.93	−1.18

Table 10.8 Demographic profile of elderly population in urban areas

Correlate	Urban male			Urban female		
	1995–96	2004	Change	1995–96	2004	Change
Marital status						
Never married	2.00	1.74	−0.26	1.65	1.23	−0.42
Currently married	81.26	81.58	0.32	38.26	38.32	0.06
Widowed	16.26	16.35	0.09	59.59	59.98	0.39
Divorced/separated	0.48	0.34	−0.14	0.50	0.47	−0.03
Educational status						
Illiterate	22.77	23.42	0.65	60.24	59.52	−0.72
Just literate	3.28	2.25	−1.03	3.22	2.11	−1.11
Below primary	11.84	11.30	−0.54	10.55	10.17	−0.38
Primary	15.96	14.03	−1.93	12.73	10.63	−2.10
Middle	14.34	13.23	−1.11	6.84	7.49	0.65
Secondary	15.68	15.34	−0.34	3.84	5.38	1.54
Higher secondary	4.69	5.10	0.41	0.95	1.67	0.72
Other educational qualification	11.44	15.34	3.90	1.62	3.03	1.41
Economic independence						
Not dependent	51.49	57.11	5.62	9.86	16.40	6.54
Partially dependent	18.11	13.55	−4.56	11.28	9.93	−1.35
Fully dependent	30.4	29.34	−1.06	78.86	73.67	−5.19
Living arrangements						
Alone and inmate of old	0.50	0.25	−0.25	0.92	0.31	−0.61
Not inmate of old age	1.75	1.98	0.23	3.29	5.25	1.96
With spouse	9.05	12.12	3.07	5.14	6.71	1.57
With spouse and others	68.62	67.11	−1.51	32.05	30.68	−1.37
Without spouse and with children	16.56	15.22	−1.34	52.18	50.69	−1.49
Other relation	2.96	2.77	−0.19	5.99	5.93	−0.06
Other non-relation	0.55	0.56	0.01	0.43	0.43	0.00
Geographical zones						
North	11.87	11.70	−0.17	10.64	11.40	0.76
Central	24.48	26.63	2.15	24.49	25.77	1.28
Northeast	4.71	6.73	2.02	4.68	5.78	1.10
East	12.55	12.52	−0.03	11.82	11.10	−0.72
West	18.78	15.33	−3.45	19.20	16.82	−2.38
South	27.62	27.08	−0.54	29.18	29.13	−0.05
Social groups						
ST	3.85	5.73	1.88	3.83	4.73	0.90
SC	9.65	10.55	0.90	9.87	11.57	1.70
General	86.50	83.72	−2.78	86.30	83.70	−2.60
Age groups						
60–69 years	61.09	62.97	1.88	62.03	62.96	0.93
70–79 years	29.95	28.21	−1.74	27.56	27.00	−0.56
80 years and above	8.96	8.82	−0.14	10.41	10.04	−0.37

Table 10.9 Results of tests for gender differences in sample proportions for inconsistent responses

Indicator of actual health	Positive response bias		Negative response bias	
	Rural	Urban	Rural	Urban
Ailment	2.01**	0.70	−3.50***	−3.32***
Mobility	−0.02	−1.17	−4.09***	−2.82***

Note ***denotes $p < 0.01$, **denotes $p < 0.05$

References

Achenbaum, W. A., & Bengston, V. L. (1994). Re-engaging the Disengagement Theory of Aging: On the History and Assessment of Theory Development in Gerontology. *Gerontologist, 34*(6), 756–763.

Agarwal, B. (1996). *A field of one's own: Gender and land rights in South Asia*. Cambridge University Press.

Alam, M. (2006). *Ageing in India: Socio-economic and health dimensions*. New Delhi: Academic Foundation.

Alam, M. (2008). *Ageing, socio-economic disparities and health outcomes: Some evidence from rural India*. Working Paper Series No. E/290/2008. Delhi: Institute of Economic Growth.

Alam, M., & Mukherjee, M. (2005). Ageing, activities of daily living disabilities and the need for public health initiatives: Some evidence from a household survey in Delhi. *Asia Pacific Population Journal, 20*(2), 47–76.

Albert, S. M., Alam, M., & Nizamuddin, M. (2005). Comparative study of functional limitation and disability in old age: Delhi and New York City. *Journal of Cross-Cultural Gerontology, 20*(3), 231–241. doi:10.1007/s10823-006-9014-2

Arifin, E. N. (2006). Living arrangements of older persons in east Java. *Asia Pacific Population Journal, 21*(3), 93–112.

Bardhan, P. K. (1974). On life and death questions. *Economic and Political Weekly, 9* (32–34), 1293–1304.

Basu, A. M. (1992). Culture, the status of women, and demographic behaviour: Illustrated with the case of India. Oxford: Clarendon Press.

Baumgartner, H., & Steenkamp, J. B. E. (2006). Response biases in marketing research. In R. Grover & M. Vriens (Eds.), *The handbook of marketing research: Uses, misuses, and future advances* (pp. 95–109). Thousand Oaks, CA: Sage.

Bengtson, V. L., Burgess, E. O., & Parrott, T. M. (1997). Theory, explanation, and a third generation of theoretical development in social gerontology. *Journal of Gerontology: Social Sciences, 52B*(2), S72–S88.

Calasanti, T. (2004). Feminist gerontology and old men. *Journal of Gerontology: Social Sciences, 59*(6), S305–S314. doi:10.1093/geronb/59.6.S305

Calasanti, T. (2009). Theorizing feminist gerontology, sexuality and beyond: An intersectional approach. In V. Bengtson, et al. (Eds.), *Handbook of theories of aging* (2nd ed., pp. 471–486). New York: Springer Publishing Company LLC.

Calasanti, T. M., & Slevin, K. F. (2001). *Gender, social inequalities, and aging*. Walnut Creek, CA: Alta Mira Press.

Chou, K. L., & Chi, I. (2002). Successful aging among the young-old, old-old and oldest-old Chinese. *International Journal of Aging and Human Development, 54*(1), 1–14. doi:10.2190/9K7T-6KXM-C0C6-3D64

Cohen, G. A. (1978). *Karl Marx's theory of history: A defence*. Oxford: Clarendon Press.

Connidis, I. (2001). *Family ties and aging*. California: Sage Publications.

Das Gupta, M. (1987). Selective discrimination against female children in rural Punjab, India. *Population and Development Review, 13*(1), 77–100.
Dyson, T., & Moore, M. (1983). On kinship structure, female autonomy, and demographic behavior in India.*Population and Development Review, 9*(1), 35–60.
Ghosh, S., & Husain, Z. (2010). Economic independence, family support and perceived health status of elderly: Recent evidences from India. *Asia Pacific Population Journal, 25*(1), 47–77.
Ginn, J., & Arber, S. (1995). Only connect: Gender relations and aging. In S. Arber & J. Ginn (Eds.), *Connecting gender and aging: A sociological approach*. Philadelphia: Open University Press.
Gupta, I., & Sankar, D. (2003). Health of the elderly in India: A multivariate analysis. *Journal of Health and Population in Developing Countries*, 24 June. Retrieved from www.longwoods.com/home.php?cat=394
Haspels, N., Majurin, E. (2008). *Work, income and gender equality: Action guide*. Bangkok: ILO.
Husain, Z., & Ghosh, S. (2011). Is health status of elderly worsening in India? A comparison of successive rounds of national sample survey data. *Journal of Biosocial Science, 43*(2), 211–231.
Jejeebhoy, S. J. (2001). Women's autonomy and reproductive behaviour in India. In Z. A. Sathar & J. F. Phillips (Eds.), *Fertility transition in South Asia* (pp. 221–441). Oxford: Clarendon Press.
Kumar, V. S. (2003). Economic security for the elderly in India: An overview. *Journal of Aging and Social Policy, 15*(2–3), 45–65.
Lall, R. B., & Seal, S. C. (1949). *General rural health survey: Singur health centre 1944*. Kolkata: All India Institute of Hygiene and Public Health.
McMullin, J. A. (2000). Diversity and the state of sociological aging theory. *The Gerontologist, 40*(5), 517–530. doi:10.1093/geront/40.5.517
Miller, B. D. (1981). *The endangered sex: Neglect of female children in rural North India*. Ithaca: Cornell University Press.
Mini, G. K. (2009). Socioeconomic and demographic diversity in the health status of elderly people in a transitional society, Kerala India. *Journal of Biosocial Science, 41*(4), 457–467. doi:10.1017/S0021932008003003
Morduch, J. J., & Stern, H. S. (1997). Using mixture models to detect sex bias in health outcomes in Bangladesh. *Journal of Econometrics, 77*(1), 259–276.
Nayyar, G. (2009). *The demand for services in India: A mirror image of Engel's Law for food?* Discussion Paper Series, No. 451, Department of Economics. Oxford: Oxford University Press.
Nyce, S. A., & Schieber, S. J. (2005). *The economic implications of aging societies: The cost of living happily ever after*. Cambridge: Cambridge University Press.
Rajan, S. I., & Kumar, S. (2003). Living arrangements among Indian elderly: New evidence from National Family Health Survey. *Economic and Political Weekly, 38*(1), 75–80.
Rajan, S. I., & Mishra, U. S. (2011). *The national policy for older persons: Critical issues in implementation*. BKPAI Working Paper No. 5. New Delhi: United Nations Population Fund.
Rajan, S. I., Sarma, P. S., & Mishra, U. S. (2003). Demography of Indian aging, 2001–2051. *Journal of Aging & Social Policy, 15*(2–3), 11–30.
Rajan, S. I. (2006). *Population ageing and health in India*. The Centre for Enquiry into Health and Allied Themes (CEHAT): Mumbai.
Rajan, S. I., Misra, U. S., & Sarma, P. S. (1999). *India's elderly—Burden or challenge?*. New Delhi: Sage Publications.
Raju, S. S. (2011). *Studies on ageing in India: A review*. BKPAI Working Paper No. 2. New Delhi: United Nations Population Fund.
Registrar General and Census Commissioner. (2006). *Population projections for India and States, 2001–2026*. New Delhi: National Commission on Population.
Rose, M. R. (1991). *Evolutionary biology of aging*. New York: Oxford University Press.
Rossi, A. S. (Ed.). (1985). *Gender and the life course*. New York: Aldine de Gruyter.
Rustagi, P. (2005). Understanding gender inequalities in wages and incomes in India. *The Indian Journal of Labour Economics, 48*(2), 319–334.

Sen, A. K. (1985). *Commodities and capabilities*. New Delhi: Oxford University Press.
Sen, A. K. (1990). Gender and cooperative conflict. In I. Tinker (Ed.), *Persistent inequalities: Women and world development* (pp. 123–149). New York: Oxford University Press.
Sen, A. K. (1993). Positional objectivity. *Philosophy and Public Affairs, 22*(2), 126–145.
Sen, A. K. (2006). The possibility of social choice. In B. Agarwal, J. Humphries, & I. Robeyns. (Eds.), *Capabilities, freedom and equality. Amartya Sen's Work from a Gender Perspective* (pp. 369–419). New Delhi: Oxford University Press.
Sen, M., & Noon, J. (2007). *Living arrangement: How does it relate to the health of the elderly in India*? Paper presented at the Annual Meeting of the Population Association of America, New York.
Shrag, C. (1967). Elements of theoretical analysis in sociology. In L. Gross (Ed.), *Sociological theory: Inquiries and paradigms* (pp. 220–253). New York: Harper and Row.
Sobieszczyk, T., Knodel, J., & Chayovan, N. (2002). *Gender and well-being among the elderly in Thailand*. PSC Research report No. 02-531. Ann Arbor, MI: Population Studies Centre, Institute of Social Research, University of Michigan.
Subrahmanya, R. K. A. (2005). *Social security for elderly*. New Delhi: Shipra Publications.
United Nations. (2002). *Political declaration and Madrid international action plan for active ageing*. Geneva: United Nations.
World Bank. (1993). *World development report: Investing in health*. Oxford: Oxford University Press.
World Bank. (1994). *Averting the old age crisis*. New York: Oxford University Press.
World Bank. (2012). *World development report: Gender equality and development*. Washington DC: The World Bank.
WHO. (1994). *Statement developed by WHO Quality of Life Working Group*. Published in the WHO Health Promotion Glossary 1998. WHO/HPR/HEP/ 98.1 Geneva: World Health Organization.
WHO. (2002). *Active aging: A policy framework*. Geneva: World Health Organization.
WHO. (2001). *Men, Ageing and Health*. Geneva: World Health Organization.

Chapter 11
The Relationship Between Spatial Activity and Wellbeing-Related Data Among Healthy Older Adults: An Exploratory Geographic and Psychological Analysis

Michal Isaacson, Hans-Werner Wahl, Noam Shoval, Frank Oswald and Gail Auslander

Abstract It has long been accepted that remaining active in later life is positively correlated with wellbeing. But is there an importance to remain spatially active in later life? This study examines data collected from participants using both questionnaires for collecting reported wellbeing indicators and GPS receivers for collecting data that describes spatial activity. An association between spatial activity, life satisfaction, perceived health and perceived physical function were found. No association between spatial activity and positive or negative affect was found. Separate analysis that was conducted for men and women revealed gender-based differences in the way that spatial activity is correlated with wellbeing. The results of this study promote a broader understanding of the importance of spatial activity in later life. Contributing to both to the theoretical aspect of activity in old age as well as having implications for considering living environments for older adults.

M. Isaacson (✉)
Department of Gerontology, University of Haifa, Haifa, Israel
e-mail: Misaacson@univ.haifa.ac.il

H.-W. Wahl
Department of Psychological Aging Research, University of Heidelberg, Heidelberg, Germany
e-mail: Hans-Werner.Wahl@psychologie.uni-heidelberg.de

N. Shoval
Department of Geography, The Hebrew University of Jerusalem, Jerusalem, Israel
e-mail: NoamShoval@huji.ac.il

F. Oswald
Interdisciplinary Ageing Research Group, Goethe University, Frankfurt, Germany
e-mail: Oswald@em.uni-frankfurt.de

G. Auslander
Paul Baerwald School of Social Work and Social Welfare, The Hebrew University of Jerusalem, Jerusalem, Israel
e-mail: Gail.Auslander@mail.huji.ac.il

T. Samanta (ed.), *Cross-Cultural and Cross-Disciplinary Perspectives in Social Gerontology*, DOI 10.1007/978-981-10-1654-7_11

Keywords Wellbeing · Spatial activity · GPS tracking · Gender · Older adults

11.1 Introduction

The Activity Theory of Aging states that a positive relationship exists between levels of activity and life satisfaction. The more active an older adult is, the more likely he or she is to be satisfied with life and to adjust to the changes associated with aging. The activity theory was inexplicitly used in gerontology research for over two decades until it was formally formulated into an explicit theory by Lemon et al. (1972) in their article on life satisfaction among in-movers to a retirement community. The existing literature has focused on older adults' engagement in activity, and by examining different types of activates has tried to quantify their contribution to wellbeing. A central theme found in the literature explaining the correlation between the level of activity and life satisfaction is the importance that older adults retain roles that they filled during their lifetime, and the understanding that the loss of such roles leads to dissatisfaction with life. *Role loss* is defined as the alteration of behaviour patterns due to the loss of a status position in society, while *role support* is the support that an individual receives from surrounding people concerning his or her fulfilment of different roles.

This study explores whether the activity theory of aging has a spatial dimension. Whether there exists an association between volumes of spatial activity and wellbeing in later life. Understanding such an association can influence the way that we build future communities, creating environments that are more effectively designed to contribute to health and wellbeing throughout the lifespan in general and in later life in particular.

11.2 Theoretical Framework

Activity was defined by Lemon et al. (1972) as "any regularized or patterned action or pursuit which is regarded beyond routine physical or personal maintenance" (513). Activities can be classified as formal or informal. In informal activities, the person interacts with family and friends. Formal activities are those which include participation in work or voluntary organizations and solitary activities such as reading a book or watching television.

In the article that presents the formalization of the activity theory by Lemon et al. (1972), not all types of activity were found to have a positive effect on life satisfaction within the sample used (older adults who moved into a retirement community in Southern California). Only social activities conducted with friends were found to have a significant contribution to life satisfaction. A decade after the publication of the article by Lemon et al. (1972), Longino and Kart (1982) published a study that aimed at replicating Lemon's work using a bigger and more diverse sample. The sample used was three times the size of that used by Lemon et al. (1209 vs. 411) and was collected in three distinctly different retirement communities. The findings of their research strengthened the previous findings, showing the positive correlation between participation in informal activities and life satisfaction. On the other hand, not only did the latter research not find a positive correlation between formal activity and life satisfaction, data collected from all three communities showed negative correlation between participation in such activities and life satisfaction. This latter finding is perhaps the most substantial finding of this research and it has great implications for the creation of formal settings of activity for older adults. These findings, as Longino and Kart explain, indicate that for people in their seventh and eighth decade of life, interaction on a daily basis within an age-segregated community (formal activity within the sample was a large activity within a day centre intended for the older adults) tends to enforce lower self-esteem and life satisfaction.

The notion that remaining active is a key to being satisfied with one's life can also be identified as a chief component in Rowe and Kahn's (1997) model of successful aging, such as Rowe and Kahn's (1997) 'engagement in life' as one of the three components that lead to successful aging. Although the activity theory is among the ten most cited theories in social gerontology (Alley et al. 2010) and is used as theoretical reasoning in many research frameworks (for a few central examples see: Cutler 1975, 1976; Schaie and Gribbin 1975; Conner et al. 1979; Hoyt et al. 1980; Ragheb and Griffith 1982; Strain and Chappell 1982; Lee and Ishiikuntz 1987; Vantilburg 1992; Everard et al. 2000; Van Willigen 2000), since the 1982 article discussed above there has not been another attempt to formally explicate the theory. That said, and although the theory has never been fully proven empirically, the theory has been used as a theoretical explanation in many studies and as the motivation behind many others.

However, from the vantage point of a gero-geographical perspective, it is interesting and at the same time limiting to see in this previous literature that activity has mostly been defined as social activity assessed via a questionnaire approach (e.g., Pinquart and Sörensen 2001). Our geographically and psychologically driven study takes a different approach to activity by focusing on the way older adults are spatially active outside of their homes rather than focussing on the type of activity that older adults are engaged in as mostly done previously. Also, we will use GPS methodology to objectively assess out-of-home mobility instead of using a questionnaire or diary device. We aim to examine whether and how both the *volumes* of out of home activity and *quality* of movement relates to wellbeing oriented indicators. Or in other words—does the activity theory of aging have a spatial dimension and if so, what does it look like?

Previous research in the area of spatial aging has predominantly examined the influence that transportation in general and access to transportation in particular has on the life satisfaction of older adults. In one of the first of these studies, Cutler (1975) concluded his study of the connection between life satisfaction and transportation with an understanding that transportation plays a crucial role in later life. In Cutler's research transportation was treated as a general and ambiguous term but has not been separated into its different components. Banister and Bowling (2004) reconfirmed the association between transportation and quality of life, adding a distinction between active and passive elements of the transport system. They have argued that examining the active part of the transport system by means of the standard transport representations such as trips and distances is not enough to capture the full extent of the influence that transportation has on quality of life in old age. MOBILATE was a European country-comparative research project that took a wider view on out of home mobility of older adults (Mollenkopf et al. 1997, 2002, 2004). Differentiating itself from the previous research done mainly in the context of transportation studies, MOBILATE concentrated on the social and functional aspects of the mobility of older adults and not only on the physical aspects of movement through space, establishing the connection between out of home mobility and social relationships. One of the important findings of this project was that the dissatisfaction with their out-of-home mobility increases with the advance of age and that the link between mobility and wellbeing becomes closer as people age (Mollenkopf et al. 1997, 2011).

At the conceptual level, out-of-home mobility can be defined as movement through space. When a person loses the ability to move freely through space, his or her mobility has been compromised and he or she can be described as having limited mobility. Recent years have seen a growing body of scholarship on the mobility of older adults, with extensive interest in how physical and cognitive functioning, the built environment and access to transport systems (including car access) affect that mobility (Schwanen and Paez 2010; Bowling 2009; Schwanen and Ziegler 2011).

We portray out-of-home mobility in later life in terms of two fundamental characteristics and that such a differentiation may be important to understand the relationship between spatial activity and wellbeing-related indicators. First, we focus the *volume* of spatial activity, which is reflected in indicators such as overall activity (average time spent out of home and average number of visited locations per day), the volume of walking movement (average time spent walking per day and average walking episodes per day), and the volume of motorized movement (average time spent driving per day and average number of driving episodes per day). Second, we focus the *quality* of movement as expressed in walking movement (average walking duration per episode, average walking distance per episode, average walking speed) and in motorized movement (average driving duration per episode, average driving distance per episode, driving speed).

In terms of wellbeing-related indicators, the more psychological part of the present study, we differentiate in accordance with respective previous research (e.g., Schilling 2005, 2006) between a cognitive component of overall wellbeing, i.e. life satisfaction, and an emotional component of overall wellbeing, i.e., positive affect. We have selected positive affect only, because previous research has found positive affect as strongly depending on external circumstances and occurrences, such as out-of-home activity, whereas negative affect is more strongly associated with personality characteristics, such as neuroticism. We added two more specific components of wellbeing, which can be expected to reveal an even closer connection with out-of-home spatial activity than overall wellbeing-related indicators, i.e. subjective health and perceived physical functioning. This is so, because previous research supports the notion that the evaluation of subjective health and particularly perceived physical functioning is substantially driven by mobility related parameters in old age.

11.3 Research Aims

The goal of this study was to explore the multi-faceted relationships between spatial activity in terms of objective GPS measures for volume and quality of out-of-home mobility and indicators of subjective wellbeing in later life. First, we expected that the association between general wellbeing indicators and the volume of spatial activity will be closer as compared to the quality of spatial activity. That may be so, because the volume component represents the more direct activity component most similar to activity assessment based on the traditional questionnaire approach. Second, we expected that the connection between spatial activity and wellbeing-related indicators is closer for specific indicators, such as subjective health and perceived physical functioning, as compared to more general wellbeing indicators, such as life satisfaction and positive affect. The main reason behind this assumption is that out-of-home mobility may serve as a major information source for evaluations of subjective health and perceived physical functioning. Third, in terms of gender differences, we expected that men may reveal an even higher connection than women between the volume characteristic of spatial activity and subjective health and perceived physical functioning. The rationale is that spatial activity may be an even stronger proxy for perceived health and perceived physical functioning for older man than for older women,

11.4 Design and Sample

The data used in this study is based on questionnaires and high-resolution time-space data that were collected as part of the SenTra project. SenTra was a bi-national, Israeli-German multidisciplinary research project that studies the connection between spatial activity and cognitive decline. Multi-disciplinary teams of researchers from both countries collaborated on designing a study that combined methodologies from social and spatial sciences, aimed at furthering the understanding of cognitive decline and mobility (for a full description of the study design see Shoval et al. 2008). Participants in the study were residents of the Greater Tel Aviv metropolitan area in Israel and the Rhine-Neckar metropolitan region in Germany. The data collection and analysis methods made it possible to present a new perspective on topics that gerontological thought has been concerned with for many years; at the same time, it also made possible the examination of new ways of studying and thinking about time-space geography in later life (Shoval et al. 2008, 2010, 2011; see also the contributions of Kaspar et al. and Wettstein et al. in this special section of the Journal of Applied Gerontology).

The Rhein-Neckar metropolitan area is located in southwestern Germany. The data for this study were collected mostly in two of the region's large cities, Heidelberg and Mannheim. The city of Heidelberg located on the River-Neckar has a historic core with its origins dating back to the fifth century and is well known for its university, the oldest university in Germany. Home to 146,000 people, the city is spread over 109 km^2. The city of Heidelberg hosts a large U.S. military presence on several bases in the city. The city has a tram system as well as an extensive local bus system. A railway station connects the city to other cities in the area and in the country via the rail network. The Old Town is located in the centre of the city on the southern banks of the River Neckar. A university town, the new university campus is located north of the river. Mannheim is located in the heart of the Rhein-Neckar metropolitan area right where the two rivers meet. The city, spread over 145 km^2, is home to 315,000 people and was declared a city in 1607 by Fredrick IV (it was previously a village). The city also has a historic core that consists of a ring road and an organic layout of roads within the historic core. Surrounding the core is a grid layout of roads that have given the city its nickname "die Quadratestadt" (city of the squares). Mannheim hosts a central railway junction and is highly connected to other parts of the country. Local transportation is based on tram lines and numerous bus routes. Unfortunately, data regarding the spatial dispersion of older adults in Heidelberg and Mannheim is not available from the German Statistics Bundesamt or from another official source.

The data presented in this paper is generated from a sub-sample consisted of 96 men and women living in the Rhein-Neckar metropolitan region, Germany, and is described in Table 11.1. While the Sentra project studied older adults with cognitive decline, the data used to write this paper was collected from the control group of cognitively healthy participants. In addition to participating in two face-to-face interviews that took place in their home, participants were asked to carry a GPS unit, which recorded their location every 10 s, with them for 28 days. The data

Table 11.1 Sample description including differentiation according to gender

Variable	Full sample	Male	Female	Test
	$N = 96$	$N = 57$	$N = 39$	
Age, years (M, SD, Range)	70.8, 4.128, 61–81	71.7, 4.14, 64–81	69.6, 3.78, 61–79	**
Married (*n*, %)	65, 67.7 %	44, 77.2 %	21, 53.8 %	*
Work or volunteer, *n* (%)	45 (46.9 %)	24 (42.1 %)	21 (53.8 %)	ns
Years of schooling (M, SD, Range)	14.63, 4.38, 2–26	15.16, 4.23, 8–26	13.86, 4.53, 2–23	ns
Satisfaction with financial situation (M, SD, Range)	7.14, 1.9, 2–10	7.11, 1.62, 4–10	7.18, 2.29, 2–10	ns

Note Differences: "ns" not significant; $*p < 0.05$; $**p < 0.01$

collected in real time was streamed by GPRS to servers that are located in The Hebrew University of Jerusalem creating very large data sets of spatial–temporal data that were later analysed to extract information describing the participants' spatial activity.

11.4.1 Assessment of Study Variables

Assessment of Spatial Activity. The spatial variables used in this research have been extracted from the GPS data collected. These variables describe different aspects of spatial activity and can be divided into two groups: First, variables that portray the *volume* of spatial activity were operationalized by means of the following GPS indicators: (a) etc. Overall activity: Average time spent out of home and Average number of visited locations per day.

(a) Volume of walking movement: Average time spent walking per day and Average walking episodes per day.
(b) Volume of motorized movement: Average time spent driving per day and Average number of driving episodes per day.

Second, variables that depict the *quality* of movement:

(c) Walking movement: Average walking duration per episode, Average walking distance per episode, Average walking speed.
(d) Motorized movement: Average driving duration per episode, Average driving distance per episode, Driving speed (Table 11.2).

Assessment of Wellbeing-Related Variables. For this study four subjective wellbeing indicators were implemented (see Table 11.3). The four wellbeing indicators were selected to cover both, cognitive as well as affective facets of general wellbeing as well as health-and function-related aspects of wellbeing. *Reported life satisfaction* was documented using a single item question asking participants to rate their life satisfaction on a scale from 1 to 10, with 1 = "not satisfied with my life" to 10 = "very satisfied with my life". This easily applicable and economic measure is used in many population-based social science surveys. Considerable methodological literature concerning its reliability and validity has been published, resulting in some consensus concluding that single-item measures provide sufficiently valid and reliable LS measurement, and do not perform markedly worse in measuring LS as multi-item questionnaires (Veenhoven 1996). Positive affect was assessed based on the positive affect subscale of the Positive and Negative Affect Schedule was implemented (PANAS) (Watson et al. 1988), yielding a score based on 10 items for positive affect. *Perceived General Health* was asked as a single item question. Answers were coded on a 5-point-ranking-scale with lower values indicating better health. In order to simplify the interpretation of the results the scores were reversed so that the meaning of a

Table 11.2 Distribution of spatial variables including differentiation according to gender

	Variable (M, SD)	Full sample	Male	Female	Test
Volume of activity	Average time spent out of home (h)	4.58, 2.31	4.62, 2.45	4.54, 2.12	ns
	Average number of visited nodes per day (count)	5, 1.06	5.02, 1.22	5.04, 1.09	ns
	Average time spent walking per day (h)	0.33, 0.37	0.31, 0.39	0.29, 0.24	ns
	Average walking tracks (count)	1.47, 1.04	1.43, 1.13	1.78, 1.2	*
	Average time Spent driving per day (h)	0.83, 0.45	0.87, 0.51	0.79, 0.39	ns
	Average number of driving tracks per day (count)	3.06, 1.35	3.26, 1.68	3.04, 1.5	ns
Quality of movement	Average driving duration per episode (h)	0.29, 0.12	0.28, 0.12	0.27, 0.08	ns
	Average walking duration per episode (h)	0.21, 0.13	0.19, 0.12	0.19, 0.12	ns
	Average walking distance per episode (km)	0.84, 0.52	0.72, 0.47	0.72, 0.57	ns
	Average driving speed (km/h)	24.38, 8.33	25.1, 8.41	23.37, 7.04	ns
	Average walking speed (km/h)	3.88, 0.69	3.75, 0.74	3.73, 0.67	ns

Note Differences: "ns" not significant; $*p < 0.05$

Table 11.3 Wellbeing indicators including differentiation according to gender

Variable (M, SD)	Full sample	Male	Female	Test
Satisfaction from life	8.23, 1.32	8.34, 1.37	7.99, 1.44	ns
Perceived general health	3.18, 0.75	3.19, 0.76	3.22, 0.74	ns
Positive affect	35.96, 5.7	35.11, 5.01	35.42, 6.64	ns
Negative affect	16.38, 5.3	15.86, 5.53	17.62, 5.1	*
Perceived physical functioning	86.4, 13.52	83.66, 17.8	81.77, 17.52	ns

Note Differences: "ns" not significant; $*p < 0.05$

higher score is better perceived health. *Perceived physical function* was documented using one of the sub-scales of the SF36 questionnaire describing how the respondent perceives his/her physical function in day-to-day life, from 1 to 5. It is important to note that people that were not able to function independently in day-to-day life as well as people with limitations to their mobility such as the inability to walk independently (without a walker or other assistive device) were excluded from participation in this research.

Background Variables. Age, gender, marital status, work, education years and satisfaction with one's financial situation were considered as background variables. Education years were measured as years of education. Satisfaction with financial situation was based on a single-item question and measured on a 1–10 scale. Participants were asked if they currently work.

11.5 Data Analysis

Given the overall exploratory nature of this study we provide an examination of our study aims at the level of bi-variant correlations. This approach better fits our rather small sample size, although out-of-home mobility has been assessed quite data-rich and intensive in our sample. But note that this "simple" approach is conceptually driven and thus nevertheless able to speak to our expectations. Because of skewed distribution in some of the variables (e.g. 90 % of the sample ranked their life satisfaction as equal or higher than 8), Spearman's non-parametric correlation was calculated and tested for statistical significance.

11.6 Results

In order to understand the connection between out of home mobility and wellbeing, each indicator of wellbeing is regarded separately (see Table 11.4). When examining the total sample strong correlations were found between all of the spatial activity volume indicators except for the average number of nodes visited per day. At the same time there was no correlation between the quality of movement indicators and life satisfaction.

Table 11.4 Spearman correlations between reported life satisfaction and spatial activity

	Life satisfaction	Full sample	Male	Female
Volume of activity	Average time spent out of home	**0.162***	**0.303****	−0.087
	Average number of visited nodes per day	0.048	**0.208***	−0.187
	Average time spent walking per day	**−0.203****	−0.173	−0.182
	Average walking tracks count	**−0.191****	−0.108	**−0.252***
	Average time spent driving per day	**0.172****	**0.192***	0.156
	Average number of driving tracks per day	**0.244*****	**0.202***	**0.262***
Quality of movement	Average driving duration per episode	−0.029	0.016	−0.147
	Average walking duration per episode	−0.019	0.006	−0.076
	Average walking speed	−0.133	−0.099	−0.187
	Average walking distance	−0.017	0.022	−0.82
	Average driving speed	0.020	0.031	−0.36

Note Differences: "ns" not significant; *$p < 0.05$; **$p < 0.01$; ***$p < 0.001$

In this study, as in previous studies as presented above, men averaged higher life satisfaction than the life satisfaction of women (see Table 11.2), this difference was not significant. Men's life satisfaction was found to be correlated with two of the volume of activity indicators while there was no correlation between life satisfaction and volume of activity for women. Men's life satisfaction was influenced by the amount of time that they spend out of their homes, men with higher life satisfaction scores spend more time out of their home. The obvious explanation for this can be that people that are absent from their home for longer times are people that still retain active roles in the work force. While this explanation fits in with the theoretical understanding of the importance of retaining roles that people have held during their lifetime and the importance that having a professional can have for men it is not supported by the findings in this study. No correlation was found between the average time spent out of home and whether or not the participant works or volunteers. The number of driving tracks and the duration of time spent driving also effects men and women differently. While men that have more driving tracks are more satisfied with their lives women's life satisfaction is adversely effected by the amount of time that they spend in an automobile. This difference can be interpreted as another dimension in which the difference between men and woman regarding being out of home is established.

11.6.1 Perceived General Health and Out of Home Mobility

When examining the full sample (see Table 11.5) there is a correlation between walking behaviour and perceived general health. Perceived health is correlated both with the volume of walking behaviour as well as with walking speed and distance. None of the other spatial activity indicators are correlated with perceived health and most interesting is the lack of correlation between perceived health and volumes of spatial activity. Surprisingly, perceived health is not associated with the amount of time a person spends out of their home.

Within the male and female sub-samples a similar correlation is found. Within these sub-samples perceived health remains correlated to walking behaviour and is found in both volume of activity indicators and quality of movement variables. Within the women, correlations were found between a higher number of walking associated variables than those found in the male sub-sample.

11.6.2 Perceived Physical Function and Out of Home Mobility

Not surprisingly perceived physical function has the strongest association with out of home mobility. Both volumes of activity as well as quality of movement

Table 11.5 Correlations between perceived health and spatial activity

	Perceived health	Full sample	Male	Female
Volume of activity	Average time spent out of home	−0.016	0.001	−0.42
	Average number of visited nodes per day	−0.058	−0.137	0.73
	Average time spent walking per day	**−0.201****	**−0.315*****	−0.44
	Average walking tracks count	−0.049	**−0.310*****	**−0.313****
	Average time spent driving per day	−0.036	0.049	−0.179
	Average number of driving tracks per day	−0.009	−0.020	−0.002
Quality of movement	Average driving duration per episode	−0.050	0.079	−0.242
	Average walking duration per episode	−0.159	−0.038	**−0.304****
	Average walking speed	**−0.291*****	**−0.233****	**−0.395*****
	Average walking distance	**−0.247*****	−0.140	**−0.365****
	Average driving speed	0.007	0.024	0.032

Note Differences: **$p < 0.01$; ***$p < 0.001$

Table 11.6 Correlations between perceived physical functioning and spatial activity

	Perceived physical functioning	Full sample	Male	Female
Volume of activity	Average time spent out of home	0.058	0.138	−0.052
	Average number of visited nodes per day	**0.269*****	**0.323*****	**0.219***
	Average time spent walking per day	**0.252*****	**0.423*****	−0.015
	Average walking tracks count	0.105	**0.376*****	**−0.239***
	Average time spent driving per day	**0.298*****	**0.314*****	**0.294****
	Average number of driving tracks per day	0.169	0.200	0.115
Quality of movement	Average walking duration per episode	**0.240****	**0.241****	**0.243***
	Average walking speed	**0.184****	0.147	**0.266***
	Average walking distance	**0.292*****	**0.278****	**0.289****
	Average driving duration per episode	**0.228****	**0.180***	**0.290****
	Average driving speed	−0.002	0.035	−0.78
	Average driving distance	0.122	**0.176***	0.052

Note Differences: *$p < 0.05$; **$p < 0.01$; ***$p < 0.001$

indicators are found to be associated with perceived physical function. This is not surprising and indicates that people who perceive themselves as more physically limited adjust their spatial activity accordingly. A finding that is of note is the lack of association between perceived physical function and the time spent out of home. There is a strong negative correlation between the amounts of time that men spend walking per day to their perceived health a correlation that does not exist within the female sub-sample (Table 11.6).

11.6.3 Affect and Out of Home Mobility

Neither positive affect nor negative affect were found to be correlated with the out of home mobility variables developed in this research (Tables 11.7 and 11.8). Most surprising is that there was no connection found between one's emotional state and the volume of out of home activity. These findings indicate that there is both no correlation between positive affect for people with higher volumes of out of home

Table 11.7 Correlations between positive affect and spatial activity

	Positive affect	Full sample	Male	Female
Volume of activity	Average time spent out of home	0.018	0.077	−0.047
	Average number of visited nodes per day	0.121	0.119	0.121
	Average time spent walking per day	−0.132	−0.164	−0.096
	Average walking tracks count	−0.059	−0.005	−0.165
	Average time spent driving per day	0.048	−0.107	0.287
	Average number of driving tracks per day	0.117	0.026	0.240
Quality of movement	Average walking duration per episode	0.054	0.047	0.054
	Average walking speed	−0.093	−0.005	−0.257
	Average walking distance	0.028	0.037	0.010
	Average driving duration per episode	−0.084	−0.190	0.090
	Average driving speed	−0.013	0.064	−0.077
	Average driving distance	−0.056	−0.088	0.005

Table 11.8 Correlations between negative affect and spatial activity

	Negative affect	Full sample	Male	Female
Volume of activity	Average time spent out of home	−0.074	−0.076	−0.067
	Average number of visited nodes per day	0.067	0.114	−0.048
	Average time spent walking per day	−0.007	−0.006	−0.015
	Average walking tracks count	0.154	0.195	0.054
	Average time spent driving per day	0.011	0.051	−0.045
	Average number of driving tracks per day	−0.010	−0.016	0.022
Quality of movement	Average walking duration per episode	−0.052	−0.136	0.076
	Average walking speed	0.005	0.009	0.149
	Average walking distance	−0.011	−0.092	0.098
	Average driving duration per episode	−0.059	0.024	−0.237
	Average driving speed	−0.005	−0.061	0.145
	Average driving distance	0.025	0.014	0.074

mobility, and no correlation between out of home mobility in people with higher negative affect.

11.7 Discussion

The most unexpected finding in this study is the lack of correlation between affect and out of home mobility in general and specifically to volumes of out of home mobility. This finding contradicts the idea that leaving home and being more active conducts the development of positive affect or that people with higher positive affect are more active out of home than people with lower positive affect, both of these notions being supported by the activity theory of aging (Lemon et al. 1972; Longino and Kart 1982). This lack of correlation can be understood in two ways: possibly the effect that out of home mobility has on affect is evident on a smaller temporal scale that would be measurable is affect was measured daily and not once for the whole month of participation. An additional way to understand this finding is by considering the role that non-spatial communication technologies play in the lives of older adults. Perhaps the social interaction that these technologies allow are more significant that spatial mobility in explaining affect. The findings in this study make it possible for us to say that on one hand, out of home mobility does not seem to improve positive affect and on the other hand, negative affect does not reduce the volume of out of home activity.

That said this research has found correlation between volumes of out of home mobility and reported life satisfaction and perceived physical limitation. Perceived general health was found to be correlated only to walking behaviours. The cause of these correlations remains speculative and will need further research to establish proof as to which factor is the factor that causes these differences, is it that people with higher quality of life feel healthier and more able to spend time out of home than those who do not, or rather does the time that older adults spend out of home raise the quality of their life. It is important to remember, as stated previously that this research had strict exclusion criteria and people with limited mobility were excluded from the sample, meaning that people that perceived their health as poor were still people that were mobile and were able to live independently.

An exciting finding that this research has established is that differences in gender are related to differences in the correlation between out of home mobility and wellbeing. For men more time spent out of home was found to be correlated with higher wellbeing while women do not display such a correlation. As discussed in the literature review men receive more role support from out of home activity while women have more role support in their roles as wives and caretakers within their homes.

Men and woman do not exhibit significant differences in the amount of time that they spend out of home or in the average number of driving tracks per day (Table 11.2). This similarity in behaviour along with the correlation between these two out of home mobility indicators with life satisfaction for males and the lack of correlation between life satisfaction and time spent out of home in the female sub-sample and the adverse effect that the average number of driving tracks can be interpreted is an indication to the different approach that men and women might have to out of home activity (Reitzes et al. 1995). In addition, this difference in correlation between time spent out of home and reported life satisfaction, raises questions as to the lack of significant differences in the average time spent out of home between people of different gender. Is the time spent out of home of older adults driven mostly by need and not desire? Perhaps the time that an older adult spends out of home reflects not only that person's preference for activity but that of his partner as partners go about their daily lives together.

The correlation between both perceived general health and perceived physical limitation and quality of movement indicators is to be expected as people adjust their out of home mobility to suit what they perceive their ability to be. This connection leads to question as to whether quality of movement data can be used as an "objective" health indicator. Perhaps like other measurements, such as blood pressure and resting heart rate the quality of movement indicators might be able to be used as indicators of one's health. This will need further research on population with greater variance in health status. The lack of correlation time spent out of home and these two measurements of wellbeing indicates that perceived health and perceived physical limitation may impact one's movement but not engagement in activity.

A possible explanation for the negative correlation between the amounts of time that men spend walking per day to their perceived health this difference can be the higher rates of access to cars that the male participants enjoy. Perhaps, when a men perceive that themselves to be physically limited they are able to use a car for more of their day-to-day activity, this would explain lower amounts of time spent walking in men who perceive themselves as physically limited while a lower rate of women are able to adjust their activity in such a way. On the other hand there is a correlation between the amount of time that women spend in driving or riding in vehicles to their perceived physical functioning, a correlation that does not exist for men.

11.8 Conclusion

The main question that this study addresses still needs to be interrogated further. Does the activity theory of aging have a spatial dimension? The importance of engagement in activity has been highlighted by many researches but is there an importance to the amount of movement that one exhibits through space? The findings in this study indicate that generally speaking, there is a connection between both volumes of out of home mobility and quality of life but these findings are not consistent throughout the mobility indicators or the same for different sub-samples. These findings illuminate the contribution that spatial data, spatial thought and discourse can have on fields of study that have not been considered spatially oriented and highlights the broader contribution that can result from the implementation of both spatial data collection methods and spatial data analysis methods.

The activity theory has been found lacking by many previous researches (Lemon et al. 1972; Longino and Kart 1982; Reitzes et al. 1995; Litwin and Shiovitz-Ezra 2006) when tested empirically. These previous studies were only able to verify the connection between activity and wellbeing in limited circumstances and were forced to reject a portion of the hypothesis that were based on the activity theory. In this light the findings in this study are not surprising. The inconsistency in the empiric evidence supporting the activity theory of aging suggests that the activity does not influence quality of life directly but rather through other and more complex mechanisms.[1]

This has been the first attempt to study the connection between out of home mobility and quality of life using high resolution time-space data. Because it is the first attempt of its kind many different aspects of mobility were examined, volume of activity, quality of movement, motorized movement, walking, length, duration speed were all examined and different aspects of spatial behaviour were found to be connected to quality of life. High resolution data, gives the opportunity to both examine activity from a different perspective than the perspective that has been considered previously, and to examine spatial activity in a resolution that has not been possible using other data collection techniques. While the data studied in this study allowed to examine the activity theory of aging from a unique angle, the data were collected as part of the SenTra project that has other research objectives and was not designed to answer the questions posed in this study.

This study implemented a straight forward or even naïve interpretation of the activity theory of aging into a spatial activity theory of aging. The results reveal that the activity theory of aging, one of the first and most intuitive theories of aging has a spatial dimension as one would intuitively assume but similar to the findings in research that examined the connection between activity and quality of life, the correlation between the two is not a simple linear relation. Further research will be needed to understand the exact mechanism through which spatial activity and quality of life are bound and to further the understanding of the way being spatially active can benefit the population of older adults.

References

Alley, D. E., Putney, N. M., Rice, M., & Bengtson, V. L. (2010). The increasing use of theory in social gerontology: 1990–2004. *The Journals of Gerontology. Series B, Psychological Sciences and Social Sciences, 65*(5), 583–590.

Banister, D., & Bowling, A. (2004). Quality of life for the elderly—The transport dimension. *Transport Policy, 11*(2), 105–115.

[1]For example, see, Litwin and Shiovitz-Ezra (2006), where activity did not directly influence wellbeing but was significant only when taking the quality of the social relationships into account.

Bowling, A. (2009). Perceptions of active ageing in Britain: Divergences between minority ethnic and whole population samples. *Age and Ageing, 38*, 703–710.

Conner, K. A., Powers, E. A., & Bultena, G. L. (1979). Social-interaction and life satisfaction: Empirical-assessment of late-life patterns. *Journals of Gerontology, 34*(1), 116–121.

Cutler, S. J. (1975). Transportation and changes in life satisfaction. *The Gerontologist, 15*(2), 155–159.

Cutler, S. J. (1976). Membership in different types of voluntary-associations and psychological well-being. *Gerontologist, 16*(4), 335–339.

Everard, K. M., Lach, H. W., Fisher, E. B., & Baum, M. C. (2000). Relationship of activity and social support to the functional health of older adults. *Journals of Gerontology Series B-Psychological Sciences and Social Sciences, 55*(4), S208–S212.

Hoyt, D. R., Kaiser, M. A., Peters, G. R., & Babchuk, N. (1980). Life satisfaction and activity theory: A multidimensional approach. *Journals of Gerontology, 35*(6), 935–941.

Lee, G. R., & Ishiikuntz, M. (1987). Social-interaction, loneliness, and emotional well-being among the elderly. *Research on Aging, 9*(4), 459–482.

Lemon, B. W., Bengtson V. L. & Peterson J. A. (1972). Exploration of activity theory of aging—Activity types and life satisfaction among in-movers to a retirement community. *Journals of Gerontology, 27*, 511–.

Litwin, H., & Shiovitz-Ezra, S. (2006). The association between activity and well-being in later-life: What really matters? *Ageing & Society, 26*, 225–242.

Longino, C. F., & Kart, C. S. (1982). Explicating activity theory: A formal replication. *Journals of Gerontology, 37*(6), 713–722.

Mollenkopf, H., Marcellini, F., Ruoppila, I., Flaschentrager, P., Gagliardi, C., & Spazzafumo, L. (1997). Outdoor mobility and social relationships of elderly people. *Archives of Gerontology and Geriatrics, 24*(3), 295–310.

Mollenkopf, H., Wahl, H., Baas, S., Rooij, R., Tacken, M., Marcellini, F., et al. (2002). Social, structural, and psychological perspectives on outdoor mobility of older people: Findings from the European Project "MOBILATE". *Gerontologist, 42*, 248–249.

Mollenkopf, H., Marcellini, F., Ruoppila, I., Szeman, Z., Tacken, M., & Wahl, H.-W. (2004). Social and behavioural science perspectives on out-of-home mobility in later life: Findings from the European project MOBILATE. *European Journal of Ageing, 1*(1), 45–53.

Mollenkopf, H., Hieber, A., & Wahl, H.-W. (2011). Continuity and change in older adults' perceptions of out-of-home mobility over ten years: a qualitative–quantitative approach. *Ageing and Society, 31*(05), 782–802.

Pinquart, M., & Sörensen, S. (2001). Gender differences in self-concept and psychological well-being in old age: A meta-analysis. *The Journals of Gerontology: Psychological Sciences, 56*(4), P195–P213.

Reitzes, D. C., Mutran, E. J., & Verrill, L. A. (1995). Activities and self-esteem: continuing the development of activity theory. *Research on Aging 17*, 260–77.

Ragheb, M. G., & Griffith, C. A. (1982). The contribution of leisure participation and leisure satisfaction to life satisfaction of older persons. *Journal of Leisure Research, 14*(4), 295–306.

Rowe, J. W., & Kahn, R. L. (1997). Successful aging. *Gerontologist, 37*(4), 433–440.

Schaie, K. W., & Gribbin, K. (1975). Adult Development and Aging. *Annual Review of Psychology, 26*(1), 65–96.

Schwanen, T., & Ziegler, F. (Eds.). (2011). Wellbeing, independence and mobility. *Ageing and Society, 31*(5).

Shoval, N., Auslander, G., Freytag, T., Landau, R., Oswald, F., Seidl, U., et al. (2008). The use of advanced tracking technologies for the analysis of mobility in Alzheimer's disease and related cognitive diseases. *BMC Geriatrics, 8*, 7.

Shoval, N., Auslander, G., Cohen, K., Isaacson, M., Landau, R., & Heinik, J. (2010). What can we learn about the mobility of the elderly in the GPS era? *Journal of Transportation Geography, 18*(5), 603–612.

Shoval, N., Wahl, H.-W., Auslander, G., Isaacson, M., Oswald, F., Edry, T., et al. (2011). Use of GPS technology for measuring timing and distance of out of home activity of elderly people with different levels of cognitive functioning. *Aging and Society., 31*(5), 849–869.

Strain, L. A., & Chappell, N. L. (1982). Confidants: Do They Make a Difference in Quality of Life? *Research on Aging, 4*(4), 479–502.

Van Willigen, M. (2000). Differential benefits of volunteering across the life course. *Journals of Gerontology Series B-Psychological Sciences and Social Sciences, 55*(5), S308–S318.

Vantilburg, T. (1992). Support networks before and after retirement. *Journal of Social and Personal Relationships, 9*(3), 433–445.

Veenhoven, R. (1996). Developments in satisfaction-research. *Social Indicators Research, 37*, 1–46.

Wahl, H.-W., Iwarsson, S., & Oswald, F. (in press). Aging well and the physical-spatial-technical environment: Toward an integrative model and a research agenda for the future. *The Gerontologist*.

Watson, D., Clark, L. A., & Tellegen, A. (1988). Development and validation of brief measures of positive and negative affect: The PANAS scales. *Journal of Personality and Social Psychology 47*, 1063–1070.

Chapter 12
The Changes of Disability-Free Life Expectancy and Inter-Generation Support for the Elderly in China: 2005–2010

Wei Guo

Abstract China, the world's most populous country, has entered a period of rapid aging. But less work has been done in systematically assessing the health status and intergenerational support resources of the elderly, a growing proportion of whom may be with inadequate government or family support. This study employs Sullivan method and SOCSIM micro-simulation analysis technique on the aggregate data of *China's 1 % National Population Sample Survey* in 2005, *the Sixth National Population Census* in 2010, and other aggregate census data in China since 1953 to estimate the changes of LE, DFLE, and the change of intergenerational support resources of Chinese elderly from gender and/or urban–rural perspectives. The results indicate that the proportion of DFLE in LE is increasing, the increasing speed of DFLE/LE is higher among rural/female elderly than that among urban/male elderly at all ages between 2005 and 2010, and the number of children of the elderly at lower ages keep decreasing, which indicate that they need more care resources in a near future. This study provides insight into imbuing community helping and pension institutions with gender and urban–rural differences in China.

Keywords Disability-free life expectancy · Intergenerational support · Sullivan method · Aging · China

12.1 Introduction

With the advanced development of Chinese economy and medical level, the fertility rate of the population has been declined continuously while the average life expectancy gets a sustained prolongation. These two factors aggravate the aging tendency in both relative and absolute quantity. It can be seen from the latest *Sixth National Population Census* in 2010, the population aged 60 and above in China is

W. Guo (✉)
Department of Social Work and Social Policy, School of Social and Behavioral Sciences, Nanjing University, Nanjing 210023, Jiangsu Province, People's Republic of China
e-mail: weiguo@nju.edu.cn

T. Samanta (ed.), *Cross-Cultural and Cross-Disciplinary Perspectives in Social Gerontology*, DOI 10.1007/978-981-10-1654-7_12

0.178 billion which accounts for the 13.26 % of the totality with the 0.119 billion people aged 65 and above (8.87 % of the totality). Compared with *the Fifth National Population Census* in 2000, the percentages increased, respectively, 2.93 % and 1.91 %.[1] It indicates that China has entered into the accelerating period of the aging population.

The economic status of the elderly, the level of the pension and the microscopic social condition will impact directly on the health of the elderly (Jiang et al. 2000; Li and Yang 2011; Tang 2005). In the research domain, on the one hand, the gender difference is a fine classification indicator in the study of the aging population, particularly the indicator of their health condition. The differences among males and females demonstrate in several aspects. In 2010, the average life expectancy of Chinese men is 72.38 years old whereas that of the women is 77.37; 4.99 years older than that of males. But the previous studies have revealed that the longer life expectancy (LE) of women does not equal to the better condition of their health, and on the contrary, the health condition of women is even worse than that of men (Liu 2001; Wang and Zeng 2001; Zeng et al. 2004). The LE is rising whereas the self-rated healthy life expectancy is declining (Qiao 2004), which indicates that the improving rate of the health condition is slower than that of the prolongation of the LE. Compared with men, the elderly women are in a comparatively inferior situation (Du and Li 2006). On the other hand, the rural population has unequal access to the rights of education, social security, consumption, etc., in comparison with the urban population, which is the result of the urban–rural dualism due to the Chinese unique household registration system (Zhang 2007). In that case, the widening gap among the rural and urban areas generates enlarging levels of the infrastructure, education and medical treatment that hinder the dual-mobility of people, goods, fund and information. That must accordingly bring about the significant variation in the aspect of the elder's health.

Under the circumstances of a weak Chinese medical security system and the influence of the traditional culture, many local governments carry out the "9073" supporting model[2] for the aged. As a result, the home-based care for the aged will remain the dominant model in the future. The modern home-based care services that China advocates now, which originated from industrialization and is supported by the social relations with the families, the society and the government as the subject liability, are essentially different from the traditional one (Ding 2013). The elderly in China have relied on their families to provide the daily cares for a long time, so, to some extent, the quality of life of the elderly has the most direct relation with the number of people in a family (Wang 2013). In China, over 90 % of aging population who have the care services at home get both living and emotional support mainly from family. Their relatives, particularly their children, play an important and main role in taking care of them (Liu and Lou 2012).

[1]2010 Main sixth national census data bulletin (No. 1), the National Bureau of Statistics of the People's Republic of China, April 28, 2011.

[2]About 90 % aged population cared in family, 7 % in community and 3 % in institutions.

According to the survey data by the Chinese Aging Scientific Research Center, the disabled aged[3] in China reached 10,843,000 in 2010 which accounts for the 6.25 % of the sum of the elderly, and the number will be up to 12,398,000 in 2015, the end of the "12th five-year plan" (Zhang et al. 2011). The challenges being faced now are a large amount of aging population and the developing trend of the core families, which were generated both from the demographic factors and housing tendency caused by the declination of the birth rates, acceleration of the population mobility and cultural transition, at the same time, the female's labour participation rate increases constantly, bringing about the shrinking of the resources the elderly can get from the family. All these contradict the requirements of care services' resources and time of the elderly with disabilities need.

Under such circumstances, the significant questions are: how to measure the health condition and its disparity among Chinese elderly and the changes of the most direct supporting resources (the average number of their children and the average number of their sons) of Chinese elderly from gender and/or rural–urban perspectives. So, based on the data from China's 1 % National Population Sample Survey in 2005, the Sixth National Population Census in 2010 and other past aggregate data of census in China, using Sullivan method featuring in using the simple data, dismissing the influence of the age structure and accordingly being compared easily and SOCSIM micro-simulation analysis technique, this paper calculated the disability-free life expectancy (DFLE) and its changing trend from gender and urban–rural perspectives, and also estimated the intergenerational support resources (the average number of their children and the average number of their sons) of Chinese elderly. These quantitative results would provide the empirical basis for establishing and refining the Chinese aging polices.

12.2 Data and Methods

12.2.1 Data Source

We used data from China's 1 % National Population Sample Survey in 2005 and the Sixth National Population Census in 2010. China's 1 % National Population Sample Survey in 2005 was the first time to investigate the health condition in a large-scale sample survey of population (the R13 item in the questionnaire was set to "health condition") in China. And the latest Sixth National Population Census in 2010 was also the first time in the population census to put such an indicator in (the R28 item in the questionnaire was "health condition" ⟨for people aged 60 and above to fill in⟩). Specifically, we used the mortality data by gender, age and rural–urban areas as the

[3]In accordance with internationally accepted standards, the disabled aged are those cannot dress, eat, bath, go to the toilet, carry out indoor sports, shop independently as the result of weakness, disability, illness, mental handicap, etc. (the aged loss of daily life self-care capacity).

basis of the life table, and used the health condition from gender, rural–urban and age perspective to calculate the disability-free life expectancy (DFLE).

What should be pointed out is that the health condition was divided into four categories in 2005 which were "health", "could basically guarantee a normal life and work", "couldn't work or live on yourself", "it depends" and we put the first and the second categories together as "could take care of themselves" and the last two categories as "could not take care of themselves".[4] In 2010, the self-rated health condition was "health", "basic health", "unhealthy, but could take care of themselves", "could not take care of themselves" four categories. In the analysis, we defined the first three as "could take care of themselves" and the last category as "could not take care of themselves". So, the self-rated health condition was divided into two categories. Though the objectivity and accuracy of the health condition may be influenced by the subjective assessment of the interviewee (the elderly and their family members), the classification criterion in this paper is comparatively objective on the ground that people can tell whether they can take care of themselves well. In addition, the highest group in 2005 survey's aggregated data was "the age of 95 and above" while that of the 2010 survey was "100 years of age and over". For the purposes of comparison, we adjusted the highest group of 2010's to "the age of 95 and above" also.

12.2.2 *Methods*

12.2.2.1 Sullivan Method

We used the life table technique to calculate the life expectancy (LE) of the elderly aged 60 and above varying in the genders and rural-city areas which provides the basis to apply the Sullivan method to calculate the disability-free life expectancy (Laditka and Wolf 1998; Sullivan 1971).

First, based on the reclassification of the health condition ("can take care of themselves" and "cannot take care of themselves"), we calculated the π_x which represents the age-ratio of people living free of disability over 60 years old (contain 60 years old):

$$\pi_x = \frac{\text{No. of people living free of disability at age x}}{\text{No. of people at age X}}$$

[4]Based on the instruction of questionnaire of the China's 1 % National Population Sample Survey in 2005, the term of "it depends" represents other situation the former three terms unable to include in. It contains the unstable health condition. So we put it into the category of "can not take care of themselves". Furthermore, the percentage of choosing this was only 1.47 %.

Then, estimate the survival person–years of those who live free of disability:

$$L_{x,\,DF} = L_x * \pi_x$$

The L_x represents the survival person–years in the life table.

Calculate the cumulative survival person–years free of disability $T_{x,\,DF}$ using the method of calculating T_x in the life table:

$$T_{x,\,DF} = \sum_{x}^{\omega} L_{x,\,DF};$$

Calculate the disability-free life expectancy $DFLE_x$:

$$DFLE_x = T_{x,\,DF}/l_x;$$

Finally, calculate the standard error of the $DFLE_x$:

$$S(DFLE_x) \approx \frac{1}{l_x^2} \sum_{x=0}^{\omega} L_x^2 \frac{\pi_x(1-\pi_x)}{N_x},$$

N_x represents the number of the aged free of the disability at age x.

12.2.2.2 SOCISM

We analysed the intergeneration support of the aged in the context of China, by the means of analysing the survival of children at different ages of the elderly.[5] According to the basic rate of lives, SOCSIM micro-simulation analysis technique was used to estimate the survival condition of children of the elderly at all ages (Hammel et al. 1976, 1981; Wachter 1987; Wachter et al. 1978). On the basis of the different fertility rates,death rates and marriage rates, we divided the monitoring period into 10 phases from 200 years before the founding of new China to get a population with stable structure which was used to emulate the relevant data after the establishment of the PRC.[6] The partition of each phase is in accordance with the mobility of fertility rates, death rates and marriage rates, and we selected the

[5]But it is difficult for us to obtain the information about surviving children according to data from previous censuses and sample surveys.

[6]It is known to all that any population will reach a stable level after experiencing a long period of steady fertility rates,death rates and marriage rates. So we emulated the data 200 years before the establishment of new China to obtain a stable population for SOCSIM micro-simulation analysis.

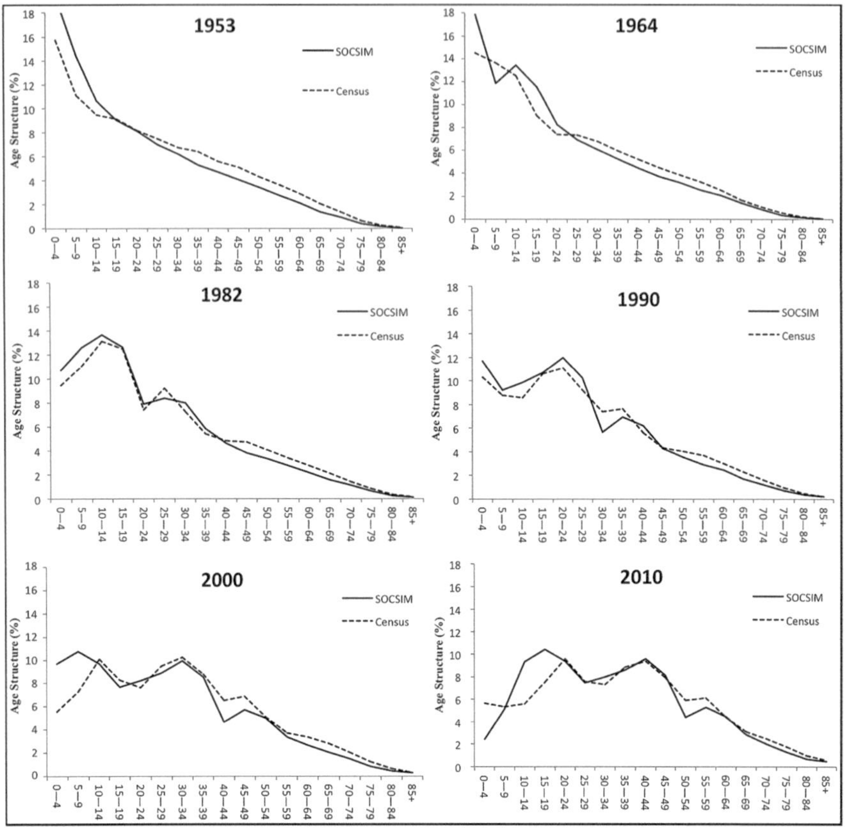

Fig. 12.1 SOCSIM simulation results and the results by the census: population age structure. *Data source:* calculations by aggregated data of previous censuses and SOCSIM micro-simulation data

average level or some specific year's data as the simulation parameters. The specific method to classify different phases and the settings of the monitoring parameters are shown in Fig. 12.1 in the chapter.

The results of the SOCSIM micro-simulation analysis cover information including the gender, the date of birth and death and the children in the monitoring periods. So we got the average survival condition of the children of male and female old people in 2005 and 2010 through the data matching. What's more, we also need to examine the results of the monitoring. Figure 12.1 shows the results of the SOCSIM micro-simulation analysis and the comparisons among six population censuses after the founding of new China since 1949. The consequences of the monitoring reach the expectations to use in the next study and analysis.

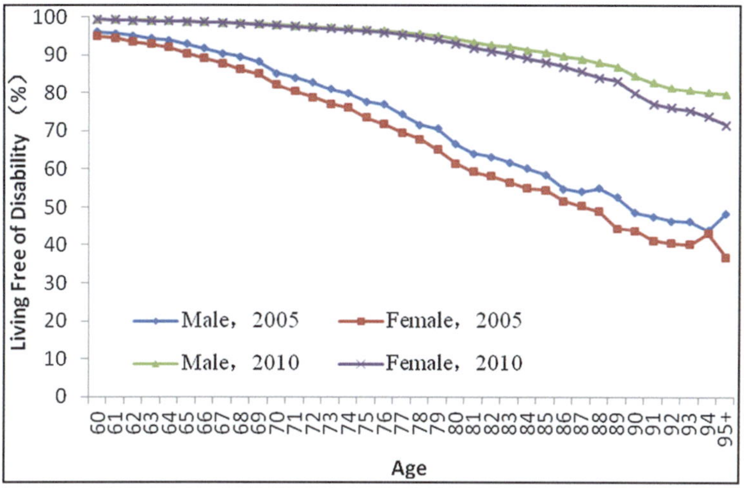

Fig. 12.2 Proportion of Chinese elderly living free of disability in 2005 and 2010 (male versus female). *Data source:* 1 % national population sample survey in 2005 and the sixth national population census in 2010 (similarly hereinafter)

12.3 Results

12.3.1 Trend of the Proportion of Chinese Elderly Living Free of Disability

Figures 12.2 and 12.3 illustrate the percentages of the aged free of disability by sexes and living areas based on the data from China's 1 % National Population Sample Survey in 2005 and the Sixth National Population Census in 2010. Overall, the percentage of the aged free of disability declined with age both in 2005 and in 2010. And, the older the age was, the bigger the descend range was. From the perspective of time changing, compared with that of data in 2005, the proportions grow at all ages and rising ranges grow with age too.

In the perspective of gender (shown in Fig. 12.2), the proportion of the male aged free of disability was higher than that of females and the disparity expanded with age in both 2005 and 2010.[7] The female's upward trend in the proportion of the elderly free of disability was higher than the male's. In the younger stage, this difference tended to increase with age, but after 80 years old it reduced with age. Therefore, compared with 2005, in 2010 the aging population differences among men and women showed a reduction in the proportion of daily self-care situation. The rise of the proportion of ability in taking care of themselves reflected the

[7]It should be noted that, due to the number of advanced aged population will reduce dramatically, the proportion of the DFLE fluctuated slightly in 2005.

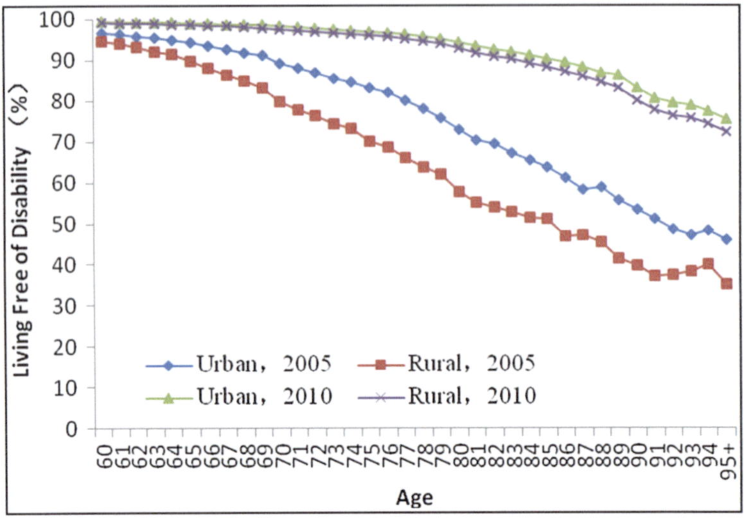

Fig. 12.3 Proportion of Chinese elderly living free of disability in 2005 and 2010 (urban versus rural)

improvements in the health condition of the aging population. However, the gender disparity of this proportion still remained which was more obvious in the senior group. Women lived longer lives, but experienced a higher and more severe disability burden than men. Thus, there were more demands of the daily caring for the elderly women in China.

Then, take a look from the rural–urban perspective (shown in Fig. 12.3). During 2005–2010, at all ages, the enhancing proportion of the elderly people free of disability living in rural areas was higher than that of urban citizens. Compared with 2005, the gap existing in the proportion of the aging population free of disability among rural and urban areas was narrowing. The difference did not visualize apparently until it showed up in groups aged 80 and older and the difference widened with age. The rising proportion of the self-care elderly population, especially those in rural areas, reflects the advanced development in social, economic, medical fields, and the narrowing trend in the gap between urban and rural areas. Meanwhile, the enhancing proportion of the self-care elderly population, especially in the elder elderly population, reflects the improvements in the health condition with not only improving the life expectancy, and that is to say the needs of daily caring will increase slowly. It reduces the burdens for family members to take care of the old people in a large scale in the China of which the family supporting is a main way in aging caring. But there still exists the differences among rural and urban areas in the proportion of self-care ability, which appears more evidently in the senior group. It is a big challenge for the refinements of rural pension system, especially for those poor ones in the Western and Central areas of China under which circumstances that most young workers migrate to make a living in cities and leave behind the old in the counties of the country alone (Guo and An 2013).

12.3.2 Changes of the Disparities of LE and DFLE among Chinese Elderly

Life expectancy (LE) is a standardized indicator which can be used to compare populations in different groups and different times, reflecting the health status of a particular group well. The disability-free life expectancy (DFLE), combining the advantages of the LE and the ratio of self-care ability, is a better measurement of the health status of the elderly population. According to the aforementioned method, the paper calculated the LE and the DELE both in rural and urban areas in 2005 and 2010 (detailed calculation results of the data can be seen in the appendix). Overall, the LE is decreasing with age. Compared to 2005's data, the LE increased at all ages in 2010 with the increasing range decrease with age.

Under the gender perspective, female's LE at all age groups were higher than men's, but the gap shrank with age. In 2010, the LE of 60-year-old male elderly population was 20.04 years and 23.14 years for women, 3.10 years older than men. The difference dropped to less than 1 year after 85 years old, and the gender difference in life expectancy fell to 0.29 year old in the group over 95 years old. In the younger stages, the increase of the women's LE was slightly higher than men's, but there were some fluctuations in the older group, embodied in the increase in male population aged between 80 and 88 years was higher than women.

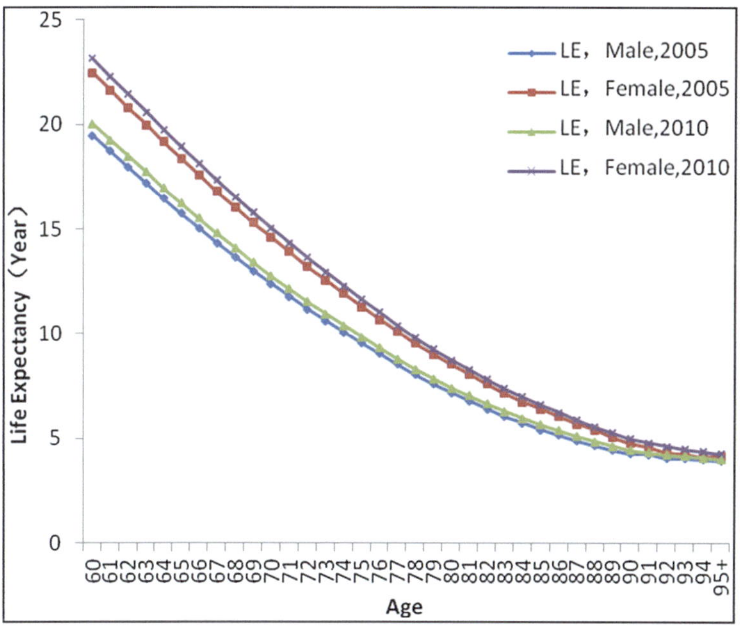

Fig. 12.4 Changes of life expectancy of Chinese elderly between 2005 and 2010 (male versus female)

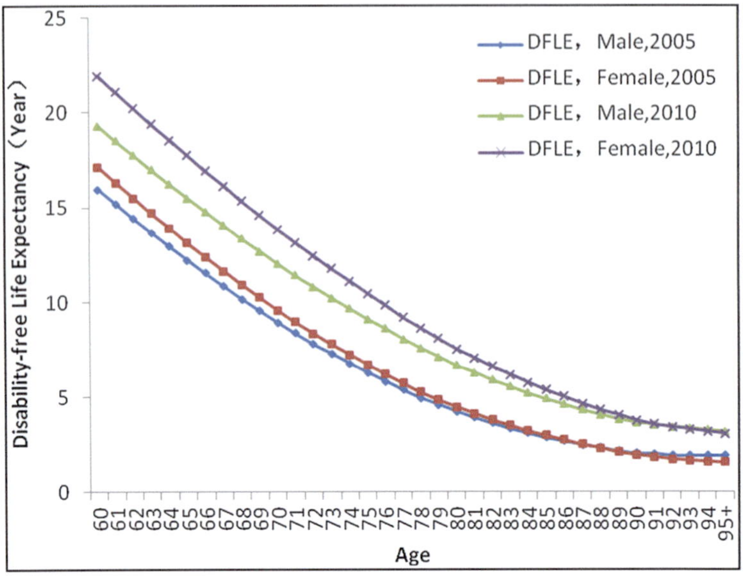

Fig. 12.5 Changes of disability-free life expectancy of Chinese elderly between 2005 and 2010 (male versus female)

The changing trend of the DFLE of different genders in 2005 and 2010 is shown in Fig. 12.5. Likewise, the DFLE reduced with age. The DFLE of the female elderly was higher than that of the male in younger stages while the situation was opposite in the senior groups. Compared with 2005, the LE was improved both of the men and the women in 2010 whereas the increasing rate was larger for women. The older the ages were, the lower the level increased. Thus, in comparison with 2005, the gap in DFLE of different genders widened in 2010 (Fig. 12.4).

Under the rural–urban perspective, the LE of urban elderly population at all ages was higher than that of rural elderly population, and the lower the age was, the more obvious the differences between urban and rural areas were. The gap reduced gradually in senior group, shown in Fig. 12.6. In 2005 and 2010, the LE in each age group improved and the increasing magnitude was relatively larger in the younger group. The improvement in the LE of the urban elderly population fluctuates more evidently with age. The increases of their LE are 0.87 years old and 0.11 years old, respectively, for 60-year-old group and 95-and-over-year-old group which showed a big disparity. However, there were comparatively slight variations in the LE of the rural elderly. The increases of the LE were 0.35 years old and 0.04 years old, respectively, for those two groups. At the same time, the magnitude of the increase of the urban elderly population was higher than that of rural areas.

Figure 12.7 shows the trend of DFLE of urban and rural elderly population from 2005 to 2010. Consistent with the differences of the LE among urban and rural populations of all age groups, each group's DFLE in the urban areas was higher than that in the rural areas, but the differences narrow with increasing age.

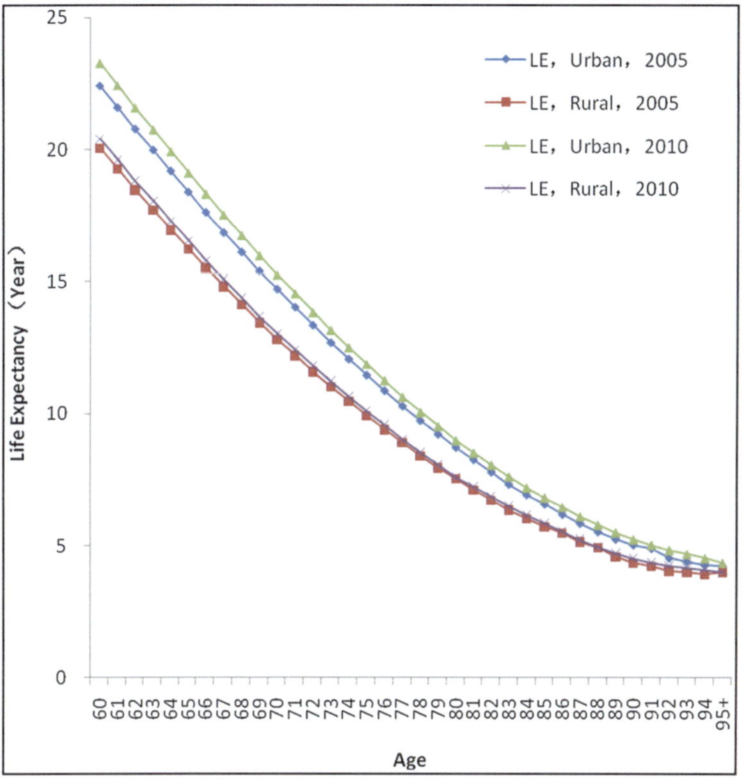

Fig. 12.6 Changes of life expectancy of Chinese elderly between 2005 and 2010 (urban versus rural)

Compared with 2005, the DFLE improved at all ages in 2010 and the increase dwindled with age. In 2005, the DFLE of the urban elderly population aged 60 was 18.63 years old which rose up to 22.28 years old in 2010, 3.65 years older than that of 2005. The DFLE of the rural ones aged 60 increased from 15.23 years old to 18.63 years old, 4.23 years older. The increase of the DFLE of the urban and rural population aged 95 and older was 1.33 and 1.51 years old, respectively. It can be concluded that the increasing magnitude of the rural DFLE was higher than the urban citizen's, and therefore, urban–rural differences of DFLE in 2010 narrowed in comparison with 2005.

After the above analysis, we found that the change of the LE and the DFLE displayed a similar increasing trend in both urban and rural areas. But the growth rate of the LE of the rural elderly population was lower than that of the urban elderly citizens, while the DFLE of the rural one is higher. The urban–rural differences characterized "one high and one low" will impact directly on the subsequent DFLE/LE of the aged.

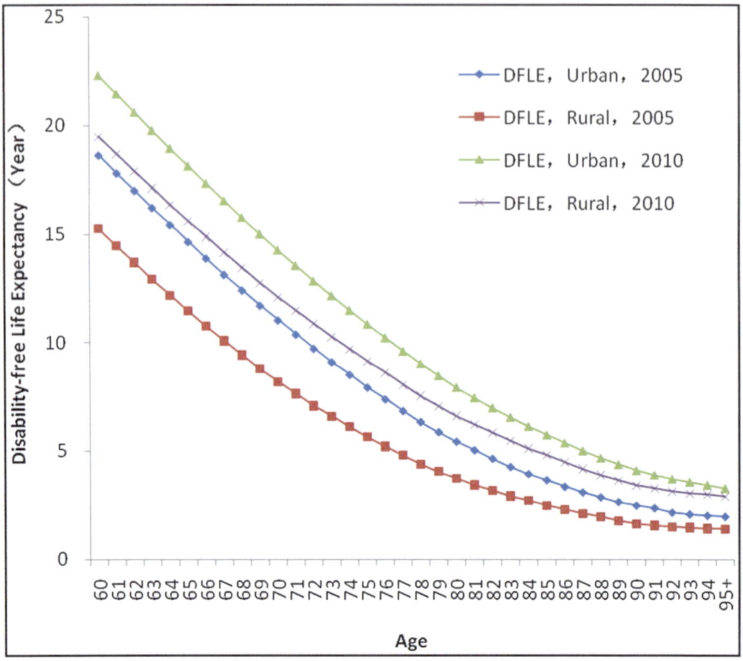

Fig. 12.7 Changes of disability-free life expectancy of Chinese elderly between 2005 and 2010 (urban versus rural)

12.3.3 Changes of the Disparity of DFLE/LE and Balanced Growth between LE and DFLE among Chinese Elderly

We analysed the changing trend of the LE and the DFLE of the elderly population in the previous section. The DFLE/LE reflects the relative speed of the increase of the LE and the DFLE. Figure 12.8 reveals the changing trend of the DFLE/LE under a gender perspective. Overall, the DFLE/LE decreased with the growing age which was in consistence with the trend of the LE. Compared with 2005, the DFLE/LE improved at all ages no matter men or women, and the rising magnitude also increased with age. The DFLE/LE of women increased larger than that of men, resulting in the shrinking gender gap of the DFLE/LE in 2010 compared to 2005.

Figure 12.9 illustrates the changing trend of the DFLE/LE under a rural–urban perspective. Compared with 2005, the DFLE/LE increased considerably by rural–urban in 2010. The older the age was, the larger the DFLE/LE increased and the magnitude of the improvement of the rural areas population was higher than that of the urban. In 2010, the disparity of the DFLE/LE among rural and urban areas tended to narrow in comparison with that in 2005. It can be concluded from the changing trend of the DFLE/LE of both rural and urban population that the expanding

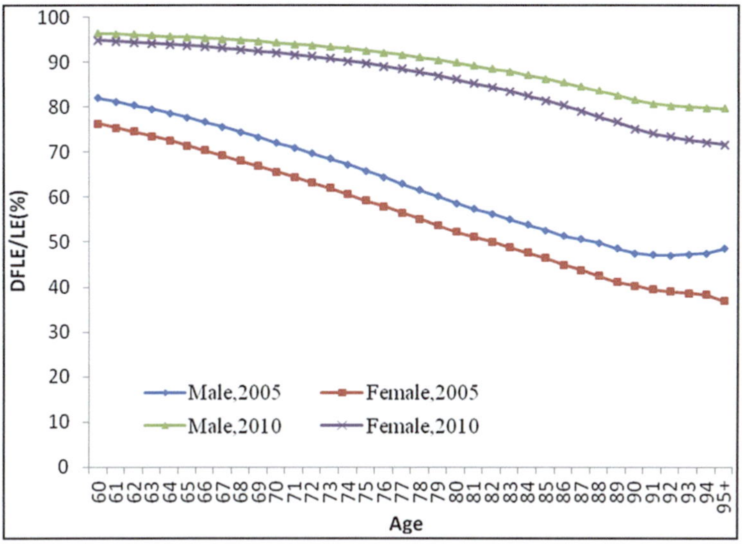

Fig. 12.8 Changes of the proportion of DFLE in LE of Chinese elderly between 2005 and 2010 (male versus female)

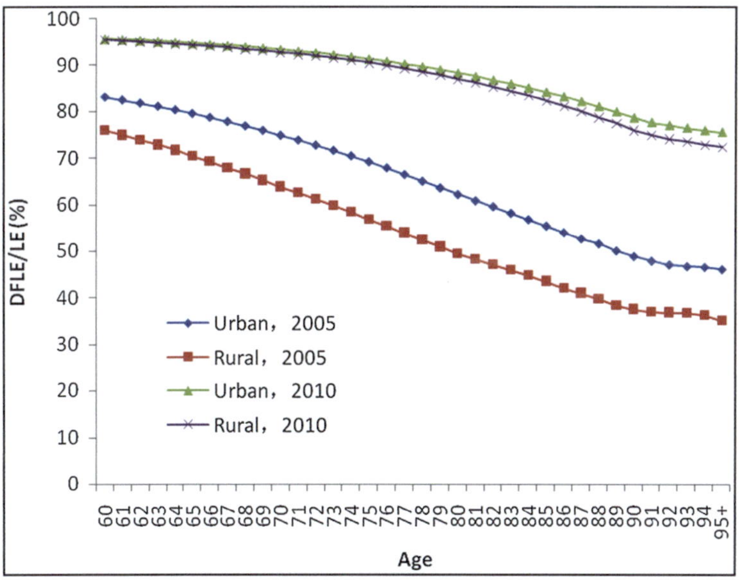

Fig. 12.9 Changes of the proportion of DFLE in LE of Chinese elderly between 2005 and 2010 (urban versus rural)

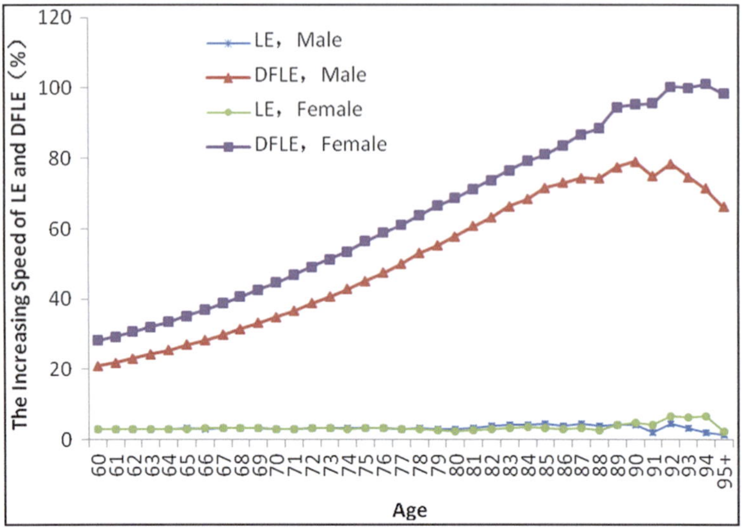

Fig. 12.10 The increasing speed of LE and DFLE of Chinese elderly between 2005 and 2010

DFLE/LE status accords to the theory of "a compression of morbidity" which is to say that not only the absolute amount of the LE gets improvement, but the DFLE/LE also enhances (Fries 1980, 1989, 2003). And it does not accord to the conclusion that the DFLE/LE of Chinese population is decreasing constantly drawn from some Chinese scholars previously (Du and Wu 2006; Zhang and Du 2009). It indicates that not only can the Chinese elderly live longer but the self-care span also gets longer even with the improved magnitude of the DFLE larger than that of the LE.

We analysed the trend of both LE and DFLE and draw a conclusion that the DFLE/LE is the reflection of comparative speed of the improvements of DFLE and LE. In the next section, we will expound the theory of "a compression of morbidity" by analysing growth proportionality of the LE and DFLE by comparing their growth rates. We used (LE2010–LE2005)/LE 2005 as the growth speed of LE and (DFLE2010–DFLE2005)/DFLE2005 as the growth speed of DFLE.

Figure 12.10 shows the growth rate of the LE and DFLE of the male and female elderly population in 2005 and 2010.[8] Overall, the growth rates of the DFLE were much faster than the LE and the increasing speeds accelerated with age. From the whole trend of the DFLE/LE of the rural and urban elderly populations, the decrease of the DFLE/LE with the growing age was consistent with that of the LE (shown in Fig. 12.11). As a whole, the increasing magnitude of the DFLE was much higher than that of LE. The unbalanced improvements of the DFLE and the LE under the gender and rural–urban perspectives illustrate the compression of

[8]The number of aged populations will decline considerably with age, thus the fluctuations of the number of the advanced aged were comparatively conspicuous.

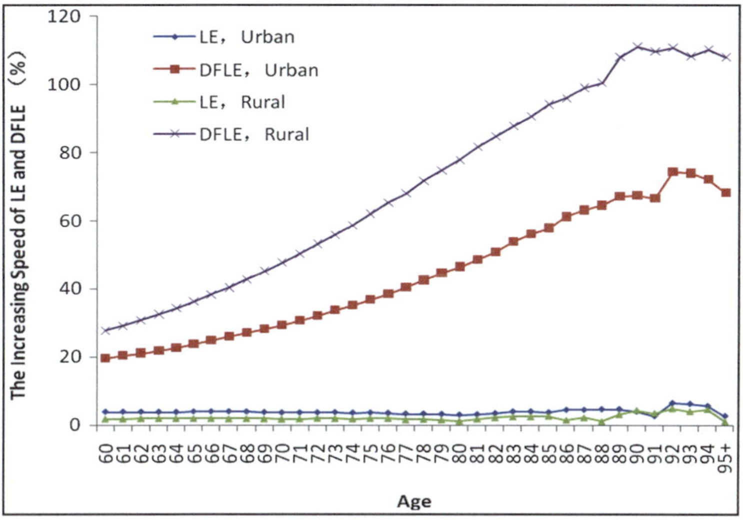

Fig. 12.11 The increasing speed of LE and DFLE of Chinese elderly between 2005 and 2010

morbidity of the Chinese old people. In addition, it is worth noticing that the LE increase of the urban population was higher than rural ones while the DFLE was on the contrary. So, the compression of morbidity is stronger for the rural elderly than the urban populations.

12.3.4 Changes of the Intergenerational Support Resources among Chinese Elderly

We have gained an overall understanding of the Chinese elderly population's health condition differentiation after analysing the heath statements and its changing trends both in the gender and rural–urban perspectives. Then, we analysed the surviving condition of the children who are the most direct life care supporting resources to the aged in the context of China. What should be declared here is that some old people, especially in the advanced ages, are cared by their grandchildren. With the limitation of data and methods, this article cannot take the condition of the grandchildren into consideration. We hope to there can be further study with the enrichment of the data and the advancement of the methods in a near future.

Figure 12.12 shows the most direct care resources of the elderly population in China, namely child survival condition of the elderly population. From the average number of surviving children, this indicator of women over 74 years old and men over 75 years old increased in 2010 compared to 2005. In contemporary Chinese society, especially in the context of the traditional concept of raising child for old age, survival status of the son is a more important indicator. So, we analysed the

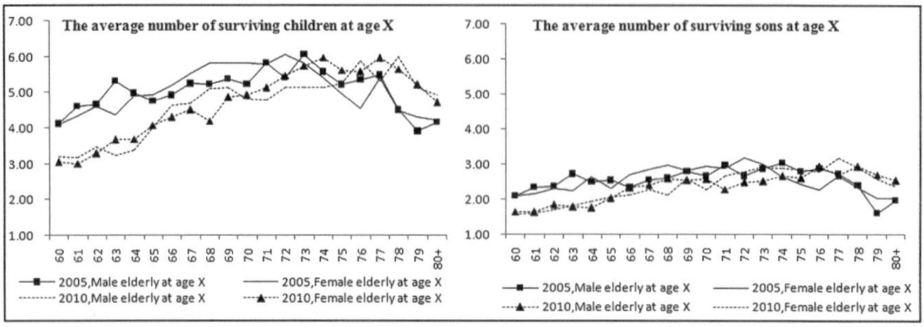

Fig. 12.12 The average number of surviving children and sons of Chinese elderly at age X. *Data source:* calculations by SOCSIM micro-simulation data

average number of surviving son of the aged and found that its changing trend is in accordance with that of surviving children. For the present situation, this is undoubtedly a good piece of news. The unhealthy time compressed in the remaining life of the elderly with the upward trend of direct care resources which indicates the care burden of the care for the Chinese elderly decreased in 2010 compared to that in 2005. However, there also exists a problem that the average number of surviving children in the younger group greater reduced considerably. With the constant decrease in the Chinese child-birth level, surviving children number will continue to decline for a long time. At the same time, these aging people gradually enter into the advanced elderly and their health status decline day by day while their care resources face a decreasing trend on the opposite. Such situation provides an implication with great significance in establishing the future of Chinese pension system.

12.4 Discussion and Conclusion

The rapid developments of Chinese society, economy, health and hygiene improve the overall health condition of the aged population. Overall, our findings are that, in present, with the constant LE prolongation, the percentage of the self-care ability increases and the improvement of DFLE is higher than LE, indicating that the health condition have achieved progress with the prolongation of LE and the needs of daily care for the old increase slower. It is undoubtedly a good piece of news for China where the family support is the main way of pension system. But there still remain a big gap in health among rural and urban areas. The conspicuous gaps of the LE and the DFLE between urban and rural areas and gender reflect the substantial difference in their health status.

Under the perspective of gender, firstly, the differences among men and women in the biological aspect result in the differences in the LE and the DFLE. In that case, the female elderly have a longer life expectancy whereas their health is at a weaker status. Second, the female proportion of morbidity is larger on the ground of the fact that, in the Chinese context, women are at a weaker position and they seldom seek medical treatment to solve comparatively minor or chronic illnesses. Third, the women's LE is longer than men's bringing about a high ratio of widow which has a negative influence on the psychological and emotional life for women elderly, and thus it may have negative effect on their health. Last but not the least, what we analysed now is the group who were born before 1950s featuring a patriarchal atmosphere which cause the condition that women's level of education and income is lower and they have more responsibility on the household. It also does harm to their health. Theses causes above can be attributed to the results that although Chinese female life expectancy is longer than men, the DFLE/LE is lower. So, the women elderly need more care. In view of this, although gender equality situation is or will be getting better in the future China, there are still some differences between them. So, we must realize that to provide better care services for women is one of the important issues need to be taken into consideration in reforming Chinese pension system.

Under the perspective of rural–urban areas, our study found that the differences between urban and rural life expectancy in 10 years increased slightly, while the DFLE showed a narrowing trend, and the degree of the morbidity compression of rural elderly population was bigger. The rapid improvements of the LE and the DFLE of the rural elderly, on the one hand, reflects the advanced development of society, economy and medical treatments and other aspects in recent years. On the other hand, the establishment and gradual refinement of medical insurance system change the former situation that rural populations have less access to the medical resources which hampered a lot in improving their health before. Furthermore, the LE and the DFLE of urban population are relatively high, so the marginal effect is relatively difficult to improve these two indicators under the circumstances of improving the standard of living and medical services. By contrast, the change of rural situation can bring a lot of marginal benefits. Thus, the degree of compression of the elderly population in rural areas is higher than that of the urban elderly population. Of course, the reduction of urban–rural gap cannot cover the disparity in an objective sense which still exists between the two groups. On the one hand, it remains the fact in China that the economic resources and health conditions in urban areas are better than rural area. On the other hand, the education and income levels of the elderly population are in a superior status which means they have stronger purchasing abilities and higher efficiency of utilization of the medical health services and they have more knowledge about keeping healthy. All these can reduce the incidence of diseases of the elderly population, especially chronic diseases. Besides, the accessibility of facilities and living environment surroundings also has great impacts on the daily living of the elderly population (Zhang and Du 2009). Facilities in urban areas are superior to those in rural areas. It reduces the difficulty of daily life, thus raising the possibility of maintaining the self-care ability of the elderly population in urban areas.

In view of this, with the rapid development of today's Chinese society, we must realize that compared to the city, the majority of rural areas, especially the western ones in context of the regional imbalances are in inferior position where the population's living standards, health status and lack of pension insurances and so on are still obstacles to the health of the elderly population. In terms of social policy settings, establishing and refining the corresponding pension system, improving the medical insurance system, strengthening the purchasing ability of health services, especially for the rural elderly population is the second issue that should be considered when reforming the current pension system in the background of aging China.

In addition, we find that the unhealthy life expectancy is shrinking for the elderly over 75 years whereas the average number of survival children and that of survival sons are increasing, which undoubtedly further reduce the pension burden at present. However, what can be concluded from our quantitative analysis, on the overall level, for younger elderly population, the unhealthy life expectancy in 2010 increased compared to 2005, which may predict the augmenting needs of health care in the future. Meanwhile, another Chinese reality is that the fertility level maintains decreasing greatly affected by the sustained and strict family planning policy. According to the latest data from the Sixth National Population Census in 2010, China's total fertility rate (TFR) was only about 1.4, and reached to 1.6 even after further amendment by professional demographers. And according to the monitoring results of SOCSIM micro-simulation analysis, we also found, compared with 2005, that the average number of survival children declined sharply in younger elderly group in 2010. And these younger elderly will enter into advanced age stage over time, bringing about an undoubted challenge for China's pension system in a near future. When we feel pleased for the improving health status of the elderly population, we cannot ignore the fact that the direct care resources are dwindling for the elderly, especially senior citizens in future. Considering the strengthening severe situation of advanced aging, the absolute amount of the unhealthy status of the aged population will be expanding, and their average number of surviving children, especially the average number of surviving son who are the main supporting resource will continue to reduce within the near future. Thus, the convergence of these two factors will lead to a more severe status of the pension system and urgent solutions must be carried out. And this may require us to go beyond the scope of the pension system itself, for example, think about the rationality of the adherence to the strict family planning policy(one-child policy).

Finally, it can be used as reference from practices of other countries, such as Germany, Japan, Singapore and many others which have put the unhealthy elderly, disabled ones in particular, into one of important public policy issues at the national level, implemented through technological, political, institutional, market-based method (Pan and Shuai 2012). And China's current pension system based mainly on family support take no consideration of the inner-group differences of the aging population. For instance, the average number of surviving children and surviving sons known as the most direct care resources show a descending trend based on the previous analysis. Under this background, the continuing existence of advanced aged women populations in rural areas, which are the intersection of the advanced aged women populations and the advanced aged populations in rural areas, should be paid more attention as vulnerable "sub-population" in political considerations, as well as their derivations which are old people without children and spouse.

Moreover, the health care problems of the disabled elderly will become more severe and prominent in China. So, at current stage of constructing the pension system in China featuring strengthening the family support while weakening institution responsibilities, it is such problems that how to develop a comprehensive community helping system covering both urban and rural areas, how to develop more adequate aging institutions, and at the same time how to enhance the aging institutions' ability to accept the special old people that should become the priorities and precautions for current Chinese government and related agencies.

Acknowledgments The author gratefully acknowledges the financial support from National Social Science Foundation of China under Grant No. 14CRK014. The Author is particularly grateful to his parents, Zhongwu Guo and Bifang Wang, who gave key advice about the implications of this analysis for policy of aging and pension reform in China.

Appendix

See Tables 12.1, 12.2, 12.3, 12.4, 12.5, 12.6, 12.7 and 12.8.

Table 12.1 Disability-free life expectancy of the Chinese elderly in 2005 (male)

Age x	l_x	L_x	π_x	$L_{x,\ DF}$	$T_{x,\ DF}$	$DFLE_x$	$S(DFLE_x)$	e_x	$DFLE_{x/e_x}$ (%)
60	84,494	83,984	0.96080	80,692	1,349,095	15.97	0.00012	19.48	81.95
65	78,636	77,864	0.92795	72,254	960,692	12.22	0.00014	15.74	77.64
70	69,818	68,664	0.85145	58,464	622,983	8.92	0.00017	12.39	72.05
75	57,212	55,717	0.77651	43,265	359,170	6.28	0.00026	9.54	65.82
80	41,404	39,673	0.66558	26,405	174,625	4.22	0.00049	7.19	58.62
85	24,594	22,948	0.58631	13,455	70,135	2.85	0.00138	5.42	52.62
90	10,952	9820	0.48603	4773	22,214	2.03	0.00697	4.27	47.45
95+	3481	13,724	0.48519	6659	6659	1.91	0.06902	3.94	48.52

Data source China 1 % national population sample survey in 2005

Table 12.2 Disability-free life expectancy of the Chinese elderly in 2005 (female)

Age x	l_x	L_x	π_x	$L_{x,\ DF}$	$T_{x,\ DF}$	$DFLE_x$	$S(DFLE_x)$	e_x	$DFLE_{x/e_x}$ (%)
60	90,650	90,308	0.94849	85,656	1,552,605	17.13	0.00000	22.45	76.28
65	86,662	86,150	0.90439	77,913	1,137,388	13.12	0.00000	18.36	71.47
70	80,341	79,496	0.82210	65,354	769,747	9.58	0.00000	14.59	65.65
75	70,368	69,112	0.73525	50,815	470,695	6.69	0.00000	11.28	59.28
80	55,817	54,106	0.61493	33,271	249,209	4.46	0.00000	8.54	52.30
85	37,699	35,804	0.54583	19,543	112,040	2.97	0.00000	6.40	46.42
90	20,403	18,741	0.43898	8227	39,354	1.93	0.00000	4.78	40.35
95+	7359	30,798	0.36947	11,379	11,379	1.55	0.00000	4.18	36.95

Data source China 1 % national population sample survey in 2005

Table 12.3 Disability-free life expectancy of the Chinese elderly in 2010 (male)

Age x	l_x	L_x	π_x	$L_{x,\ DF}$	$T_{x,\ DF}$	$DFLE_x$	$S(DFLE_x)$	e_x	$DFLE_{x/e_x}$ (%)
60	87,621	87,148	0.99337	86,570	1,692,762	19.32	0.00000	20.04	96.41
65	82,004	81,287	0.98767	80,285	1,271,443	15.50	0.00000	16.23	95.54
70	73,672	72,537	0.97894	71,010	886,593	12.03	0.00001	12.76	94.30
75	61,125	59,619	0.96642	57,617	556,826	9.11	0.00001	9.84	92.55
80	45,180	43,291	0.94113	40,742	300,537	6.65	0.00001	7.41	89.83
85	27,204	25,553	0.90732	23,185	133,179	4.90	0.00004	5.67	86.32
90	12,682	11,517	0.84433	9724	46,068	3.63	0.00017	4.46	81.50
95+	4300	17,174	0.79631	13,675	13,675	3.18	0.00151	3.99	79.63

Data source China's sixth national population census in 2010

Table 12.4 Disability-free life expectancy of the Chinese elderly in 2010 (female)

Age x	l_x	L_x	π_x	$L_{x,\ DF}$	$T_{x,\ DF}$	$DFLE_x$	$S(DFLE_x)$	e_x	$DFLE_{x/e_x}$ (%)
60	93,309	93,026	0.99313	92,387	2,048,047	21.95	0.00001	23.14	94.85
65	89,817	89,340	0.98750	88,224	1,593,703	17.74	0.00001	18.93	93.71
70	84,096	83,270	0.97801	81,439	1,164,537	13.85	0.00001	15.04	92.08
75	74,401	73,182	0.96296	70,471	777,729	10.45	0.00001	11.65	89.72
80	60,367	58,512	0.93013	54,424	454,450	7.53	0.00002	8.74	86.12
85	41,294	39,354	0.88052	34,652	222,126	5.38	0.00004	6.61	81.42
90	22,683	20,978	0.79888	16,759	85,449	3.77	0.00012	5.01	75.16
95+	8911	38,156	0.71637	27,334	27,334	3.07	0.00076	4.28	71.64

Data source China's sixth national population census in 2010

Table 12.5 Disability-free life expectancy of the Chinese elderly in 2005 (urban)

Age x	l_x	L_x	π_x	$L_{x,\ DF}$	$T_{x,\ DF}$	$DFLE_x$	$S(DFLE_x)$	e_x	$DFLE_{x/e_x}$ (%)
60	90,609	90,250	0.96639	87,217	1,687,968	18.63	0.00022	22.41	83.12
65	86,340	85,789	0.94316	80,912	1,263,139	14.63	0.00025	18.39	79.55
70	79,624	78,752	0.89180	70,231	877,066	11.02	0.00029	14.71	74.87
75	69,631	68,347	0.83136	56,821	551,272	7.92	0.00038	11.44	69.19
80	55,350	53,653	0.72766	39,041	300,007	5.42	0.00060	8.72	62.17
85	37,816	36,100	0.63775	23,023	137,626	3.64	0.00128	6.57	55.39
90	20,497	18,844	0.53343	10,052	50,302	2.45	0.00436	5.02	48.86
95+	8056	34,064	0.46071	15,693	15,693	1.95	0.02825	4.23	46.07

Data source China 1 % national population sample survey in 2005

Table 12.6 Disability-free life expectancy of the Chinese elderly in 2005 (rural)

Age x	l_x	L_x	π_x	$L_{x, DF}$	$T_{x, DF}$	$DFLE_x$	$S(DFLE_x)$	e_x	$DFLE_{x/e_x}$ (%)
60	85,057	84,579	0.94675	80,075	1,295,730	15.23	0.00000	20.04	76.00
65	79,602	78,885	0.89662	70,730	911,369	11.45	0.00000	16.23	70.53
70	71,309	70,213	0.79666	55,936	583,485	8.18	0.00000	12.81	63.88
75	59,098	57,666	0.70137	40,445	333,325	5.64	0.00000	9.91	56.90
80	43,569	41,880	0.57768	24,193	162,116	3.72	0.00000	7.52	49.47
85	26,711	25,002	0.51150	12,789	66,235	2.48	0.00000	5.69	43.55
90	12,874	11,660	0.39828	4644	20,912	1.62	0.00000	4.33	37.52
95+	4068	16,186	0.35145	5689	5689	1.40	0.00000	3.98	35.14

Data source China 1 % national population sample survey in 2005

Table 12.7 Disability-free life expectancy of the Chinese elderly in 2010 (urban)

Age x	l_x	L_x	π_x	$L_{x, DF}$	$T_{x, DF}$	$DFLE_x$	$S(DFLE_x)$	e_x	$DFLE_{x/e_x}$ (%)
60	93,262	92,966	0.99450	92,455	2,078,235	22.28	0.00001	23.28	95.73
65	89,627	89,136	0.99007	88,252	1,623,536	18.11	0.00001	19.11	94.77
70	83,830	83,015	0.98365	81,658	1,193,982	14.24	0.00001	15.25	93.41
75	74,324	73,108	0.97063	70,961	805,415	10.84	0.00002	11.86	91.39
80	60,405	58,619	0.94476	55380	479,251	7.93	0.00002	8.98	88.35
85	41,930	40,014	0.90289	36,129	240,765	5.74	0.00005	6.82	84.24
90	23,542	21,837	0.83260	18,181	96,697	4.11	0.00015	5.22	78.66
95+	9869	42,849	0.75515	32,357	32,357	3.28	0.00085	4.34	75.51

Data source China's sixth national population census in 2010

Table 12.8 Disability-free life expectancy of the Chinese elderly in 2010 (rural)

Age x	l_x	L_x	π_x	$L_{x, DF}$	$T_{x, DF}$	$DFLE_x$	$S(DFLE_x)$	e_x	$DFLE_{x/e_x}$ (%)
60	87,563	87,113	0.99226	86,439	1,704,635	19.47	0.00000	20.39	95.46
65	82,264	81,585	0.98576	80,423	1,283,560	15.60	0.00000	16.53	94.36
70	74,274	73,161	0.97448	71,294	897,620	12.09	0.00000	13.02	92.79
75	61,938	60,483	0.95997	58,062	566,300	9.14	0.00001	10.10	90.55
80	46,412	44,530	0.92811	41,329	307,194	6.62	0.00001	7.61	87.01
85	28,494	26,855	0.88337	23,723	137,098	4.81	0.00003	5.84	82.36
90	13,810	12,603	0.80201	10,108	47,356	3.43	0.00013	4.52	75.95
95+	4722	18,984	0.72408	13,746	13,746	2.91	0.00115	4.02	72.41

Data source China's sixth national population census in 2010

References

Ding, J. D. (2013). The home care service: Recognizing mistakes, rational principles and perfecting measures. *Journal of Renmin University of China, 2*, 20–26. [in Chinese].

Du, P., & Li, Q. (2006). Disability-free life expectancy of Chinese elderly and its change between 1994 and 2004. *Population Research, 5*, 9–16. [in Chinese].

Du, P., & Wu, C. (2006). Ability of daily life of the Chinese elderly: Status and change. *Population Research, 1*, 50–56. [in Chinese].

Fries, J. F. (1980). Aging, natural death and the compression of morbidity. *New England Journal of Medicine, 3*, 130–134.

Fries, J. F. (2003). Measuring and monitoring success in compressing morbidity. *Annals of Internal Medicine, 5*, 455–459.

Fries, J. F., Green, L. W., & Levine, S. (1989). Health promotion and the compression of morbidity. *The Lancet, 3*, 481–483.

Guo, W., & An, S. X. (2013). Unhealthy life expectancy and inter-generation support of Chinese elderly: A gender perspective. *Journal of Nanjing Agricultural University (Social Sciences Edition), 13*(6), 93–100. [in Chinese].

Hammel, E. A., Hutchinson, D. W., Wachter, K. W., Lundy, R. T., & Deuel, R. Z. (1976). *The SOCSIM demographic-sociological microsimulation program operating manual* (Vol. 27). Institute of International Studies Research Series Berkeley: University of California.

Jiang, J. M., Han, S. M., Sun, G. Q., & Zhang, C. X. (2000). Analysis of health of the urban elderly among different areas in China. *Chinese Journal of Health Statistics, 17*(1), 26–27. [in Chinese].

Laditka, S. B., & Wolf, D. A. (1998). New methods for analyzing active life expectancy. *Journal of Aging and Health, 10*(2), 214–241.

Li, Sh, & Yang, S. (2011). The impact of pension and inequality on health of older adults in China. *Chinese Journal of Population Science, 3*, 26–34. [in Chinese].

Liu, Y. Z. (2001). Concerned about gender issues among Chinese people of advanced age-project research introduction of factors influencing the health and long life of Chinese people of advanced age. *Collection of Women's Studies, 4*, 47–51. [in Chinese].

Liu, J., & Lou, W. Q. (2012). Improving the social support system for family caregivers of the disabled old in Shanghai. *Journal of East China Normal University (Philosophy and Social Sciences), 1*, 19–25. [in Chinese].

Pan, J. H., & Shuai, Y. L. (2012). Analysis of the disability rate and prevalence of elder population in China: Based on 6th national census. *Journal of Nanjing College for Population Programme Management, 4*, 3–6. [in Chinese].

Qiao, X. C. (2004). *Health changes of the elderly in china from* 1987, 1992 *to* 2000, *by gender and age* (*extended abstract*). Paper Presented in 2004 PAA.

Sullivan, D. F. (1971). A single index of mortality and morbidity. *HSMHA Health Reports, 86*(4), 347–354.

Tang, Z. (2005). Study on the active life expectancy of the elderly and its longitudinal transition in Beijing. *Chinese Journal of Epidemiology, 12*, 939–942. [in Chinese].

Wachter, K. W. (1987). Microsimulation of household cycles. In J. Bongaarts, T. K. Burch, & K. W. Wachter (Eds.), *Family demography: Methods and their applications*. Oxford: Clarendon Press.

Wachter, K. W., Hammel, E. A., & Laslett, P. (1978). *Statistical studies of historical social structure*. New York: Academic Press.

Wang, L. L. (2013). A study on the demand, supply and utilization of home-based care services for the elderly based on the theory of "services chain". *Population Journal, 2*, 49–59. [in Chinese].

Wang, S. X., & Zeng, X. X. (2001). Gender differences in the activities of daily living of the Chinese oldest old. *Chinese Journal of Population Science, S1*, 1–5. [in Chinese].

Zeng, Y., Liu, Y Zh, Xiao, Zh Y, & Zhang, C. Y. (2004). The socio-demographic and healthy status of the oldest-old in China. *Chinese Journal of Population Science, S1*, 4–13. [in Chinese].
Zhang, L. Y. (2007). The analysis and policy research on consumption difference between urban and rural areas under the dualistic structure. *China Soft Science, 2*, 11–17. [in Chinese].
Zhang, W. J., & Du, P. (2009). Regional disparity in changes of healthy life expectancy of Chinese elderly-expansion or compression. *Population Research, 5*, 68–76. [in Chinese].
Zhang, K. D., Sun, L. J., Mou, X. Y., Wang, H. T., & Li, M. Z. (2011). Research on situation of urban and rural disabled elderly. *Disability Research, 2*, 11–16. [in Chinese].

Chapter 13
Elderly Inpatient Care Utilization and Financing in India: Is There a Gender Difference?

William Joe, Abhishek Kumar and Udaya Shankar Mishra

Abstract This paper examines utilization and financing of elderly inpatient care in India with a specific objective to unravel intersecting inequalities in distressed financing. For this purpose, we draw insights from three broad theoretical perspectives namely, unitary, collective and intersectionality framework that are well-recognized in the literature on intra-household welfare and its distribution. The analysis reveals significant gender differentials as well as income gradient in both utilization and financing of elderly inpatient care in India. The econometric inferences are consistent with the theoretical inferences and unravel significant disadvantages for elderly females in accessing distressed financing. It is also observed that households are more likely to resort to means such as borrowings, asset selling and contributions from friends and relatives to support hospitalization of elderly male. Given the disutility associated with distressed financing, even richer households are less likely to incur debt for inpatient care of female elderly. In view of such intricacies, this paper argues for universal healthcare coverage and calls for policymaking to strengthen community-based elderly health care.

Keywords Gender differentials · Intersectionality framework · Elderly healthcare utilization · Hospitalization · Out-of-pocket expenditure · India

W. Joe (✉) · A. Kumar
Institute of Economic Growth (IEG), New Delhi, Delhi, India
e-mail: william@iegindia.org

A. Kumar
e-mail: abhishekkmr48@gmail.com

U.S. Mishra
Centre for Development Studies, Trivandrum, Kerala, India
e-mail: mishra@cds.ac.in

T. Samanta (ed.), *Cross-Cultural and Cross-Disciplinary Perspectives in Social Gerontology*, DOI 10.1007/978-981-10-1654-7_13

13.1 Introduction

Out-of-pocket (OOP) healthcare payments account for around 70 % of the total healthcare expenditure in India (Government of India 2009). Predominance of such regressive mode of healthcare financing has severe impoverishing effect on numerous households across the country (O'Donnell et al. 2008; Van Doorslaer et al. 2007; Flores et al. 2008; Garg and Karan 2009; Selvaraj and Karan 2009; Berman et al. 2010; Gupta and Joe 2013). In particular, financing of inpatient care via distressed means such as borrowings and sale of household assets are known to jeopardize customary living standards of several households (Krishna 2004, 2006). Although events such as hospitalization can be unpredictable at times but households with elderly members are exposed to an elevated risk of such adversities. In fact, with health system deficiencies and rising medical costs, such households have to decide on two key issues: (a) whether to support hospitalization of the elderly member and (b) whether to resort to distressed financing to cover hospitalization costs, if necessary? Clearly, these decisions are influenced by various determinants and an improved understanding can be critical for ensuring healthy aging in India.

Against this backdrop, this paper aims to examine elderly inpatient care utilization in India with a specific emphasis to comprehend gender-related inequalities in healthcare financing. Out-of-pocket health expenditure is met through households' own income and savings or via distressed financing which usually refers to borrowing (with or without interest) and selling of household assets for healthcare utilization (Kruk et al. 2009; Leive and Xu 2008; Dilip and Duggal 2002; Binnendijk et al. 2012). However, it is plausible that for several households even the process of arranging contributions or interest-free borrowings could be rather cumbersome and may involve repayments. Therefore, it is desirable to adopt a broader definition of 'distressed financing' to include: (i) borrowings (with or without interest), (ii) sale of assets and (iii) contributions from friends and relatives (with or without repayments). Clearly, composition of distressed financing can significantly affect household wellbeing hence households are expected to be cautious about decisions regarding hospitalization and financing of elderly inpatient care.

Household decision-making can be complex and is well recognized in the economic literature on intra-household welfare and its distribution (Iversen 2002; Haddad et al. 1997; Agarwal 1997a, b; Sen 1990). It is noted that household decisions are contingent both upon basic material foundations including household income, wealth and educational background as well as sociocultural aspects such as gender, intra-household relations, and other forms of bargaining power. Decision-making for elderly health care is also influenced by contextual factors such as widespread income deprivations, profound gender inequities and dysfunctional public health system. Under such circumstances, several elderly members may perhaps forego treatment and face the worst of the adverse consequences. Nevertheless, households availing inpatient care face a basic trade-off regarding how much to invest and whether to use distressed means for this purpose. Given such relevance, it is important to understand intricacies related to household behaviour towards elderly inpatient care.

To comprehend empirical regularities in the data, the paper draws insights from three broad theoretical perspectives, namely unitary, collective and intersectionality framework (Becker 1965; Sen 1990; Agarwal 1997a, b; Haddad et al. 1997; Iversen 2002; Crenshaw 1991; Springer et al. 2012). As such, the unitary approach is among the earliest attempts to understand intra-household allocations and assumes that household resources and preferences are smoothly aggregated to yield the household welfare function. The collective models emphasize that each household member may have varied preferences and that intra-household resource allocations are largely determined by the bargaining power of the household members. Interestingly, bargaining among members can be based on either perfect information (cooperative model) or asymmetric information (non-cooperative model). In the cooperative model, household members agree to the rules of the bargaining while enforcing the contract, whereas in the latter the rules should be self-enforcing at each stage to achieve any agreed allocation. A natural extension of such bargaining framework is to recognize each elderly member as one bearing multiple identities (gender or relation) and owning different resources (occupation, skills or wealth). The interactions of these identities can provide vital insights regarding intra-household resource allocations for elderly care. In fact, socially dominant identities (such as gender advantage for men) are known to have influential leveraging effect while deciding upon hospitalization care utilization and financing (Sen and Iyer 2012). These issues form the crux of the intersectional framework whereby multiple axes of power such as gender, family relations and economic status can be combined to discern the impact on health, healthcare utilization and its financing.

These alternative frameworks display some continuity as they gradually shift focus from households as the unit of analysis to individuality of household members. Hence, from an empirical perspective, it is rather desirable to approach the analysis and interpret the results in a unified framework. In fact, Haddad et al. (1997: p. 11) note that "*that much observed [household] behavior lends itself to interpretations consistent with more than one household model*". Besides they also observe that it is difficult to distinguish between "*endowment*" effects and "*bargaining*" effects and suggest that "*an interaction between the unitary approach (in which the identification of endowments is central) and collective models may prove fruitful*". Nevertheless, for a consistent verification of the theory it is critical to approach the analysis with full information regarding the severity of illness and also about details regarding intra-household distribution and ownership of resources. But due to data limitations on this front, we could only discuss certain broad inferences regarding influence of key socioeconomic factors on elderly inpatient care utilization and its financing in India. With this backdrop, the rest of the paper is organized as follows. Section 13.2 presents a brief review of alternative theoretical approaches with specific focus on models of intra-household allocation. Section 13.3 informs regarding the data and methods used for the analysis whereas Sect. 13.4 presents the major results. Section 13.5 concludes with a discussion of the key findings.

13.2 Inequalities in Elderly Hospitalization and Financing: Some Theory

Conventional economic theory, particularly household economics, identifies households as a unit of production and consumption and assumes that households act '*as one*'—hence the *unitary approach*—to maximize household welfare or utility (Haddad et al. 1997). It is supposed that household members are able to combine their set of preferences (work, time, leisure, etc.) to generate utility for the household (Becker 1965). The reason for aggregating utility of different household members is based on a rather implicit assumption that households can easily arrive at consensus (Samuelson 1956). For example, it is plausible that due to shared thinking or social norms, households can be expected to provide adequate health care to the elderly members. In an unconstrained situation this implies that household will continue to invest in elderly health until household marginal utility diminishes to zero. Interestingly, a number of possibilities emerge once we introduce household constraints. For instance, households face various time and resource constraints and aim to maximize the overall household welfare subject to these constraints. When subjected to constraints, it is plausible that households may assign differing weights to each household member including the elderly. Under such circumstances, the household utility from treatment of elderly member can be expressed as the weighted sum of the net utility of all members. Such an aggregation of preferences and household utility is referred to as the cooperative model of the household (Sen 1966).

Although in the cooperative models 'individuality' of the household members is explicit but this model works effectively only when members display 'symmetric sympathy' towards the elderly or when the household utility is significantly dependent on the utility of the altruistic member in the household (Haddad et al. 1997). The altruistic member is expected to have a favourable attitude towards elderly health care and should display effective control over utility function of non-conforming household members such that the household's utility function is synonymous with the altruists' function. For example, the altruistic member can be the head of the household who can seek cooperation from all other members for elderly health care and can bargain effectively with other non-conforming members. This aspect of preference aggregation in the cooperative model is brought out by the popular 'rotten-kid theorem' (Becker 1974a, b, 1981). In general, the acts of the altruistic member can be based on two broad motivations. For instance, the altruistic head may aim to equalize health (utility) of the elderly member with that of other members. Alternatively, the head may prefer to invest the limited resources more efficiently to enhance overall household welfare even after accounting for any welfare losses due to ill health of the elderly. In the literature, the former type of concern can be regarded as the *equity concern* whereas the latter reflects the *efficiency concern* (Haddad et al. 1997). This trade-off between equity and efficiency concerns is an important aspect of elderly healthcare utilization in resource poor settings. The equity concern demands that all the elderly who are in need are provided with adequate health care. On the contrary, the efficiency concern may introduce systematic inequalities in elderly health care whereby the disadvantaged elderly may be neglected in provisioning and financing of health care. Although this form of cooperative model is useful to aggregate preferences of household members but divergent behaviour of household members (including counter-response of the altruist) may be an issue in aggregation. As an alternative, a broad class of collective models is therefore employed to emphasize upon the individuality of the household members while aggregating household preference. In fact, unitary or cooperative approach can be regarded as special cases of collective models whereby all members display or share common preferences.

This brief review suggests that the issue of elderly healthcare utilization can be well-represented through collective models of household members whereby it is plausible that (a) benefits from cooperation by investing in elderly health is always greater for *all* the members; (b) benefits from cooperation at least improves household welfare in the Pareto sense, and; (c) benefits from cooperation gained either through bargaining or through threat of non-cooperation at least enhances elderly welfare. These mutually exclusive grouping has far-reaching consequences on the distribution of elderly healthcare utilization and its financing. For instance, if households belong to the first category then we can expect that all the elderly members of the household who are in need of health care receive timely and adequate care. This also implies that there will not be any intra-household inequities (such as gender inequity) in inpatient care utilization or its financing. However, if households belong to the second category then we may expect endowment-based inequalities across elderly members in utilization and also its financing. This is because households may invest in elderly health as long as it does not lead to a decline in overall household welfare. Importantly, if households belong to the third type of classification then we can expect both endowment-based as well as bargaining-based inequalities in healthcare utilization and financing. The bargaining power of an elderly can emanate from a more general social identity such as gender or biological relationship with members to more specific sources of bargaining power such as asset ownership or bequests.

At this juncture, it is important to draw attention towards the intersectional nature of these endowments and the sources of bargaining power. In particular, intersections of social and material factors can create complex synergy of disadvantages and engender health inequities (McGibbon and McPherson 2011; Crenshaw 1991; Agarwal 1997a, b; Davis 2008; Hankivsky and Christoffersen 2008; Hankivsky 2012; Sen and Iyer 2012; Springer et al. 2012). For example, the household response may differ between elderly with assets and those who are *assetless*. Hence by intersecting multiple axes of power it is feasible to discern power relations across heterogeneous social groups and to inform policy regarding the unique multiplicative effect of identifying people with more than one social identity (Kelly 2009). These intersections across multiple identities apparently creates population subgroups whereby the extremes, i.e. the least advantaged and the most advantaged, can at times easily be identified but the large section of middle groups lack any consistent hierarchy (Griffin et al. 2002). Also, because of varying relative impact of intersectional groups there is no a priori reason to presume superiority of any specific dimension. It also implies that positioning of middle groups is determined by *leveraging* of the dominant characteristics to compensate for disadvantageous features (Sen and Iyer 2012). In fact, leveraging is a critical feature while analyzing elderly healthcare utilization and emerges as a natural extension within the collective models. The leveraging power can be drawn from multiple sources (such as gender, financial status, relationship to the head of the household) and can be often discerned empirically.

As a synthesis of the theoretical approach, we now present a simple stylized model to describe the link between elderly healthcare utilization and household welfare. In particular, the emphasis here is to illustrate how inequalities get manifested in elderly healthcare utilization and how it can determine use of various sources of financing. This stylized model assumes that all elderly have identical illness and severity and face identical cost functions. In Fig. 12.1, healthcare expenditure (X) is depicted on the horizontal axis and marginal cost of treatment and household marginal utility from elderly health care is presented on the vertical axis. The health production function $H = f(X)$ is also well-behaved and offers diminishing returns to healthcare investments. The figure shows the marginal cost (MC) and marginal utility (MU) curves for households seeking hospitalization care for the elderly. The downward sloping marginal utility curve signifies that each additional unit spent on health expenditure leads to diminishing marginal increments in total health. The positive slope of marginal cost indicates increasing incremental costs for marginal health improvements at higher levels.

The intersection of marginal cost and marginal utility curve shows optimal level of investment (X^*) that maximizes household utility by investing in the elderly health care. At this level of investment, the elderly gains a health level denoted H, based on the production function $H = f(X)$. It may be noted that the level H may or may not be the optimal level because it does not refer to a situation where marginal cost is equated to marginal health benefits. In other words, we assume that household utility function and elderly health benefit function differ from each other. In particular, the former is a function of household members' endowments and bargaining power whereas the latter is a function of health expenditure. As shown in Fig. 13.1a, households may also choose to invest until a level such as $X_{\max}$ is reached where household marginal utility approaches zero (i.e. total health gains are the maximum). However, in resource-poor settings it is more likely that households may have incomes X_Y which is lower than the required level of optimal expenditure X^*. Under such circumstances, households may either forego treatment or may try to avail free or limited care. Also, several households may attempt to raise resources through distressed means such as borrowing and asset sales to achieve desirable level of health expenditure (X^*) that would equate household marginal utility with marginal cost. In such cases, the total health expenditure (OX^*) can be expressed as a sum of own resources (OX_Y) and distressed means (X_YX^*).

Figure 13.1a also depicts a case where there are no systematic inequalities in elderly inpatient care utilization and that household invests similar resources (including distressed means). However, households may have member-specific marginal utility schedule which depends on social, material as well as other intrinsic characteristics associated with the elderly. This implies that households seeking to optimize healthcare investments will differentiate between elderly members and this can lead to systematic inequities and discrimination in health, health care and healthcare financing. Such a situation is depicted in Fig. 13.1b. Here, we consider case of two elderly members—a male and a female—needing inpatient care. The marginal utility schedule for the members is given by $\mathrm{MU_M}$ and $\mathrm{MU_F}$, respectively. It is assumed that households draw greater (lower) utility from investment in health care of the male (female) member. The optimality conditions suggest that households invest greater resources in health care of males than compared to females thus engendering inequalities in health and health care. The absolute inequality in health outcomes can be given by $H_\mathrm{M} - H_\mathrm{F}$ whereas the relative inequality can be expressed as $H_\mathrm{M}/H_\mathrm{F}$.

It may be noted that such health inequities may not be entirely to gender bias but rather due to possible endowment benefits (such as regular income) associated with the elderly man. This endowment enters the utility function of the household and leads to a dominant marginal utility schedule for the man. Under such circumstances, it is even likely that the amount of distressed financing for male will be higher than female, which here is associated with endowment effects. Alternatively, it is possible that female member may have certain advantage in terms of asset or property ownership and that may influence the household marginal utility. In Fig. 13.1c, the dotted curve $\mathrm{MU_F}$, represents such a situation where female member can have effective bargaining power within the household to ensure higher household utility than endowment effects associated with males. This outweighs the utility schedule for males and offers a relative advantage for female in terms of health.

Figure 13.1d, depicts the case when an elderly member is recognized to possess multiple identities. For example, the elderly can be male or female, employed or unemployed, head of the household or a distant relative. Under such circumstances, the marginal utility of household varies as per the individual and hence health, healthcare utilization and healthcare financing can be largely determined by effective leveraging power of one identity over others. The intersectional approach also suggests that the gap in health, health care and healthcare financing is the widest among those who are advantaged across all identities and those who are disadvantaged in every sense. Although, sometimes it is relatively easier to identify extreme groups (such as elderly male, employed and head of the household) but at times the hierarchy may be obscured. Figure 13.1d depicts such a case based on intersectionality framework whereby member-specific household marginal utility schedule is largely determined by both endowment and leveraging effect of the group identity. In the figure, non-working female's health is assumed to be the most disadvantaged (H_FO) whereas the employed male (H_MW) has better health status. In particular, the figure shows that the household marginal utility for the female member can be dependent on leveraging effect in terms of income, working status or relationship to household members. Thus households can accordingly allocate greater or smaller resources for her health care and this shift is denoted by the

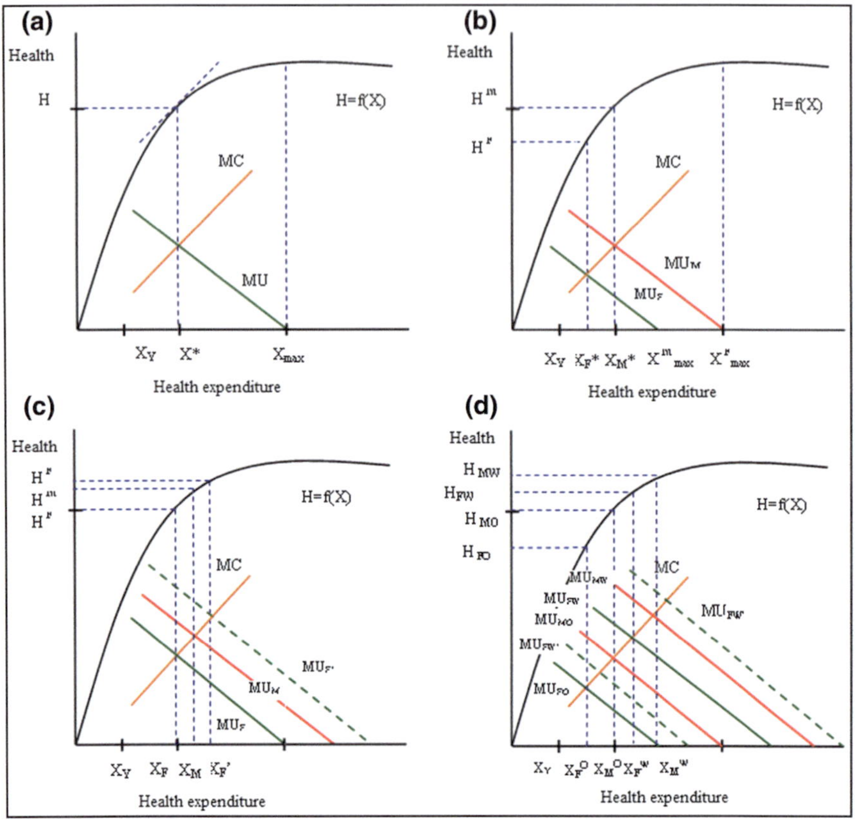

Fig. 13.1 A stylized model of elderly health care and household utility

dotted lines for MU_{FW}. These issues are now empirically explored in the context of elderly inpatient care in India.

13.3 Data and Methods

Nationally representative data from Morbidity and Health Care Survey (60th round) of India is used for the analysis (NSSO 2006). This survey was conducted in 2004 by the National Sample Survey Organisation (NSSO), Ministry of Statistics and Programme Implementation, Government of India and covered a sample of over 73 thousand households (around 47 thousand in rural areas and 26 thousand in urban areas). The survey adopted a two-stage stratified design, with census villages and urban blocks as the first-stage units for the rural and urban areas, respectively, and households as the second-stage units. The survey was conducted during

January–June 2004 and was split up into two rounds of three months each (NSSO 2006). Key information on aspects of hospitalization (inpatient) and ambulatory (outpatient) care for a reference period of 365 days and 15 days, respectively, was collected through this survey. This paper analyses the data on about 34,831 elderly (aged 60 and above) persons with focus on inpatient care utilization during past one year from the date of survey.

The survey provides information about out-of-pocket healthcare expenditures on inpatient care and its financing at a household level. Specifically, the sources of financing are classified as follows: (a) own income and savings, (b) borrowings (with or without interests), (c) resources from sale of assets and (d) contributions or assistance from friends and relatives (with or without repayment). As such, the survey manual does not clarify whether the borrowings are with or without interest (and contributions are with or without repayments) hence it is assumed that the former may be mostly with repayment and interests whereas the latter reflects the part which need not involve either interest or repayment. Distressed financing is defined to comprise of: (i) borrowings (with or without interest), (ii) sale of assets and (iii) contributions from friends and relatives (with or without repayments).

Huge gender disadvantages for females have been a fundamental developmental and policy concern for India. This study therefore hypothesizes that there are gender differentials in elderly inpatient care utilization and its financing. However, before presenting econometric inferences regarding the same, we present certain broad characteristics of elderly inpatient care and its sources of financing. In particular, we use concentration curve (CC) and concentration index (CI) to discern the socioeconomic gradient in use of elderly inpatient care (Wagstaff et al. 1991). The CC plots the cumulative proportions of the population (ranked by socioeconomic status) on the *x-axis* against the cumulative proportions of elderly using inpatient care on *y-axis*. For interpretative purposes, if the incidence of hospitalization is evenly distributed across socio-economic spectrum then the concentration curve would coincide with the diagonal (line of equality); if it is concentrated among higher (lower) income groups, then CC lies below (above) the diagonal; and farther the CC from the diagonal, greater would be the incidence among the poorer households. The CI could be derived from the CC and is defined as twice the area between the CC and the diagonal. Following Erreygers (2009), the CI could be computed as follows:

$$C(h) = \frac{2}{n^2 \mu_h} \sum_{i=1}^{n} z_i h_i; \quad z_i = \frac{n+1}{2} - \rho_i,$$

where i: (i = 1, 2, …, n) represents a given population; ρ_i is the socio-economic rank of the person with the best well-off individual ranked first and the least well-off ranked last. In the case of ties, each member of the tied group is assigned the average rank of the group. The CI ranges between +1 and −1 with zero depicting equal distribution across socio-economic spectrum and large positive (negative) values indicating greater incidence of a particular source of financing among the richer (poorer) households.

Econometric analysis of association between gender and sources of healthcare financing involves at least two steps: one, to decide whether to utilize care and second, to decide about the sources of financing (Asfaw et al. 2007). The first decision itself depends on whether households select to visit a doctor and follow the advice for hospitalization whereas the second decision is contingent upon hospitalization and also depends on a combination of social, economic and intra-household factors. If the first decision is biassed then any analysis of sources of financing without adjusting for such selection bias can yield biassed and inconsistent estimates. Therefore, we employ Heckman probit model for the analysis (Heckman 1979; van de Ven and van Praag 1981; Greene 2003). The model captures the two-stage decision-making process for healthcare financing and presents estimates adjusted for selection bias. We also use this method to infer the relationship of prominent socio-economic correlates with private sector hospitalization and also the magnitude of inpatient care expenditure.

Estimation of probit selection model requires instrumental variables that are expected to influence elderly hospitalization but have no direct influence on decisions regarding distressed financing. In our analysis, we identify age, education and mobility status of the elderly as along with availability of tap water and number of elderly members in the household as instruments that directly affect hospitalization but are less likely to influence distress financing. As such, elderly with much higher age, better education and poor mobility are more likely to avail hospitalization care but this may not directly influence financing decisions. More number of elderly members in the household implies increased probability of witnessing hospitalization but it is not correlated with source of financing. The variable of access to tap water is used as a proxy to indicate availability of hospital care in the vicinity but is unrelated to financing decisions.

The analysis is adjusted for information on key socio-economic and demographic variables. Since in Indian context it is difficult to collect reliable income data (NSSO 2006), therefore information on household consumption is used to proxy the socio-economic status of the households. For analytical purposes, we classify households in the upper two quintiles of monthly per capita household consumption expenditure as richer households and the rest as poorer households. Gender and elderly members' relation to household head are used to examine their association with the sources of financing. Also, the analysis classifies households into three broad social groups, namely Scheduled Caste and Scheduled Tribe (SC and ST) and others. The SC and ST households are socially and economically vulnerable population subgroups that are accorded special status by the Constitution of India. Two important non-communicable diseases, namely cancer and cardiovascular diseases (CVD) are considered to examine their associations with distressed financing. Age of the hospitalized person, medical expenditure and financial protection (insurance) information are used to adjust the probability of accessing various sources of financing across individuals.

An important objective of the paper is to unravel how gender related differences tend to interact with endowment and bargaining power to yield differing outcomes for elderly bearing multiple identities. After a theoretical review of intersectionality, it is now expected that elderly with endowment and bargaining advantage will fare better than those without any endowment or bargaining power. In this context, family relations are considered as an integral part of this analysis therefore aspects such as status of economic dependence and source of financial support for the elderly are also included. The analysis views family as a group of people related by blood or marriage, who live together and financially support each other. While adjusting for these covariates, the aim of the analysis is to examine how various intersectional groups are advantaged (or disadvantaged) in terms of distressed financing. These groups are formed by intersecting three prominent determinants viz. poverty (poor and non-poor), gender (male and female) and relation with household head (self head or others). The intersection of these determinants yields eight mutually exclusive subgroups, namely: (1) poorer-female-head (PFH); (2) poorer-female-other (PFO); (3) richer-female-head (RFH); (4) richer-female-others (RFO); (5) poorer-male-head (PMH); (6) poorer-male-others (PMO); (7) richer-male-head (RMH); (8) richer-male-others (RMO).

Finally, it may be noted that the analysis is restricted to only those households which have reported hospitalization of only one elderly person. This is because the survey obtains source of finance information at the aggregate household level and inclusion of households with more than one hospitalized persons would disallow any inference regarding association of sources of financing with prominent factors such as gender and age of the hospitalized person. Altogether, we have information regarding 34,831 elderly persons of which 4970 have been hospitalized during the reference period of last 365 days. However, when we restrict our analysis to only cases with only one hospitalisation then it yields a sample of 33,785 elderly with 4391 hospitalization cases. All the analysis is performed in statistical software Stata 12 (StataCorp 2007) and the estimates are presented using recommended household weights to reflect survey design (NSSO 2006).

13.4 Results

This section reports certain broad descriptive patterns observed in the context of elderly inpatient care utilization and its financing in India. As shown in Table 13.1, inpatient care utilization is higher among males (6.8 %) than females (5.4 %). Moreover, the gap in utilization is greater in urban areas. Also, higher utilization is noted among the richer households (8.7 %) as compared to the poorer ones (4.6 %). Leveraging on account of being the head of the household is also significant. On average higher number of elderly (7 %) who are head of household use inpatient care as compared to elderly who are not the head of their household. Comparing the groups formed by intersecting gender and wealth, the table shows that elderly male belonging to the richer group enjoy a distinct advantage whereas elderly female belonging to the poorer group are the least advantaged. While comparing the groups formed by intersection of gender, household headship and income class it is observed that the elderly male (head of household) in upper wealth group have a

Table 13.1 Percentage elderly reporting inpatient care utilization, India 2004

Groups	Rural	Urban	All India
Male (M)	5.8	9.6	6.8
Female (F)	4.7	7.7	5.5
Poorer 60 % households (P)	4.4	6.2	4.6
Richer 40 % households (R)	8.1	9.5	8.8
Head of the household (H)	6.1	9.7	7.0
Household member (O)	4.5	7.4	5.2
Poorer male (PM)	5.0	6.9	5.2
Poorer female (PF)	3.8	5.7	4.0
Richer male (RM)	8.7	10.5	9.5
Richer female (RF)	7.5	8.5	8.0
Richer male and household head (RMH)	9.0	10.7	9.8
Richer female and household head (RFH)	8.7	9.8	9.1
Poorer male and household member (PMO)	4.1	3.4	4.0
Poorer female and household member (PFO)	3.7	5.1	3.9
Poorer male and household head (PMH)	5.2	7.8	5.4
Poorer female and household head (PFH)	4.0	7.5	4.6
Richer male and household member (RMO)	6.1	9.8	8.0
Richer female and household member (RFO)	7.1	8.2	7.6
All	5.4	8.6	6.1

clear advantage and female (not head of household) belonging to the lower income group are the most disadvantaged. Observing the trend for rural and urban areas it can be observed that utilization of inpatient care and ranking of group also varies.

There is a significant income gradient in utilization of elderly inpatient care in India. This can be discerned through the concentration curves that exhibit income-related inequalities in utilization of inpatient care. As shown in Fig. 13.2, all these concentration curves are below the line of equity indicating that hospitalization is concentrated among the richer sections of the society. The concentration curve for females falls below the concentration curve for males signifying that the magnitude of such income-related inequality is higher for females. A comparative glance further informs that these income-related inequalities are higher in rural areas. Estimates of income-related inequality in elderly hospitalization are obtained from the concentration index. These index values (and 95 % confidence intervals) are as follows: Rural male 0.164 [0.122–0.206], Rural female 0.234 [0.185–0.283], urban male 0.074 [0.029–0.118] and Urban female 0.108 [0.058–0.157]. All the concentration indices are positive and significant thus confirming greater inpatient care among richer sections. Also, the non-overlapping confidence intervals for rural areas indicate significant gender-related difference in distribution of hospitalization across socioeconomic rank.

In Table 13.2 we report the results of the probit selection model to understand the association of hospitalization with various socioeconomic correlates of

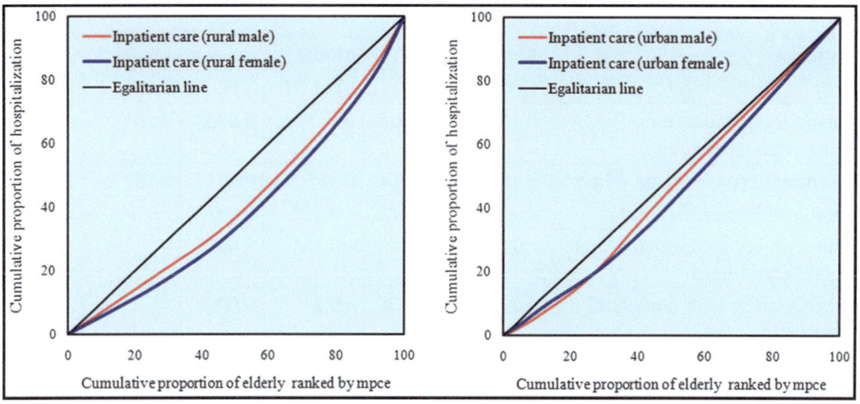

Fig. 13.2 Concentration curve for elderly inpatient care, India 2004

Table 13.2 Probit selection model results for elderly hospitalization: selection equation

Variables	Model 1		Model 2		Model 3	
	Coef.	S.E	Coef.	S.E	Coef.	S.E
Richer male and household head (RMH) (Reference)						
Richer female and household head (RFH)	–	–	–	–	−0.02*	*0.0013*
Poorer female and household member (PFO)	–	–	–	–	−0.43*	*0.0011*
Poorer male and household member (PMO)	–	–	–	–	−0.51*	*0.0015*
Poorer female and household head (PFH)	–	–	–	–	−0.28*	*0.0015*
Poorer male and household head (PMH)	–	–	–	–	−0.21*	*0.0008*
Richer female and household member (RFO)	–	–	–	–	−0.21*	*0.0011*
Richer male and household member (RMO)	–	–	–	–	−0.3*	*0.0017*
Poorer female (PF) (Reference)						
Poorer male (PM)	–	–	0.02*	*0.0009*	–	–
Richer female (RF)	–	–	0.23*	*0.0008*	–	–
Richer male (RM)	–	–	0.23*	*0.0010*	–	–
Female (F)	−0.01*	*0.0008*	–	–	–	–
Poorer 60 % households (P)	−0.22*	*0.0006*	–	–	–	–
Rural	−0.05*	*0.0007*	−0.05*	*0.0007*	−0.06*	*0.0007*
Scheduled caste	−0.02*	*0.0008*	−0.02*	*0.0008*	−0.02*	*0.0008*
Scheduled tribe	−0.15*	*0.0013*	−0.15*	*0.0013*	−0.15*	*0.0013*
Financial dependence on spouse	0.22*	*0.0011*	0.22*	*0.0012*	0.23*	*0.0012*

(continued)

Table 13.2 (continued)

Variables	Model 1		Model 2		Model 3	
	Coef.	S.E	Coef.	S.E	Coef.	S.E
Financial dependence on children/grandchildren	0.15*	*0.0007*	0.15*	*0.0007*	0.16*	*0.0007*
Financial dependence on others	0.15*	*0.0014*	0.15*	*0.0014*	0.23*	*0.0014*
Spouse of household head	−0.22*	*0.0012*	−0.22*	*0.0012*	−0.04*	*0.0012*
Father (in-law)/Mother (in-law) of household head	−0.16*	*0.0053*	−0.16*	*0.0053*	–	–
Other relation with household head	−0.2*	*0.0008*	−0.2*	*0.0008*	–	–
Number of elderly in household	0***	*0.0006*	0***	*0.0006*	0.01*	*0.0006*
Logarithm of age of elderly	0.42*	*0.0030*	0.41*	*0.0030*	0.44*	*0.0030*
Education—up to primary education	0.23*	*0.0007*	0.23*	*0.0007*	0.22*	*0.0007*
Education—up to secondary education	0.15*	*0.0009*	0.15*	*0.0009*	0.15*	*0.0009*
Education—higher education	0.13*	*0.0013*	0.13*	*0.0013*	0.13*	*0.0013*
Physical mobility restricted	0.58*	*0.0008*	0.58*	*0.0008*	0.58*	*0.0008*
Availability of tap water for drinking	0.11*	*0.0006*	0.11*	*0.0006*	0.11*	*0.0006*
Rho	−0.156*		−0.148*		−0.149*	

Source Authors using NSSO (2006)
*Indicates significance at 1%

household and the elderly person. In order to demonstrate intersectional inequalities, we use three different models. Model 1 does not include any interactions of key variables (gender, household income and members' relationship with head), model 2 interacts gender and household income status, whereas model 3 interacts gender, household income status and relationship to household head. In Model 1, the variables that are turning out to be significant and in the expected directions are elderly female, lower wealth, rural area, schedule caste, schedule tribe, financially dependent on spouse, financially dependent on grand children or children; other relatives are financially dependent, spouse of head, father-mother-in laws of head, other relative of head, total number of elderly in household, age, education and clean water. The chances of an elderly female of being hospitalized are lower than an elderly male, indicating that elderly females are less likely to be hospitalized. Also, income and caste based inequalities are apparent as elderly belonging to lower income group and those from scheduled caste and scheduled tribe community are less likely to have received inpatient care in the given one-year reference period. Elderly persons from rural areas are also less likely to be hospitalized. The chances of hospitalization are higher if the elderly is financially dependent on spouse, children or grandchildren. Compared to an elderly who is the household head, the chances of a non-head elderly to receive inpatient care is much lower. The instrumental variables used in the selection model also behave in expected direction. For example, higher is the age, greater are the chances of being hospitalized. Elderly reporting poor physical mobility are more likely to utilize inpatient care. Also chances of elderly utilizing inpatient care increases with education level. The availability of clean water is a proxy for health facilities in the vicinity and it also has a favourable impact on elderly inpatient care utilization.

In model 2 gender and income variables are interacted to examine whether income advantage can outweigh gender disadvantage for elderly females in utilization of inpatient care. The inference here conforms with the theoretical premise presented earlier that endowments are important aspect mediating gender-related healthcare utilization in India. For instance, there is evidence that elderly from richer households would have greater probability of hospitalization whereas within same income groups we witness greater utilization by males than females. Thus endowment effects as well as leveraging are apparent in utilization of elderly inpatient care in India. However, leveraging effect can be complex and is expected to be associated with factors such as relationship of the concerned elderly with the household head. This is demonstrated by further interacting gender and household income status with information on household headship and members' relationship with the head. For convenience we have created eight subgroups based on intersections of gender (male, female), household income (poorer, richer) and relationship (household head, other relation). For analytical purposes, the base category here is *elderly male being head of household and belonging to upper class*. It is observed that compared to the elderly in the reference category all other elderly persons have less chances of receiving inpatient care. Also, it is apparent that elderly males and females from poorer households have the lowest likelihood of hospitalization. The influence of household headship is apparent because even in lower income categories, household head are more likely to have received inpatient care. The spouse of household head also has much better chances of hospitalization compared to other relatives. Thus it is emerging that within households, the elderly person's relationship with the household head is an important determining factor in matters of inpatient care utilization.

Table 13.3 presents the marginal effect of different socio-economic variables on utilization of private facility by elderly which can also be considered as a proxy for higher healthcare expenditure. Here again three models have been formulated for examining the marginal effects of various socioeconomic variables on private sector hospitalization for elderly. The dependent variable utilization of private facility is subject to selection equation for hospitalization and assumes two values either zero (not hospitalized) or one (hospitalized). The significant and negative rho values indicate that there is a possible selection bias as the errors of the selection and outcome equation are negatively correlated. This also suggests that unobservable factors that may make hospitalization more likely are perhaps less likely to lead to private sector hospitalization. The coefficients of the different variables in the selection equation are consistent with the findings on gender differences in hospitalization. In fact, it is observed that elderly females have about 1.5 % lower chances of accessing private healthcare facilities than compared to elderly males. Elderly from lower income households have 14 % less chances of using a private

Table 13.3 Probit selection model results for private sector hospitalization and hospitalization expenditure for elderly: marginal effects from outcome equation

Variables	Model 1		Model 2		Model 3	
	dy/d*x*	S.E.	dy/d*x*	S.E.	dy/d*x*	S.E.
Richer male and household head (RMH) (Reference)						
Richer female and household head (RFH)	–	–	–	–	0.016*	*0.0011*
Poorer female and household member (PFO)	–	–	–	–	−0.132*	*0.0014*
Poorer male and household member (PMO)	–	–	–	–	−0.101*	*0.0018*
Poorer female and household head (PFH)	–	–	–	–	−0.184*	*0.0017*
Poorer male and household head (PMH)	–	–	–	–	−0.139*	*0.0010*
Richer female and household member (RFO)	–	–	–	–	0.039*	*0.0009*
Richer male and household member (RMO)	–	–	–	–	0.064*	*0.0014*
Poorer female (PF) (Reference)		–		*0.0008*	–	–
Poorer male (PM)	–	–	0.032*	*0.0012*	–	–
Richer female (RF)	–	–	0.155*	*0.0013*	–	–
Richer male (RM)	–	*0.0007*	0.154*	–	–	–
Female (F)	−0.015*	*0.0010*	–	–	–	–
Poorer 60 % households (P)	−0.145*	*0.0006*	–	*0.0006*	–	*0.0006*
Rural	−0.001	*0.0010*	−0.002*	*0.0010*	−0.002*	*0.0010*
Scheduled caste	−0.165*	*0.0016*	−0.167*	*0.0016*	−0.166*	*0.0016*
Scheduled tribe	−0.155*	*0.0011*	−0.158*	*0.0011*	−0.158*	*0.0011*
Financial dependence on spouse	−0.025*	*0.0006*	−0.027*	*0.0006*	−0.028*	*0.0006*
Financial dependence on children/grandchildren	−0.031*	*0.0012*	−0.032*	*0.0012*	−0.034*	*0.0013*
Financial dependence on others	0.076*	*0.0009*	0.075*	*0.0009*	0.062*	*0.0009*
Spouse of household head	0.063*	*0.0040*	0.063*	*0.0042*	0.029*	*0.0011*
Father (in-law)/Mother (in-law) of household head	0.142*	*0.0007*	0.142*	*0.0007*	–	–
Other relation with household head	0.039*	–	0.04*	–	–	–

Source Authors using NSSO (2006)
Note dy/d*x* represents marginal effects for a discrete change in the explanatory variables
*Indicates significance at 1%

facility, implying that endowments are an important aspect in determining elderly care. The chances of elderly belonging to schedule caste and schedule tribe being treated at a private facility are 16 and 15 % lower than the remaining population. In case the elderly is financially dependent on the spouse, than the chances of using a private facility are 2.5 % lower. The chances of using private facility are 3 % lower in case the elderly is financially dependent on grandchildren or children. Interestingly, spouse of the household head has 6.3 % higher chances of using a private facility. The financial dependency of other relatives implies that the chances of elderly using the private facility are 7 % more. In case the head of the household has father, mother, and father-in-law and mother-in law than the chances of elderly being admitted to a private hospital are 14 % higher. In case there are other relatives than the chances of elderly using a private facility are 3.9 % lower. This implies those elderly members who are financially supported by immediate family also have better chances of utilizing care from private sector. The results based on interaction of gender and income informs that elderly male from lower income households have 3 % higher chances of using private healthcare facility compared to his female counterpart. However, the impact of economic wellbeing on private sector care utilization is much stronger as both male and female from better income households. The intersections further confirm the influence of income on deciding upon sector of inpatient care utilization thus restricting the choice between sectors and healthcare providers for several poor households. However, even within the low income households, the female disadvantage is apparent as males are relatively more likely to use private sector care.

Table 13.4 depicts the incidence of various sources of financing for inpatient care utilization of elderly members. Overall, 31 % households incurred borrowings

Table 13.4 Sources of financing used by households for elderly inpatient care, India 2004

Groups	Own income/ savings	Borrowings	Sale of assets	Contributions from friends and relatives
Male (M)	84.1	34.8	4.8	20.3
Female (F)	85.8	25.1	5.2	20.6
Poorer 60 % households (P)	82.1	39.1	6.1	22.2
Richer 40 % households (R)	87.4	23.0	4.0	18.8
Head of the household (H)	81.9	31.8	5.9	23.7
Household member (O)	89.4	28.6	3.6	15.4
Poorer male (PM)	80.6	43.1	6.9	23.4
Poorer female (PF)	84.0	33.5	4.9	20.7
Richer male (RM)	87.5	27.0	2.9	17.4
Richer female (RF)	87.2	18.2	5.4	20.5
Richer male and household head (RMH)	87.5	26.8	3.0	17.9
Richer female and household head (RFH)	71.4	15.3	12.5	32.5
Poorer male and household member (PMO)	82.7	49.9	4.0	12.3

(continued)

Table 13.4 (continued)

Groups	Own income/ savings	Borrowings	Sale of assets	Contributions from friends and relatives
Poorer female and household member (PFO)	86.8	34.4	4.8	15.9
Poorer male and household head (PMH)	80.3	42.2	7.3	24.8
Poorer female and household head (PFH)	73.2	29.9	5.4	39.3
Richer male and household member (RMO)	86.7	28.6	2.2	13.3
Richer female and household member (RFO)	93.4	19.4	2.6	15.9
All	84.9	30.5	5.0	20.4

Source Authors using NSSO (2006)

to finance elderly hospitalization whereas 20 % households have used contributions from friends and relatives for inpatient care. Around 5 % households had to sell household assets to support hospitalization of elderly member. In Table 13.2, the emphasis is to highlight gender, wealth and household headship related differentials in financing the hospitalization services among the elderly. In this regard, certain systematic differences are apparent. Households incur greater borrowings for males or for the household head. Incidence of borrowings is relatively lower for female household members whereas financing through sale of assets and contributions is slightly more for females. Elderly belonging to the poorer households (bottom 60 % in terms of monthly per capita consumption expenditure) are more dependent on distressed financing mechanisms. Around 39, 6 and 22 % of elderly from poorer households have healthcare financed through borrowings, sale of assets and contributions respectively as compared to 23, 4 and 19 % in the richer group.

Further, if the hospitalized elderly happens to be the head of the household then there is a higher chance of using distressed financing options; borrowing (32 %), sale of assets (6 %) and contribution from friends and relatives (24 %). Comparing the groups formed by intersecting gender and wealth, the table shows that poorer households are more willing to borrow resources for healthcare of elderly male (43 %) whereas it is much lower for elderly female (33 %). Similar gender differential is observed even among higher income households. Finally, we compare groups by intersecting three dimensions namely gender, household headship and income class. Here gender difference is prominent across all groups belonging to similar income and household headship status. In fact, hospitalization of poorer elderly males is heavily dependent on borrowings whereas females have generally higher reliance on contributions from friends and relatives.

Finally, we estimate alternative models to understand the effects of different socio-economic variables on financing of health care of elderly through distressed means like borrowings, sale of assets and contribution from friends and relatives.

The dependent variable here is binary in nature and indicates whether or not the household has relied on the concerned source of financing. The analysis also controls for selection problem but due to space considerations, in Table 13.5 we report only the marginal effects for key variables only. These variables are the intersectional groups that are in particular important to understand the relative

Table 13.5 Probit selection model results for sources of financing for elderly hospitalization: marginal effects from outcome equations

Variables	Coef. for use of sources of financing					
	Borrowing		Sale of assets		Contributions from friends/relatives	
	dy/dx	S.E.	dy/dx	S.E.	dy/dx	S.E.
Model 1						
Male (M) (Reference)						
Female (F)	−0.086*	*0.0010*	−0.016	*0.0130*	−0.047*	*0.0012*
Richer 40 % households (R) (Reference)						
Poorer 60 % households (P)	0.113*	*0.0011*	0.021	*0.0179*	0.028*	*0.0007*
Model 2						
Poorer female (PF) (Reference)						
Poorer male (PM)	0.076*	*0.0011*	0.023	*0.0216*	0.063*	*0.0013*
Richer female (RF)	−0.123*	*0.0013*	−0.009	*0.0094*	−0.009*	*0.0009*
Richer male (RM)	−0.033*	*0.0011*	−0.007	*0.0070*	0.019*	*0.0014*
Model 3						
Richer male and household head (RMH) (Reference)						
Richer female and household head (RFH)	−0.107*	*0.0011*	0.069*	*0.0015*	0.138*	*0.0013*
Poorer female and household member (PFO)	0.034*	*0.0012*	0.005*	*0.0005*	−0.096*	*0.0010*
Poorer male and household member (PMO)	0.138*	*0.0021*	−0.008*	*0.0004*	−0.123*	*0.0014*
Poorer female and household head (PFH)	0.043*	*0.0015*	0.026*	*0.0011*	0.198*	*0.0017*
Poorer male and household head (PMH)	0.109*	*0.0011*	0.032*	*0.0008*	0.050*	*0.0009*
Richer female and household member (RFO)	−0.088*	*0.0008*	−0.016*	*0.0004*	−0.086*	*0.0009*
Richer male and household member (RMO)	−0.016*	*0.0013*	−0.015*	*0.0005*	−0.087*	*0.0016*

Source Authors using NSSO (2006)
Note The regression estimates are adjusted for key socioeconomic variables, including place of residence, social group, financial dependency status of elderly, insurance. Model 1 and 2 also adjust for relationship with household head

importance of gender, household income and household relationship in financing health care. Model 1 is the base model and it finds significant disadvantages for elderly females in accessing distressed financing. It implies that households are more likely to resort to means such as borrowings, asset selling and contributions from friends and relatives to support hospitalization of elderly male. In fact, the probability of borrowings for hospitalization is almost nine percent lower for female elderly than compared to elderly male. Income disadvantage is also apparent as households with lower incomes are more likely to resort to distressed financing. Model 2 presents the estimates based on intersections of gender and household income. Since the relationship between gender and use of distressed financing is mediated by household income, therefore, it is observed that elderly male and female from richer households are less likely to use distressed financing. However, within same income category the gender advantage is apparent. For example, compared to poor females, the poor males are almost 8 % more likely to use borrowings and 6 % more likely to use contributions from friends and relatives. On the other hand, females from richer households also face a gender disadvantage as households are more likely to use these distressed financing means for the males.

Model 3 unravels the impact of intersection of gender, household income and elderly person's relationship with household head on use of distressed financing mechanisms. The results unravel an interesting hierarchy in use of different sources of financing. For example, when compared to elderly males from richer households, the elderly females from richer households are less likely to use borrowings to finance inpatient care. However, these females are more likely to use contributions from friends and relatives and sale of household assets for health care. This suggests that even richer households are less likely to incur debt for health care of female elderly. All the elderly groups belonging to low income households have relatively higher chances of using borrowings for inpatient care. Nevertheless, gender advantage is apparent as the probability of borrowings for males is significantly higher then for females. This finding holds irrespective of household headship status thus highlighting important gender differentials in utilization of inpatient care among Indian elderly.

13.5 Discussion

Higher incidence and magnitude of out-of-pocket healthcare payments continues to be a major health financing concern in India (HLEG 2011). Such regressive means of healthcare financing forces several households to resort to distressed means of healthcare financing. The problem is much severe for households while accessing inpatient care (Krishna 2004, 2006). In fact, in mid-2000s about 60 and 40 % households in rural and urban India with hospitalized cases had to borrow, sell household assets and use contributions from friends and relatives to cope up with the hospitalization costs (Joe 2014). Given the disutility associated with distressed financing, it is likely that households will be selective and cautious while using such options. In fact, the selective nature of households is apparent from the fact that households are more likely to resort to borrowings or sale of assets to finance inpatient care of non-elderly members than compared to elderly members (Joe 2014).

This paper finds that inpatient care of about one-half of the elderly members is financed through distressed means. In particular, about 30 % households have borrowed resources and 5 % households have sold assets; and 20 % of households have received contribution from friends and relatives to pay for hospitalization of elderly member. Although, a number of key insights have emerged from the analysis but it should be noted that there are groups who are in dire need for finance but do not have any options. The condition of such groups may not get reflected in the discussion as our analysis is confined only to groups that have actually opted for hospitalization. Clearly, the true need for financial protection may be even greater than what is revealed and is apparent from the distribution of hospitalisation across socioeconomic categories. In particular, elderly disadvantaged in terms of gender, household income or biological relations are also less likely to utilize inpatient care. However, those who utilize inpatient care are more likely to receive it through public health facilities where poor quality may be a concern. Also, households tend to incur lower medical expenditure on females (results not reported here) and the result remains significant even after adjusting for key socioeconomic correlates. The disadvantages also manifest in the form of significantly lower use of distressed financing for health care compared to other elderly. Thus, it emerges that such elderly are disadvantaged at each step including likelihood of hospitalization, quality of care and use of distressed financing and are vulnerable to poor health care and health outcomes.

From our analysis it can be observed that a number of socio-economic aspects have a significant effect on the households decision to seek inpatient care for the elderly. The economic aspects include household income, wealth and educational background; and the socio-cultural aspects include gender, intra-household relations and other forms of bargaining power. To guide the analysis these factors were initially discussed in a theoretical framework. Although, the above list of socio-economic factors is not exhaustive but the leveraging effect emanating from the interaction of these multiple axes of power proves to be more potent than the individual effects. Apart, from descriptive analysis, the econometric analysis is based on Heckman probit model that controls for selection bias and yields robust econometric inferences.

We also present inferences regarding determinants of elderly inpatient care utilization from private sector. Here it is observed that the dependency of the elderly on others as well as dependency of others (close and distant relatives) on the household head can alter the chances of elderly seeking care at a private facility. The place of residence and social group also explains variation in the utilization of private sector facility. Financial protection offered through health insurance (state, private or employer provided) can affect the chances of inpatient care. Apart from these demand side factors (characteristics of the households), a number of supply side factors are also at work. Also, the state of affairs does not support universal health care and such circumstances can only lead to deteriorating healthcare environment for the elderly. A few environmental determinants that have a negative impact on elderly healthcare utilization are poverty, rural area disadvantage, backward social group affiliations and lack of social insurance and social security.

Theoretically, it is inferred that the interaction of gender and wealth is advantageous to certain sections though the hierarchy of groups can vary. For instance, the male elderly that belong to richer group have a clear advantage in availing inpatient care as compared to other groups, and female elderly belonging to the poorer group are the most disadvantaged. Similarly, the interaction effect of gender, wealth and head of household strongly leverage against weaker identities. These aspects are also verified through the econometric analysis for explaining the determinants of utilizing different sources of distressed financing such as borrowing, sale of assets and contribution from friends and relatives. The results confirm of significant gender discrimination in distressed financing. On average, households are less likely to borrow to meet inpatient care expenditure of elderly female than compared to males. Similar results are also observed in the context of child healthcare utilization in India (Asfaw et al. 2007). This reflects that households may derive lower utility from treatment of female elderly and this could in part be related with income status of the household or status of the female in the concerned household. In recognition of this fact, the theoretical framework also introduces wealth as an enabling determinant that has an important bearing on the ability of the household to seek inpatient care for the elderly. This theoretical inference is also supported as elderly from the richer households are more likely to receive inpatient care as compared to those belonging to the poorer ones. Elders who are head of the household also have considerable bargaining power.

As shown in Fig. 13.3, the inferences from the models are consistent with the theoretical model and suggest that inequalities can manifest in different forms and could be captured effectively when an intersectional approach is adopted. Figure 13.3, in fact helps us to understand three important concerns: (i) whether gender disadvantage diminish with improvements in household income; (ii) whether household head have a distinct advantage in matters of healthcare utilization; and (iii) whether distressed financing is a key feature of financing for the poor. As regards the first concern, Fig. 13.3a shows that when gender and income are viewed as two different variables then disadvantages are obvious for the vulnerable groups (females and poor households). However, interaction of gender and income reveals that gender disadvantages are present among both poorer and richer households. This also suggests that gender disadvantage is not easily eliminated despite improvements in household income. Second, it is inferred that household headship is associated with distressed financing but this effect is not significant for the poorer households because elderly males, irrespective of headship status have higher probability of using distressed financing compared to females. Finally, it is also evident from Fig. 13.3b that there is a consistent hierarchy in use of borrowings whereby all the poorer groups display relatively higher chances of borrowings than compared to the richer groups including the reference group (elderly male household head from richer section). Apparently, the formation of extreme

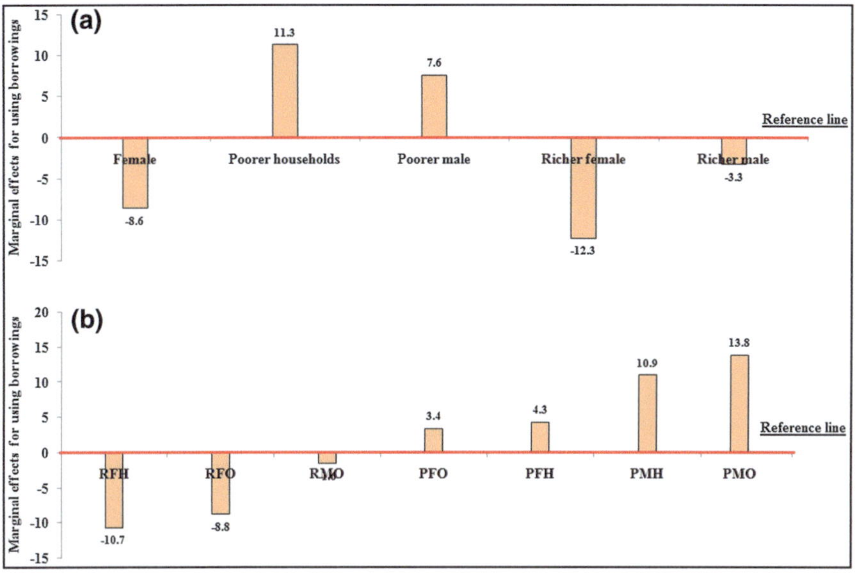

Fig. 13.3 Marginal effects for using borrowings by intersectional groups (*Note* the marginal effects are based on econometric results reported in Table 12.5. The groups are formed by intersecting three prominent determinants viz. household income (poorer and richer households), gender (male and female) and relation with household head (self head or others). The intersection of these determinants yields eight mutually exclusive subgroups, namely: (1) poorer-female-head (PFH); (2) poorer-female-other (PFO); (3) richer-female-head (RFH); (4) richer-female-others (RFO); (5) poorer-male-head (PMH); (6) poorer-male-others (PMO); (7) richer-male-head (RMH); (8) richer-male-others (RMO)

groups reveals implicit social hierarchy whereby households are most likely to use borrowings for poorer males and are least likely to use it for females from richer households.

13.6 Conclusion

Most of the hospitalization in India is financed via distressed means such as borrowings, sale of household assets or through contributions from friends and relatives. These worrying trends become further dismal when the fact is considered that the disutility associated with distressed financing can force the households to be selective and cautious while using distressed financing options. This motivates us to understand the selective nature of the households by testing the premises: if households are more likely to incur borrowings or sale of assets to finance inpatient care of non-elderly members than compared to elderly members. Also, the study identifies the socio-economic determinants of utilization of major sources of distressed financing by households for elderly health care. Although, the analysis is based on data for the year 2004 but arguably social relations are not likely to change drastically over a short period of time and this perhaps compensates for this limitation.

Household decision-making is a complex task but we adopt three dominant theoretical frameworks: unitary, collective and intersectionality framework to examine the various determinants of elderly inpatient care utilization and its financing. The key variables that influence the decision of the households to seek treatment for elderly have been identified as wealth (endowment), gender and relationship with the head (bargaining power). Also, the interaction of these factors also gives rise to leveraging effects that explains the patterns observed in using different sources of distressed financing. It is worth noting that gender disadvantage is a very consistent and significant decisive factor behind explaining intra-household variations in hospitalization and distressed financing. But, nothing could be further from the fact that it is really difficult to formulate an effective policy for dealing with intra household dynamics. Therefore, policymakers should be sensitive towards intra household dynamics. In fact, further research is necessary to understand the influence of endowment and bargaining power on intra-household resource allocation for health care.

In concluding, it is worthwhile to discuss some important policy concerns. India has rather dismal social security provisions for the elderly including social health insurance. Although, there is an increasing willingness of the government to expand health care for the elderly but despite prominent initiatives (such as Rashtriya Swasthya Bima Yojana or Arogyashree Scheme) the prevalence of very high burden of out-of-pocket healthcare expenditure in the private sector remains a formidable concern. Moreover, inpatient care utilization can be a catastrophe for several poorer households but amidst highly privatized tertiary care sector these issues have not received adequate policy attention. In fact, the analysis shows that households with cancer and heart patients are more likely to use distressed means to finance health care. Clearly, major financial and infrastructure investments are critical to boost the dysfunctional public health system and to gear it to meet the healthcare requirements of the elderly. In view of such intricacies, universal healthcare coverage through direct public provision is perhaps a pragmatic way forward to enhance elderly wellbeing.

These issues assume further relevance because of ongoing demographic and epidemiological transition in India. In fact, with growing burden of non-communicable diseases, improvements in geriatric care and appropriate human resources (specialists, doctors and community health workers) would also constitute one component of the solutions. Finally, it is also important to strengthen mechanisms for community-based health care of elderly. For example, India—with second largest elderly (aged 60+) population in the world—is certainly looking forward to integrate family in policy discourse via the Maintenance and Welfare of Senior Citizens Act, 2007 of India that mandates effective provisions for the maintenance (includes provision for food, clothing, residence and medical attendance and treatment) and welfare of parents and senior citizens (Government of India 2007). Legal provisions notwithstanding, community in general should be motivated or incentivized to provide social, economic and emotional support to the elderly. Likewise family has an equally critical role in encouraging healthcare utilization by elderly.

References

Agarwal, B. (1997a). Environmental action, gender equity and women's participation. *Development and Change, 28*(1), 1–44.

Agarwal, B. (1997b). Bargaining and gender relations: Within and beyond the household. *Feminist Economics, 3*(1), 1–51.

Asfaw, A., Klasen, S., & Lamanna, L. (2007). Intrahousehold Gender Disparities in Childrens Medical Care before Death in India. IZA Discussion Paper No. 2586.

Becker, G. (1981). A treatise on the family. Cambridge, USA: Harvard University Press.

Becker, G. (1965). A theory of the allocation of time. *Economic Journal, 75*, 493–517.

Becker, G. (1974a). A theory of marriage: Part II. *Journal of Political Economy, 82*(2), S11–S26.

Becker, G. (1974b). A theory of social interactions. *Journal of Political Economy, 82*(6), 1063–1093.

Berman, P., Ahuja, R., Bhandari, L. (2010). The impoverishing effect of healthcare payments in India: New methodology and findings. *Economic and Political Weekly XLV*, 65–71.

Binnendijk, E., Koren, R., & Dror, D. M. (2012). Hardship financing of healthcare among rural poor in Orissa, India. *BMC Health Services Research, 12*, 23.

Crenshaw, K. (1991). Mapping the margins: Intersectionality, identity politics, and violence against women of color. *Stanford Law Review, 43*(6), 1241–1299.

Davis, K. (2008). Intersectionality as buzzword: A sociology of science perspective on what makes a feminist theory successful. *Feminist Theory, 9*(1), 67–85.

Dilip, T. R., Duggal, R. (2002). *Incidence of non-fatal health outcomes and debt in urban India.* Draft paper presented for urban research symposium, December 9–11, 2002. Washington DC: World Bank.

Erreygers, G. (2009). Correcting the concentration index. *Journal of Health Economics, 28*, 504–515.

Flores, G., et al. (2008). Coping with health-care costs: Implications for the measurement of catastrophic expenditures and poverty. *Health Economics, 17*, 1393–1412.

Garg, C. C., & Karan, A. A. (2009). Reducing out-of-pocket expenditures to reduce poverty: A disaggregate analysis at rural urban and state level in India. *Health Policy and Planning, 24*, 116–128.

Government of India. (2007). *Maintenance and Welfare of Parents and Senior Citizens Act, 2007.* The Gazette of India, December 2007. http://socialjustice.nic.in/oldageact.php.

Government of India. (2009). *National Health Accounts—India (2004–2005).* New Delhi: Ministry of Health & Family Welfare.

Greene, W. (2003). *Econometric analysis*. New Jersey: Prentice Hall.

Griffin, J. M., Fuhrer, R., Stansfeld, S. A., & Marmot, M. (2002). The importance of low control at work and home on depression and anxiety: Do these effects vary by gender and social class? *Social Science & Medicine, 54*(5), 783–798.

Gupta, I., & Joe, W. (2013). Refining estimates of catastrophic healthcare expenditure: An application in the Indian context. *International Journal of Health Care Finance and Economics, 13*, 157–172.

Haddad, L., Hoddinott, J., & Alderman, H. (1997). *Intrahousehold resource allocation in developing countries—Models, methods and policy*. Baltimore, MD: Johns Hopkins University Press.

Hankivsky, O. (2012). Women's health, men's health, and gender and health: Implications of intersectionality. *Social Science and Medicine, 74*, 1712–1720.
Hankivsky, O., & Christoffersen, A. (2008). Intersectionality and the determinants of health: A Canadian perspective. *Critical Public Health, 18*(3), 271–283.
Heckman, J. (1979). Sample selection bias as a specification error. *Econometrica, 47*(1), 153–161.
High Level Expert Group (HLEG) Report 2011. (2011). High level expert group report on universal health coverage in India. Submitted to the Planning Commission of India. New Delhi: Public Health Foundation of India.
Iversen, V. (2002). Autonomy in child labor migrants. *World Development, 30*(5), 817–834.
Joe, W. (2014). Distressed financing of household out-of-pocket health care payments in India: Incidence and correlates. *Health Policy and Planning*, 1–14. doi:10.1093/heapol/czu050
Kelly, U. A. (2009). Integrating intersectionality and biomedicine in health disparities research. *Advances in Nursing Science, 32*(2), E42–E56.
Krishna, A. (2004). Escaping poverty and becoming poor: Who gains, who loses and why? *World Development, 32*, 121–136.
Krishna, A. (2006). Pathways out of and into poverty in 36 villages of Andhra Pradesh. *India World Development, 34*, 271–288.
Kruk, M. E., Goldmann, E., & Galea, S. (2009). Borrowing and selling to pay for health care in low- and middle-income countries. *Health Affairs, 28*, 1056–1066.
Leive, A., & Xu, K. (2008). Coping with out-of-pocket health payments: Empirical evidence from 15 African countries. *Bulletin of the World Health Organization, 86*, 849–856.
McGibbon, E., & McPherson, C. (2011). Applying intersectionality and complexity theory to address the social determinants of women's health. *Women's Health and Urban Life, 10*, 59–86.
National Sample Survey Organization (NSSO). (2006). Morbidity health care and the condition of aged. Report No 507. New Delhi: Ministry of Statistics and Programme Implementation.
O'Donnell, O., van Doorslaer, E., Rannan-Eliya, R. P., et al. (2008). Who pays for health care in Asia? *Journal of Health Economics, 27*, 460–475.
Samuelson, P. (1956). Social indifference curves. *Quarterly Journal of Economics, 70*, 1–22.
Selvaraj, S., & Karan, A. (2009). Deepening health insecurity in India: Evidence from national sample surveys since 1980s. *Economic and Political Weekly, 44*, 55–60.
Sen, A. (1966). Labour allocation in a cooperative enterprise. *Review of Economic Studies, 33*, 361–371.
Sen, A. (1990). Gender and cooperative conflicts. In I. Tinker (Ed.), *Persistent inequalities: Women and world development*. New York: Oxford University Press.
Sen, G., & Iyer, A. (2012). Who gains, who loses and how: Leveraging gender and class intersections to secure health entitlements. *Social Science and Medicine, 74*, 1802–1811.
Springer, K. W., Hankivsky, O., & Bates, L. M. (2012). Introduction—Gender and health: Relational, intersectional and biosocial approaches. *Social Science and Medicine, 74*, 1661–1666.
StataCorp. (2007). *Stata statistical software: Release 10*. College Station, TX: StataCorp LP.
van de Ven, W. P. M. M., & van Praag, B. M. S. (1981). The demand for deductibles in private health insurance: A probit model with sample selection. *Journal of Econometrics, 17*, 229–252.
van Doorslaer, E., et al. (2007). Catastrophic payments for health care in Asia. *Health Economics, 16*, 1159–1184.
Wagstaff, A., Paci, P., & Van Doorslaer, E. (1991). On the measurement of inequalities in health. *Social Science and Medicine, 33*, 545–557.

Index

A
Abuse, 15, 104
Active aging, 14
Activities of daily living (ADL), 76
Activity Theory of Aging, 204
Affect, 214
Age categorization, 53
Age discrimination, 48
Ageism, 12, 50, 86
Agency, 74, 110
Age prejudice, 50
Age stereotypes, 50
Aging society crisis, 14, 126
Arogyashree Scheme, 268

B
Baby boom, 137
Bargaining framework, 247
Bereavement, 74
Burden, 73, 127

C
Caregiver, 25
Caregiving, 12, 25
Care-prevention, 128
Care work, 33
Cognitive, 207
Communal ownership, 97
Community, 131, 268
Community ties, 79
Conceptual models, 5
Convoy model, 32
Cooperation, 113
Critical theory, 7
Cross-cultural, 11
Cultural context of aging, 10
Cultural model (, 72
Cultural norms, 28, 70
Cultural scripts, 88, 144
Cultural theories, 69
Cumulative advantage/disadvantage theory, 9, 27
Cybernetics theory, 128

D
Data rich, theory poor, 6
Dependence, 73
Dependency, 99
Dependency ratio, 35
Dependent, 33, 35, 149
Depression, 24
Depressive symptoms, 171
Disability free life expectancy (DFLE), 17, 223
Discrimination, 86, 90, 120, 192, 266
Distressed financing, 246

E
Empowerment, 192
Epistemology, 6
Ethnography, 74
European Social Survey, 54
Expectations, 149, 183
Expenditure, 194
Experience, 74

F
Familial, 165
Familial relations, 89
Family, 74, 196
Family structure, 25, 144
Feedback systems, 128
Feminization, 178
Filial piety, 28, 156
Frailty, 133
Future, 14

T. Samanta (ed.), *Cross-Cultural and Cross-Disciplinary Perspectives in Social Gerontology*, DOI 10.1007/978-981-10-1654-7

G
Gender, 16, 71, 96, 165, 179, 227, 247, 254
Geriatric depression scale (GDS-15), 164
Geriatric hospital, 128
Gero-geographical perspective, 205
Gerontological imagination, 8, 13
Globalization, 70, 148
Global south, 6, 11
Good old age, 10
GPS, 17, 205
Great Depression, 17, 49
Group in the individual, 49

H
Happiness, 144
Health care utilization, 16, 250
Health outcomes, 57
History, 17
Hospitalization, 180, 246, 253
Household resource allocation, 16
Household utility, 250
Household welfare, 250
Human Development Index, 144

I
Identity, 10, 59, 71
Individual in the group, 48
Individualism, 90, 97, 113
Individuality, 248
Inequality, 113
Informal or at-home care, 24
Instrumental variables, 254
Interaction, 215
Interactionist approach, 14, 104
Interdependence, 80, 110
Intergenerational, 147
Intergenerational co-residence, 25
Intergenerational relationships, 94
Interpretive, 7
Intersectionality framework, 247
Intervention, 59
Intra-household resource allocations, 247
Intra-household welfare, 246

J
Japan, 71
Joint family, 26

K
Kerala, 144

L
Language, 138
Liberalization, 90
Life course, 17, 32, 121, 160
Life course perspective, 27, 161
Life expectancy, 221
Life histories, 74
Life satisfaction, 17, 57, 212
Life stage, 30, 160
Living arrangements, 25, 113, 146
Locomotive disease, 134
Logit model, 187
Loneliness, 10
Longitudinal data, 9
Long Term Care Insurance (LTCI), 14, 75, 126

M
Market, 110
Maya, 119
Meaningful decline, 11
Meaning in life, 71
Measurement, 160
Memories, 48, 70
Methodological collectivism, 50
Middle-range theories, 8
Migration, 178
Mobility, 17, 180, 207
Moral, 114
Moral-cultural scripts, 110
Morality, 29
Multigenerational, 12
Multigenerational families, 113
Multilevel modelling, 50

N
Narrative, 74
National Sample Survey Organisation (NSSO), 252
Neglect, 15, 92
Neoliberalism, 70, 87
Network dynamics, 32
Non-familial, 111, 121
Normative, 103, 155, 161
NSSO, 184

O
Objectivity, 183
Off-timedness, 30
Off-time events, 161
Old age homes, 15, 113

Old age security, 120
On-time role transitions, 162
Ontogenesis, 14, 126
Out of pocket, 264

P
Paradigm shift, 6
Participant observation, 74
Patriarchal, 196
Pentecostalism, 98
Perceived, 207
Perceived ageism, 94
Perceived and actual health, 16
Perceived health, 182
Perceived social status, 58
Phenomenology, 10
Policy-Making, 59
Political, 71
Positional objectivity, 16, 182
Positivistic, 7
Post-conflict communities, 91
Post-conflict regime, 15
Poverty, 15, 24, 266
Probit selection model, 254
Psychological, 161
Purposive sampling approach, 93

R
Rashtriya Swasthya Bima Yojana, 268
Reciprocal, 114
Reciprocity, 155
Reflexivity, 11
Resilience, 86
Retirement, 23, 90, 204
Role-enrichment hypothesis, 31
Role loss, 204
Role occupancy theory, 31
Role support, 204
Role theory, 27

S
Selection bias, 259
Self, 138
Self-reported health, 180
Sense of self, 72
Seva, 26, 113
Sexuality, 10, 71
Social capital, 15, 58, 110
Social class, 96
Social clock theory, 161
Social constructionist, 7
Social constructionist perspective, 110
Social control, 121
Social exclusion, 95
Social Identity Approach, 50
Socialization, 129
Social networks, 144
Social norms, 75
Social psychology, 15, 48
Social security, 179, 266
Social support, 58
Social ties, 58, 110
Solidarity, 102, 121, 147, 148
South West Nigeria, 87
Spatial, 204
Stereotypes, 12
Stereotyping, 86
Stressors, 57, 162
Stress process model, 27
Structure of scientific revolutions,, 6
Subjective, 160, 179, 207
Subjective experience, 53
Subjective wellbeing, 71, 89, 144
Subjectivity, 11, 70
Successful aging, 11, 15, 70, 113
Sullivan method, 17
Symbolic interactionists, 51

T
Terror management theory, 54
Time-space data, 207
Time-use surveys, 35
Transitions, 30, 161

U
Unitary approach, 248
Urbanization, 148

W
Wellbeing, 58, 161
Widow, 120
Widowhood, 179

Y
Yoruba, 86

MIX
Papier aus verantwortungsvollen Quellen
Paper from responsible sources
FSC® C105338

If you have any concerns about our products,
you can contact us on
ProductSafety@springernature.com

In case Publisher is established outside the EU,
the EU authorized representative is:
Springer Nature Customer Service Center GmbH
Europaplatz 3, 69115 Heidelberg, Germany

Printed by Libri Plureos GmbH
in Hamburg, Germany